Macroscale and Microscale Organic Experiments

6th Edition

Kenneth L. Williamson | Katherine M. Masters

CENGAGE
Learning™

Australia • Brazil • Japan • Korea • Mexico • Singapore • Spain • United Kingdom • United States

Macroscale and Microscale Organic Experiments, 6th Edition

Kenneth L. Williamson | Katherine M. Masters

Executive Editors:
Michele Baird
Maureen Staudt
Michael Stranz

Project Development Manager:
Linda deStefano

Senior Marketing Coordinators:
Sara Mercurio
Lindsay Shapiro

Senior Production / Manufacturing Manager:
Donna M. Brown

PreMedia Services Supervisor:
Rebecca A. Walker

Rights & Permissions Specialist:
Kalina Hintz

Cover Image:
Getty Images*

* Unless otherwise noted, all cover images used by Custom Solutions, a part of Cengage Learning, have been supplied courtesy of Getty Images with the exception of the Earthview cover image, which has been supplied by the National Aeronautics and Space Administration (NASA).

© 2011 Kenneth L. Williamson | Katherine M. Masters

ALL RIGHTS RESERVED. No part of this work covered by the copyright herein may be reproduced, transmitted, stored or used in any form or by any means graphic, electronic, or mechanical, including but not limited to photocopying, recording, scanning, digitizing, taping, Web distribution, information networks, or information storage and retrieval systems, except as permitted under Section 107 or 108 of the 1976 United States Copyright Act, without the prior written permission of **Kenneth L. Williamson and Katherine M. Masters.**

For product information and technology assistance, contact us at
Cengage Learning Customer & Sales Support, 1-800-354-9706
For permission to use material from this text or product, submit all requests online at **cengage.com/permissions**
Further permissions questions can be emailed to
permissionrequest@cengage.com

ISBN-13: 978-1-133-44537-1
ISBN-10: 1-133-44537-3

Cengage Learning
5191 Natorp Boulevard
Mason, Ohio 45040
USA

Cengage Learning is a leading provider of customized learning solutions with office locations around the globe, including Singapore, the United Kingdom, Australia, Mexico, Brazil, and Japan. Locate your local office at:
international.cengage.com/region

Cengage Learning products are represented in Canada by Nelson Education, Ltd.

For your lifelong learning solutions, visit **custom.cengage.com**

Visit our corporate website at **cengage.com**

Printed in the United States of America

Contents

Chapter 1	Introduction	1
Chapter 2	Laboratory Safety, Courtesy, and Waste Disposal	26
Chapter 3	Melting Points and Boiling Points	41
Chapter 4	Recrystallization	61
Chapter 5	Distillation	86
Chapter 7	Extraction	131
Chapter 8	Thin-Layer Chromatography: Analyzing Analgesics and Isolating Lycopene from Tomato Paste	164
Chapter 9	Column Chromatography: Fluorenone, Cholesteryl Acetate, Acetylferrocene, and Plant Pigments	185
Chapter 10	Gas Chromatography: Analyzing Alkene Isomers	205
Chapter 11	Infrared Spectroscopy	220
Chapter 12	Nuclear Magnetic Resonance Spectroscopy	239
Chapter 14	Ultraviolet Spectroscopy, Refractive Indices, and Qualitative Instrumental Organic Analysis	279
Chapter 19	Alkenes from Alcohols: Cyclohexene from Cyclohexanol	334
Chapter 22	Oxidation: Cyclohexanol to Cyclohexanone; Cyclohexanone to Adipic Acid	356
Chapter 24	Oxidative Coupling of Alkynes: 2,7-Dimethyl-3,5-octadiyn-2,7-diol	372
Chapter 25	Catalytic Hydrogenation	377
Chapter 29	Friedel–Crafts Alkylation of Benzene and Dimethoxybenzene; Host-Guest Chemistry	406
Chapter 36	Aldehydes and Ketones	467
Chapter 40	Esterification and Hydrolysis	515
Chapter 53	The Benzoin Condensation: Catalysis by the Cyanide Ion and Thiamine	655
Chapter 54	Nitric Acid Oxidation; Preparation of Benzil from Benzoin; and Synthesis of a Heterocycle: Diphenylquinoxaline	661
Chapter 55	The Borohydride Reduction of a Ketone: Hydrobenzoin from Benzil	668
Chapter 58	The Synthesis of an Alkyne from an Alkene; Bromination and Dehydrobromination: Stilbene and Diphenylacetylene	680
Appendices	Experiment 4 - Molecular Shapes: Structural Isomerism, Conformational Analysis, and Stereoisomerism	A-1

The S$_N$1 Reaction: Synthesis of 2-Chloro-2-methylpropane (*tert*-butyl chloride)	A10
S$_N$2 Reaction: Synthesis of 1-Chlorotetradecane from 1-Tetradecanol	A13
Saponification: Synthesis of Salicylic Acid from Methyl Salicylate (Oil of Wintergreen)	A16
Thin Layer Chromatography - Analysis of Analgesics	A20
Aldol Condensation Reaction: Synthesis and Identification of an Aldol Reaction Product	A23
General Procedures for Identifying Unknowns	A26
Test for Phenols	A30
Unknown Alcohols: Functional Group Characterization[1]	A31
Unknown Alcohols: Derivative Preparation[1]	A34
Unknown Aldehydes and Ketones: Functional Group Characterization[1]	A36
Unknown Aldehydes and Ketones: Derivative Formation[1]	A39
Unknown Carboxylic Acids: Revised Procedures for Preparation of Derivatives	A41
Unknown Amines - Test Tube Tests[1]	A44
Unknown Amine Derivatives	A48

CHAPTER 1

Introduction

When you see this icon, sign in at this book's premium website at www.cengage.com/login to access videos, Pre-Lab Exercises, and other online resources.

PRELAB EXERCISE: Study the glassware diagrams presented in this chapter and be prepared to identify the reaction tube, the fractionating column, the distilling head, the filter adapter, and the Hirsch funnel.

Welcome to the organic chemistry laboratory! Here, the reactions that you learned in your organic lectures and studied in your textbook will come to life. The main goal of the laboratory course is for you to learn and carry out techniques for the synthesis, isolation, purification, and analysis of organic compounds, thus experiencing the experimental nature of organic chemistry. We want you to enjoy your laboratory experience and ask you to remember that safety always comes first.

EXPERIMENTAL ORGANIC CHEMISTRY

You are probably not a chemistry major. The vast majority of students in this laboratory course are majoring in the life sciences. Although you may never use the exact same techniques taught in this course, you will undoubtedly apply the skills taught here to whatever problem or question your ultimate career may present. Application of the scientific method involves the following steps:

1. Designing an experiment, therapy, or approach to solve a problem.
2. Executing the plan or experiment.
3. Observing the outcome to verify that you obtained the desired results.
4. Recording the findings to communicate them both orally and in writing.

The teaching lab is more controlled than the real world. In this laboratory environment, you will be guided more than you would be on the job. Nevertheless, the experiments in this text are designed to be sufficiently challenging to give you a taste of experimental problem-solving methods practiced by professional scientists. We earnestly hope that you will find the techniques, the apparatus, and the experiments to be of just the right complexity, not too easy but not too hard, so that you can learn at a satisfying pace.

Macroscale and Microscale Experiments

This laboratory text presents a unique approach for carrying out organic experiments; they can be conducted on either a *macroscale* or a *microscale*. Macroscale was the traditional way of teaching the principles of experimental organic chemistry and is the basis for all the experiments in this book, a book that traces its history to

1934 when the late Louis Fieser, an outstanding organic chemist and professor at Harvard University, was its author. Macroscale experiments typically involve the use of a few grams of *starting material*, the chief reagent used in the reaction. Most teaching institutions are equipped to carry out traditional macroscale experiments. Instructors are familiar with these techniques and experiments, and much research in industry and academe is carried out on this scale. For these reasons, this book has macroscale versions of most experiments.

For reasons primarily related to safety and cost, there is a growing trend toward carrying out microscale laboratory work, on a scale one-tenth to one-thousandth of that previously used. Using smaller quantities of chemicals exposes the laboratory worker to smaller amounts of toxic, flammable, explosive, carcinogenic, and teratogenic material. Microscale experiments can be carried out more rapidly than macroscale experiments because of rapid heat transfer, filtration, and drying. Because the apparatus advocated by the authors is inexpensive, more than one reaction may be set up at once. The cost of chemicals is, of course, greatly reduced. A principal advantage of microscale experimentation is that the quantity of waste is one-tenth to one-thousandth of that formerly produced. To allow maximum flexibility in the conduct of organic experiments, this book presents both macroscale and microscale procedures for the vast majority of the experiments. As will be seen, some of the equipment and techniques differ. A careful reading of both the microscale and macroscale procedures will reveal which changes and precautions must be employed in going from one scale to the other.

Synthesis and Analysis

The typical sequence of activity in synthetic organic chemistry involves the following steps:

1. Designing the experiment based on knowledge of chemical reactivity, the equipment and techniques available, and full awareness of all safety issues.
2. Setting up and running the reaction.
3. Isolating the reaction product.
4. Purifying the crude product, if necessary.
5. Analyzing the product using chromatography or spectroscopy to verify purity and structure.
6. Disposing of unwanted chemicals in a safe manner.

1. Designing the Experiment

Because the first step of experimental design often requires considerable experience, this part has already been done for you for most of the experiments in this introductory level book. Synthetic experimental design becomes increasingly important in an advanced course and in graduate research programs. Safety is paramount, and therefore it is important to be aware of all possible personal and environmental hazards before running any reaction.

2. Running the Reaction

The rational synthesis of an organic compound, whether it involves the transformation of one functional group into another or a carbon-carbon bond-forming reaction, starts with a *reaction*. Organic reactions usually take place in the liquid phase and are *homogeneous*—the reactants are entirely in one phase. The reactants can be solids and/or liquids dissolved in an appropriate solvent to mediate the reaction. Some reactions are *heterogeneous*—that is, one of the reactants is a solid and requires stirring or shaking to bring it in contact with another reactant. A few heterogeneous reactions involve the reaction of a gas, such as oxygen, carbon dioxide, or hydrogen, with material in solution.

An *exothermic* reaction evolves heat. If it is highly exothermic with a low activation energy, one reactant is added slowly to the other, and heat is removed by external cooling. Most organic reactions are, however, mildly *endothermic*, which means the reaction mixture must be heated to overcome the activation barrier and to increase the rate of the reaction. A very useful rule of thumb is that *the rate of an organic reaction doubles with a 10°C rise in temperature*. Louis Fieser introduced the idea of changing the traditional solvents of many reactions to high-boiling solvents to reduce reaction times. Throughout this book we will use solvents such as triethylene glycol, with a boiling point (bp) of 290°C, to replace ethanol (bp 78°C), and triethylene glycol dimethyl ether (bp 222°C) to replace dimethoxyethane (bp 85°C). Using these high-boiling solvents can greatly increase the rates of many reactions.

Effect of temperature

The progress of a reaction can be followed by observing: a change in color or pH, the evolution of a gas, or the separation of a solid product or a liquid layer. Quite often, the extent of the reaction can be determined by withdrawing tiny samples at certain time intervals and analyzing them by *thin-layer chromatography* or *gas chromatography* to measure the amount of starting material remaining and/or the amount of product formed.

Chapters 8–10: **Chromatography**

The next step, product isolation, should not be carried out until one is confident that the desired amount of product has been formed.

3. Product Isolation: Workup of the Reaction

Running an organic reaction is usually the easiest part of a synthesis. The real challenge lies in isolating and purifying the product from the reaction because organic reactions seldom give quantitative yields of a single pure substance.

In some cases the solvent and concentrations of reactants are chosen so that after the reaction mixture has been cooled, the product will *crystallize* or *precipitate* if it is a solid. The product is then collected by *filtration*, and the crystals are washed with an appropriate solvent. If sufficiently pure at that point, the product is dried and collected; otherwise, it is purified by the process of *recrystallization* or, less commonly, by *sublimation*.

Chapter 4: **Recrystallization**

More typically, the product of a reaction does not crystallize from the reaction mixture and is often isolated by the process of *liquid/liquid extraction*.

Chapter 7: **Liquid/Liquid Extraction**

This process involves two liquids, a water-insoluble organic liquid such as dichloromethane and a neutral, acidic, or basic aqueous solution. The two liquids do not mix, but when shaken together, the organic materials and inorganic byproducts go into the liquid layer that they are the most soluble in, either organic or aqueous. After shaking, two layers again form and can be separated. Most organic products remain in the organic liquid and can be isolated by evaporation of the organic solvent.

Chapter 5: **Distillation**

If the product is a liquid, it is isolated by *distillation*, usually after extraction. Occasionally, an extraction is not necessary and the product can be isolated by the process of *steam distillation* from the reaction mixture.

Chapter 6: **Steam Distillation and Vacuum Distillation**

4. Purification

When an organic product is first isolated, it will often contain significant impurities. This impure or crude product will need to be further purified or cleaned up before it can be analyzed or used in other reactions. Solids may be purified by recrystallization or sublimation, and liquids by distillation or steam distillation. Small amounts of solids and liquids can also be purified by *chromatography*.

Chapters 11–14: Structure Analysis

5. Analysis to Verify Purity and Structure

The purity of the product can be determined by melting point analysis for solids, boiling point analysis or, less often, refractive index for liquids, and chromatographic analysis for either solids or liquids. Once the purity of the product has been verified, structure determination can be accomplished by using one of the various spectroscopic methods, such as ^{1}H and ^{13}C nuclear magnetic resonance (NMR), infrared (IR), and ultraviolet/visible (UV/Vis) spectroscopies. Mass spectrometry (MS) is another tool that can aid in the identification of a structure.

⚠ Never smell chemicals in an attempt to identify them.

6. Chemical Waste Disposal

All waste chemicals must be disposed of in their proper waste containers. Instructions on chemical disposal will appear at the end of each experiment. It is recommended that nothing be disposed of until you are sure of your product identity and purity; you do not want to accidentally throw out your product before the analysis is complete. Proper disposal of chemicals is essential for protecting the environment in accordance with local, state, and federal regulations.

EQUIPMENT FOR EXPERIMENTAL ORGANIC CHEMISTRY

A. Equipment for Running Reactions

Organic reactions are usually carried out by dissolving the reactants in a solvent and then heating the mixture to its boiling point, thus maintaining the reaction at that elevated temperature for as long as is necessary to complete the reaction. To keep the solvent from boiling away, the vapor is condensed to a liquid, which is allowed to run back into the boiling solvent.

Microscale reactions with volumes up to 4 mL can be carried out in a *reaction tube* (Fig. 1.1a). The mass of the reaction tube is so small and heat transfer is so rapid that 1 mL of nitrobenzene (bp 210°C) will boil in 10 seconds, and 1 mL of benzene (mp 5°C) will crystallize in the same period of time. Cooling is effected by simply agitating the tube in a small beaker of ice water, and heating is effected by immersing the reaction tube to an appropriate depth in an electrically heated sand bath. This sand bath usually consists of an electric 100-mL flask heater or heating mantle half filled with sand. The temperature is controlled by the setting on a variable voltage controller, but it heats slowly and changes temperature slowly.

Turn on the sand bath about 20 minutes before you intend to use it. The sand heats slowly and changes temperature slowly.

The air above the heater is not hot. It is possible to hold a reaction tube containing refluxing solvents between the thumb and forefinger without the need for forceps or other protective devices. Because sand is a fairly poor conductor of heat, there can be a very large variation in temperature in the sand bath depending on its depth. The temperature of a 5-mL flask can be regulated by using a spatula to pile up or remove sand from near the flask's base. The heater is easily capable of producing temperatures in excess of 300°C; therefore, never leave the controller at its maximum setting. Ordinarily, it is set at 20%–40% of maximum.

 Never put a mercury thermometer in a sand bath! It will break, releasing highly toxic mercury vapor.

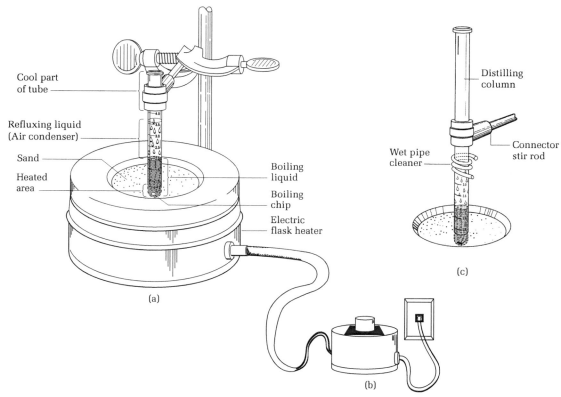

FIG. 1.1
**(a) A reaction tube being heated on a hot sand bath in a flask heater. The area of the tube exposed to the heat is small. The liquid boils and condenses on the cool upper portion of the tube, which functions as an air condenser.
(b) A variable voltage controller used to control the temperature of the sand bath. (c) The condensing area can be increased by adding a distilling column as an air condenser.**

Photos: *Williamson Microscale Kit, Refluxing a Liquid in a Reaction Tube on a Sand Bath*; Video: *The Reaction Tube in Use*

Because the area of the tube exposed to heat is fairly small, it is difficult to transfer enough heat to the contents of the tube to cause the solvents to boil away. The reaction tube is 100 mm long, so the upper part of the tube can function as an efficient *air condenser* (Fig. 1.1a) because the area of glass is large and the volume of vapor is comparatively small. The air condenser can be made even longer by attaching the empty *distilling column* (Fig. 1.1c and 1.13o) to the reaction tube using the *connector with support rod* (Fig. 1.1c and Fig. 1.13m). The black connector is made of Viton, which is resistant to high-boiling aromatic solvents. The cream-colored connector is made of Santoprene, which is resistant to all but high-boiling aromatic solvents. As solvents such as water and ethanol boil, the hot vapor ascends to the upper part of the tube. These condense and run back down the tube. This process is called *refluxing* and is the most common method for conducting a reaction at a constant temperature, the boiling point of the solvent. For very low-boiling solvents such as diethyl ether (bp 35°C), a pipe cleaner dampened with water makes an efficient cooling device. A water-cooled condenser is also available (Fig. 1.2) but is seldom needed for microscale experiments.

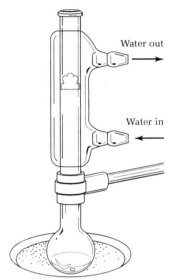

FIG. 1.2
Refluxing solvent in a 5-mL round-bottomed flask fitted with a water-cooled condenser.

Video: *How to Assemble Apparatus*

Organic reactions should be conducted in a fume hood with the sash lowered.

A Petri dish containing sand and heated on a hot plate is not recommended for microscale experiments. It is too easy to burn oneself on the hot plate; too much heat wells up from the sand, so air condensers do not function well; the glass dishes will break from thermal shock; and the ceramic coating on some hot plates will chip and come off.

Larger scale (macroscale) reactions involving volumes of tens to thousands of milliliters are usually carried out in large, round-bottom flasks that fit snugly (without sand!) into the appropriately sized flask heater or heating mantle (Figure 16.2). The round shape can be heated more evenly than a flat-bottom flask or beaker. Heat transfer is slower than in microscale because of the smaller ratio of surface area to volume in a round-bottomed flask. Cooling is again conducted using an ice bath, but heating is sometimes done in a steam bath or hot water bath for low-boiling liquids. The narrow neck is necessary for connection via a *standard-taper ground glass joint* to a water-cooled *reflux condenser*, where the water flows in a jacket around the central tube.

The high heat capacity of water makes it possible to remove a large amount of heat in the larger volume of refluxing vapor (Fig. 1.2).

Heating and Stirring

In modern organic laboratories, electric flask heaters (heating mantles), used alone or as sand baths, are used exclusively for heating. Bunsen burners are almost never used because of the danger of igniting flammable organic vapors. For solvents that boil below 90°C, the most common method for heating macroscale flasks is the *steam bath* or *hot water bath*.

Reactions are often stirred using a *magnetic stirrer* to help mix reagents and to promote smooth boiling. A Teflon-coated bar magnet (*stirring bar*) is placed in the reaction flask, and a magnetic stirrer is placed under the flask and flask heater. The stirrer contains a large, horizontally rotating bar magnet just underneath its metal surface that attracts the Teflon-coated stirring bar magnet through the glass of the flask and causes it to turn. The speed of stirring can be adjusted on the front of the magnetic stirrer.

B. Equipment for the Isolation of Products

Filtration

If the product of a reaction crystallizes from the reaction mixture on cooling, the solid crystals are isolated by *filtration*. This can be done in several ways when using microscale techniques. If the crystals are large enough and in a reaction tube, expel the air from a *Pasteur pipette*, insert it to the bottom of the tube and withdraw the solvent (Fig. 1.3). Highly effective filtration occurs between the square, flat tip of the pipette and the bottom of the tube. This method of filtration has several advantages over the alternatives. The mixture of crystals and solvent can be kept on ice during the entire process. This minimizes the solubility of the crystals in the solvent. There are no transfer losses of material because an external filtration device is not used. This technique allows several recrystallizations to be carried out in the same tube with final drying of the product under vacuum. If you know the *tare* (the weight of the empty tube), the weight of the product can be determined without removing it from the tube. In this manner a compound can be synthesized, purified by crystallization, and dried all in the same reaction tube. After removal of material for analysis, the compound in the tube can then be used for the next

FIG. 1.3
Filtration using a Pasteur pipette and a reaction tube.

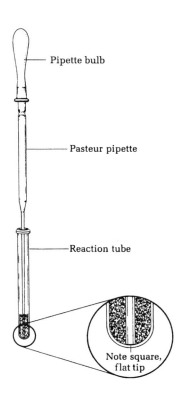

FIG. 1.4
A Hirsch funnel with an integral adapter, a polyethylene frit, and a 25-mL filter flask.

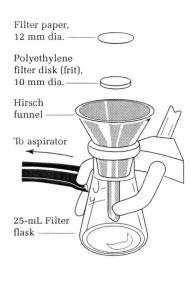

Videos: *The Reaction Tube in Use; Filtration of Crystals Using the Pasteur Pipette*

Video: *Macroscale Crystallization*

Chapter 7: Extraction

reaction. This technique is used in many of this book's microscale experiments. When the crystals are dry, they are easily removed from the reaction tube. When they are wet, it is difficult to scrape them out. If the crystals are in more than about 2 mL of solvent, they can be isolated by filtration with a *Hirsch funnel*. The one that is in the microscale apparatus kit is particularly easy to use because the funnel fits into the *filter flask* with no adapter and is equipped with a *polyethylene frit* for the capture of the filtered crystals (Fig. 1.4). The Wilfilter is especially good for collecting small quantities of crystals (Fig. 1.5).

Macroscale quantities of material can be recrystallized in conical *Erlenmeyer flasks* of the appropriate size. The crystals are collected in porcelain or plastic *Büchner funnels* fit with pieces of filter paper covering the holes in the bottom of the funnel (Fig. 1.6). A rubber *filter adapter* (*Filtervac*) is used to form a vacuum tight seal between the flask and the funnel.

Extraction

The product of a reaction will often not crystallize. It may be a liquid or a viscous oil, it may be a mixture of compounds, or it may be too soluble in the reaction solvent being used. In this case, an immiscible solvent is added, the two layers are shaken to effect *extraction*, and after the layers separate, one layer is removed. On a microscale, this can be done with a Pasteur pipette. The extraction process is repeated if necessary. A tall, thin column of liquid, such as that produced in a reaction tube, makes it easy to selectively remove one layer by pipette. This is more difficult to do in the usual test tube because the height/diameter ratio is small.

FIG. 1.5
A Wilfilter is placed upside down in a centrifuge tube and spun in a centrifuge.

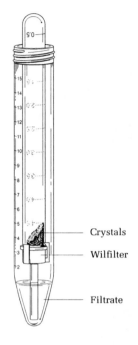

FIG. 1.6
A suction filter assembly.

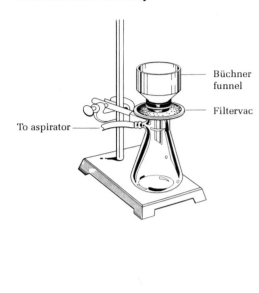

Photos: *Extraction with Ether and Extraction with Dichloromethane*; Videos: *Extraction with Ether, Extraction with Dichloromethane*

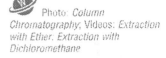
Photos: *Sublimation Apparatus*

Photo: *Column Chromatography*; Videos: *Extraction with Ether, Extraction with Dichloromethane*

Chapter 9: Column Chromatography

On a larger scale, a *separatory funnel* is used for extraction (Fig. 1.7a). The mixture can be shaken in the funnel and then the lower layer removed through the stopcock after the stopper is removed. These funnels are available in sizes from 10 mL to 5000 mL. The chromatography column in the apparatus kit is also a *micro separatory funnel* (Fig. 1.7b). Remember to remove the frit at the column base of the micro Büchner funnel and to close the valve before adding liquid.

C. Equipment for Purification

Many solids can be purified by the process of *sublimation*. The solid is heated, and the vapor of the solid condenses on a cold surface to form crystals in an apparatus constructed from a *centrifuge tube* fitted with a rubber adapter and pushed into a *filter flask* (Fig. 1.8). Caffeine can be purified in this manner. This is primarily a microscale technique, although sublimers holding several grams of solid are available.

Mixtures of solids and, occasionally, of liquids can be separated and purified by *column chromatography*. The *chromatography column* for both microscale and macroscale work is very similar (Fig. 1.9).

Some of the compounds to be synthesized in these experiments are liquids. On a very small scale, the best way to separate and purify a mixture of liquids is by *gas chromatography*, but this technique is limited to less than 100 mg of material for the usual gas chromatograph. For larger quantities of material, *distillation* is used. For this purpose, small distilling flasks are used. These flasks have a large surface area to allow sufficient heat input to cause the liquid to vaporize rapidly so that it can be distilled and then condensed for collection in a *receiver*. The complete apparatus (Fig. 1.10) consists of a *distilling flask*, a *distilling adapter* (which also functions as an air condenser on a microscale), a *thermometer adapter*, and a *thermometer*; for macroscale, a water-cooled *condenser* and *distilling adapter* are added to the

FIG. 1.7
**(a) A separatory funnel with a Teflon stopcock.
(b) A microscale separatory funnel. Remove the polyethylene frit from the micro Büchner funnel before using.**

FIG. 1.8
A small-scale sublimation apparatus.

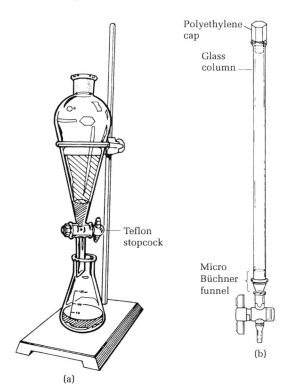

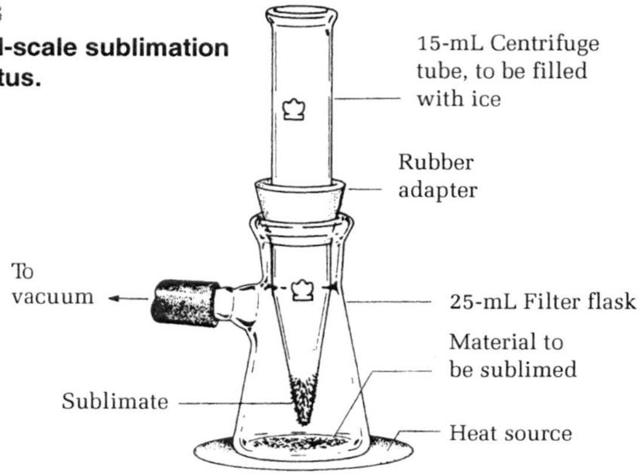

FIG. 1.9
A chromatography column consisting of a funnel, a tube, a base fitted with a polyethylene frit, and a Leur valve.

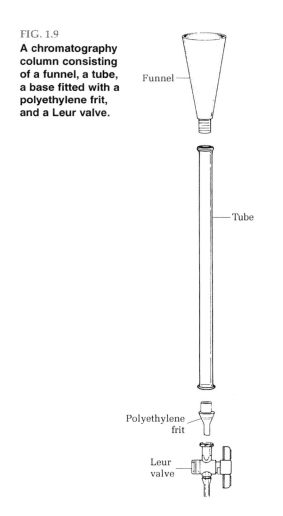

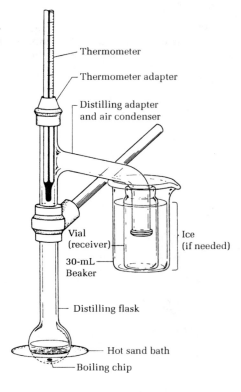

FIG. 1.10

A small-scale simple distillation apparatus. Note that the entire thermometer bulb is below the side arm of the distilling adapter.

apparatus (Fig. 1.11). *Fractional distillation* is carried out using a small, packed *fractionating column* (Fig. 1.12). The apparatus for fractional distillation is very similar for both microscale and macroscale. On a microscale, 2 mL to 4 mL of a liquid can be fractionally distilled, and 1 mL or more can be simply distilled. The usual scale in these experiments for macroscale distillation is about 25 mL.

Some liquids with a relatively high vapor pressure can be isolated and purified by *steam distillation*, a process in which the organic compound codistills with water at the boiling point of water and is then further purified and concentrated. The microscale and macroscale apparatus for this process are shown in Chapter 6.

The collection of typical equipment used for microscale experimentation is shown in Figure 1.13 and for macroscale experimentation in Figure 1.14. Other equipment commonly used in the organic laboratory is shown in Figure 1.15.

Use mercury-free thermometers whenever possible.

Photos: *Simple and Fractional Distillation Apparatus*; Video: *How to Assemble the Apparatus*

CHECK-IN OF LAB EQUIPMENT

Your first duty will be to check in to your assigned lab desk. The identity of much of the apparatus should already be apparent from the preceding outline of the experimental processes used in the organic laboratory.

Check to see that your thermometer reads about 22–25°C (71.6–77°F), which is normal room temperature. Examine the fluid column to see that it is unbroken and continuous from the bulb up. Replace any flasks that have star-shaped cracks.

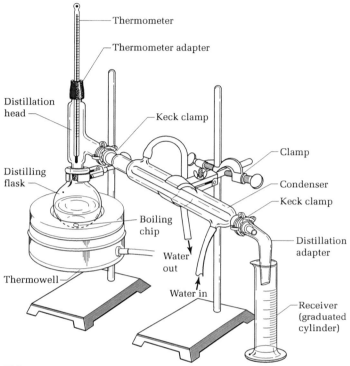

FIG. 1.11
An apparatus for simple distillation.

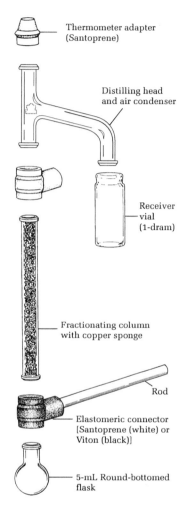

FIG. 1.12
A microscale fractional distillation apparatus. The thermometer adapter is to be fitted with a thermometer.

Remember that apparatus with graduations, stopcocks, or ground glass joints and anything porcelain are expensive. Erlenmeyer flasks, beakers, and test tubes are, by comparison, fairly cheap.

TRANSFER OF LIQUIDS AND SOLIDS

Borosilicate Pasteur pipettes (Fig. 1.16) are very useful for transferring small quantities of liquid, adding reagents dropwise, and carrying out recrystallizations. Discard used Pasteur pipettes in the special disposal container for waste glass. Surprisingly, the acetone used to wash out a dirty Pasteur pipette usually costs more than the pipette itself.

A plastic funnel that fits on the top of the reaction tube is very convenient for the transfer of solids to reaction tubes or small Erlenmeyer flasks for

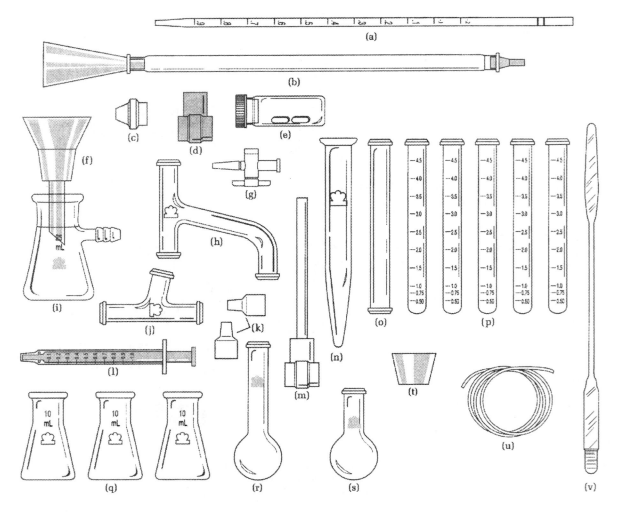

FIG. 1.13
Microscale apparatus kit.
(a) Pipette (1 mL), graduated in 1/1000ths.
(b) Chromatography column (glass) with a polypropylene funnel and 20-μm polyethylene frit in the base, which doubles as a micro Büchner funnel. The column, base, and stopcock can also be used as a separatory funnel.
(c) Thermometer adapter.
(d) Connector only (Viton).
(e) Magnetic stirring bars (4 × 12 mm) in a distillation receiver vial.
(f) Hirsch funnel (polypropylene) with a 20-μm fritted polyethylene disk.
(g) Stopcock for a chromatography column or separatory funnel.
(h) Claisen adapter/distillation head with an air condenser.
(i) Filter flask, 25 mL.
(j) Distillation head with a 105° connecting adapter.
(k) Rubber septa/sleeve stoppers, 8 mm.
(l) Syringe (polypropylene).
(m) Connector with a support rod.
(n) Centrifuge tube (15 mL)/sublimation receiver, with cap.
(o) Distillation column/air condenser.
(p) Reaction tube, calibrated, 10 × 100 mm.
(q) Erlenmeyer flasks, 10 mL.
(r) Long-necked flask, 5 mL.
(s) Short-necked flask, 5 mL.
(t) Rubber adapter for sublimation apparatus.
(u) Tubing (polyethylene), 1/16-in. diameter.
(v) Spatula (stainless steel) with scoop end.

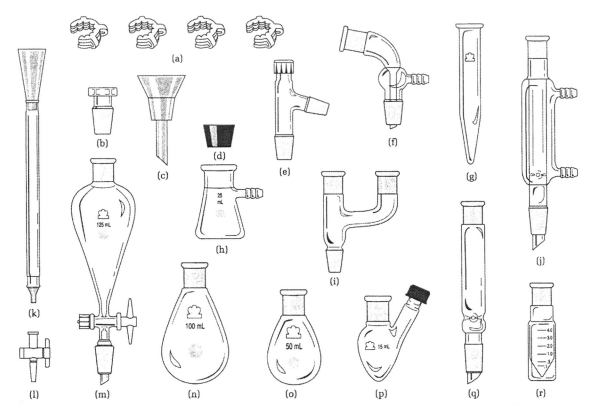

FIG. 1.14
Macroscale apparatus kit with 14/20 standard-taper ground-glass joints.

(a) Polyacetal Keck clamps, size 14.
(b) Hex-head glass stopper, 14/20 standard taper.
(c) Hirsch funnel (polypropylene) with a 20-μm fritted polyethylene disk.
(d) Filter adapter for use with sublimation apparatus.
(e) Distilling head with O-ring thermometer adapter.
(f) Vacuum adapter.
(g) Centrifuge tube (15 mL)/sublimation receiver.
(h) Filter flask, 25 mL.
(i) Claisen adapter.
(j) Water-jacketed condenser.
(k) Chromatography column (glass) with a polypropylene funnel and 20-μm polyethylene frit in the base, which doubles as a micro Büchner funnel.
(l) Stopcock for a chromatography column.
(m) Separatory funnel, 125 mL.
(n) Pear-shaped flask, 100 mL.
(o) Pear-shaped flask, 50 mL.
(p) Conical flask (15 mL) with a side arm for an inlet tube.
(q) Distilling column/air condenser.
(r) Conical reaction vial (5 mL)/distillation receiver.

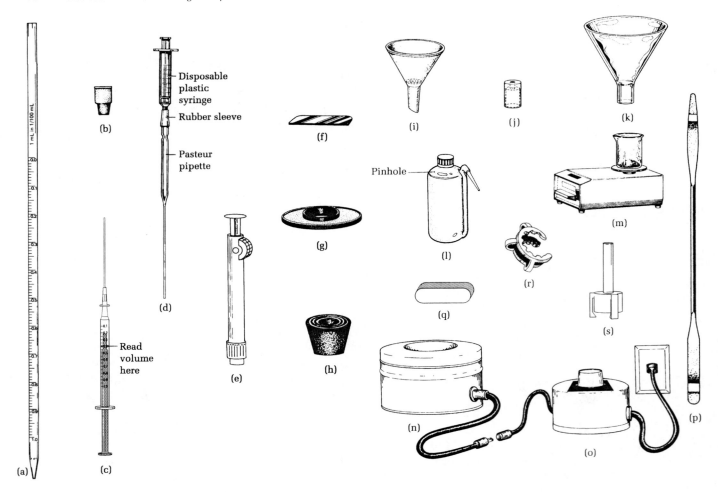

FIG. 1.15
Miscellaneous apparatus.
(a) 1.0 ± 0.01 mL graduated pipette.
(b) Septum.
(c) 1.0-mL syringe with a blunt needle.
(d) Calibrated Pasteur pipette.
(e) Pipette pump.
(f) Glass scorer.
(g) Filtervac.
(h) Set of neoprene filter adapters.
(i) Hirsch funnel with a perforated plate in place.
(j) Rubber thermometer adapter.
(k) Powder funnel.
(l) Polyethylene wash bottle.
(m) Single-pan electronic balance with automatic zeroing and 0.001 g digital readout; 100 g capacity.
(n) Electric flask heater.
(o) Solid-state control for electric flask heater.
(p) Stainless steel spatula.
(q) Stirring bar.
(r) Keck clamp.
(s) Wilfilter.

microscale experiments (Fig. 1.17). It can also function as the top of a chromatography column (Fig. 1.9). A special spatula with a scoop end (Fig. 1.13v) is used to remove solid material from the reaction tube. On a large scale, a powder funnel is useful for adding solids to a flask (Fig. 1.15k). A funnel can also be fashioned from a sheet of weighing paper for transferring lightweight solids.

WEIGHING AND MEASURING

The single-pan electronic balance (Fig. 1.15m), which is capable of weighing to ±0.001 g and having a capacity of at least 100 g, is the single most important instrument that makes microscale organic experiments possible. Most of the weighing measurements made in microscale experiments will use this type of balance. Weighing is fast and accurate with these balances as compared to mechanical balances. There should be one electronic balance for every 12 students. For macroscale experiments, a balance of such high accuracy is not necessary. Here, a balance with ±0.01 g accuracy would be satisfactory.

A container such as a reaction tube standing in a beaker or flask is placed on the balance pan. Set the digital readout to register zero and then add the desired quantity of the reagent to the reaction tube as the weight is measured periodically to the nearest milligram. Even liquids are weighed when accuracy is needed. It is much easier to weigh a liquid to 0.001 g than it is to measure it volumetrically to 0.001 mL.

It is often convenient to weigh reagents on glossy weighing paper and then transfer the chemical to the reaction container. The success of an experiment often depends on using just the right amount of starting materials and reagents. Inexperienced workers might think that if 1 mL of a reagent will do the job, then 2 mL will do the job twice as well. Such assumptions are usually erroneous.

Liquids can be measured by either volume or weight according to the following relationship:

$$\text{Volume (mL)} = \frac{\text{Weight (g)}}{\text{Density (g/mL)}}$$

Modern Erlenmeyer flasks and beakers have approximate volume calibrations fused into the glass, but these are *very* approximate. Better graduations are found on the microscale *reaction tube*. Somewhat more accurate volumetric measurements are made in 10-mL graduated cylinders. For volumes less than 4 mL, use a graduated pipette. **Never** apply suction to a pipette by mouth. Use a rubber bulb, a pipette pump, or fit the pipette with a small plastic syringe using appropriately sized rubber tubing. A Pasteur pipette can be converted into a calibrated pipette with the addition of a plastic syringe (Fig. 1.15d). Figure 1.16 also shows the calibration marks for a 9-in. Pasteur pipette. You will find among your equipment a 1-mL pipette, calibrated in hundredths of a milliliter (Fig. 1.15a). Determine whether it is designed to *deliver* 1 mL or *contain* 1 mL between the top and bottom calibration marks. For our purposes, the latter is the better pipette.

Because the viscosity, surface tension, vapor pressure, and wetting characteristics of organic liquids are different from those of water, the so-called automatic pipette (designed for aqueous solutions) gives poor accuracy in measuring organic

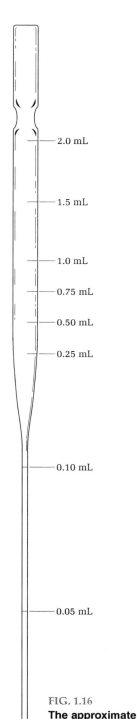

FIG. 1.16

The approximate calibration of a Kimble 9" Pasteur pipette.

FIG. 1.17
A funnel for adding solids and liquids to a reaction tube.

FIG. 1.18
Using a pipette pump to measure liquids to ±0.01 mL.

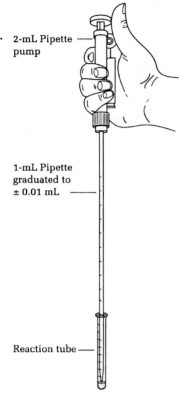

Never pipette by mouth!

liquids. Syringes (Fig. 1.15c and Fig. 1.15d) and pipette pumps (Fig. 1.18), on the other hand, are quite useful, and these will be used frequently. Do not use a syringe that is equipped with a metal needle to measure corrosive reagents because these reagents will dissolve the metal in the needle. Because many reactions are "killed" by traces of moisture, many students' experiments are ruined by damp or wet apparatus. Several reactions that require especially dry or oxygen-free atmospheres will be run in systems sealed with a rubber septum (Fig. 1.15b). Reagents can be added to the system via syringe through this septum to minimize exposure to oxygen or atmospheric moisture.

Careful measurements of weights and volumes take more time than less accurate measurements. Think carefully about which measurements need to be made with accuracy and which do not.

TARES

Tare = weight of empty container

The tare of a container is its weight when empty. Throughout this laboratory course, it will be necessary to know the tares of containers so that the weights of the compounds within can be calculated. If identifying marks can be placed on the

WASHING AND DRYING LABORATORY EQUIPMENT

Washing

Clean apparatus immediately.

⚠️ Both ethanol and acetone are very flammable.

Wash acetone is disposed in an organic solvents waste container; halogenated solvents go in the halogenated solvents waste container.

Considerable time may be saved by cleaning each piece of equipment soon after use, for you will know at that point which contaminant is present and be able to select the proper method for its removal. A residue is easier to remove before it has dried and hardened. A small amount of organic residue can usually be dissolved with a few milliliters of an appropriate organic solvent. Acetone (bp 56.1°C) has great solvent power and is often effective, but it is extremely flammable and somewhat expensive. Because it is miscible with water and vaporizes readily, it is easy to remove. Detergent and water can also be used to clean dirty glassware if an appropriate solvent cannot be found. Cleaning after an operation may often be carried out while another experiment is in process.

A *polyethylene bottle* (Fig. 1.15l) is a convenient wash bottle for acetone. Be careful not to store solvent bottles in the vicinity of a reaction where they can provide additional fuel for an accidental fire. The name, symbol, and formula of a solvent should be written on a bottle with a marker or a wax pencil. For macroscale crystallizations, extractions, and quick cleaning of apparatus, it is convenient to have a bottle of each frequently used solvent—95% ethanol, ligroin or hexanes, dichloromethane, ether, and ethyl acetate. A pinhole opposite the spout, which is covered with the finger when in use, will prevent the spout from dribbling the solvent. For microscale work, these solvents are best dispensed from 25-mL or 50-mL bottles with an attached test tube containing a graduated (1-mL) polypropylene pipette (Fig. 1.19). Be aware of any potential hazards stemming from the reactivity of these wash solvents with chemical residues in flasks. Also, be sure to dispose of wash solvents in the proper container. Acetone and most other organic solvents do not contain halogens and can therefore go in the regular organic solvents waste container. However, if dichloromethane or another halogen-containing solvent is used, it must be disposed of in the halogenated solvents waste container.

Sometimes a flask will not be clean after a washing with detergent and acetone. At that point, try an abrasive household cleaner. If still no success, try adding dilute acid or base to the dirty glassware, let it soak for a few minutes, and rinse with plenty of water and acetone.

FIG. 1.19
A recrystallization solvent bottle and dispenser.

Drying

To dry a piece of apparatus rapidly, rinse with a few milliliters of acetone and invert over a beaker to drain. **Do not use compressed air**, which contains droplets of oil, water, and particles of rust. Instead, draw a slow stream of air through the apparatus using the suction of your water aspirator or house vacuum line.

MISCELLANEOUS CLEANUP

If a glass tube or thermometer becomes stuck to a rubber connector, it can be removed by painting on glycerol and forcing the pointed tip of a small spatula between the rubber and the glass. Another method is to select a cork borer that fits snugly over the glass tube, moisten it with glycerol, and slowly work it through the connector. If the stuck object is valuable, such as a thermometer, the best policy is to cut the rubber with a sharp knife. Care should be taken to avoid force that could potentially cause a thermometer to break, causing injury and the release of mercury.

THE LABORATORY NOTEBOOK

The Laboratory Notebook
  What you did.
  How you did it.
  What you observed.
  Your conclusions.

A complete, accurate record is an essential part of laboratory work. Failure to keep such a record means laboratory labor lost. An adequate record includes the procedure (what was done), observations (what happened), and conclusions (what the results mean).

Never record anything on scraps of paper. Use a lined, 8.5" × 11" paperbound notebook, and record all data in ink. Allow space at the front for a table of contents, number the pages throughout, and date each day's work. Reserve the left-hand page for calculations and numerical data, and use the right-hand page for notes. Never record **anything** on scraps of paper to be recorded later in the notebook. Do not erase, remove, or obliterate notes; simply draw a single line through incorrect entries.

The notebook should contain a statement or title for each experiment and its purpose, followed by balanced equations for all principal and side reactions and, where relevant, mechanisms of the reactions. Consult your textbook for supplementary information on the class of compounds or type of reaction involved. Give a reference to the procedure used; do not copy verbatim the procedure in the laboratory manual. Make particular note of safety precautions and the procedures for cleaning up at the end of the experiment.

Before coming to the lab to do preparative experiments, prepare a table (in your notebook) of reagents to be used and the products expected, with their physical properties. Use the molar ratios of reactions from your table to determine the limiting reagent, and then calculate the theoretical yield (in grams) of the desired product. Begin each experiment on a new page.

Include an outline of the procedure and the method of purification of the product in a flow sheet if this is the best way to organize the experiment (for example, an extraction; *see* Chapter 8). The flow sheet should list all possible products, by-products, unused reagents, solvents, and so on that are expected to appear in the crude reaction mixture. On the flow sheet diagram indicate how each of these is removed (e.g., by extraction, various washing procedures, distillation, or crystallization). With this information entered in your notebook before coming to the laboratory, you will be ready to carry out the experiments with the utmost efficiency. Plan your time before the laboratory period. Often two or three experiments can be run simultaneously.

When working in the laboratory, record everything you do and everything you observe **as it happens.** The recorded observations constitute the most important part of the laboratory record, since they form the basis for the conclusions you will draw at the end of each experiment. One way to do this is in a narrative form.

Alternatively, the procedure can be written in outline form on the left-hand side of the page and the observations recorded on the right-hand side.

In some colleges and universities, you will be expected to have all the relevant information about the running of an experiment entered in your notebook *before coming to the laboratory* so that your textbook will not be needed when you are conducting experiments. In industrial laboratories, your notebook may be designed so that carbon copies of all entries are kept. These are signed and dated by your supervisor and removed from your notebook each day. Your notebook becomes a legal document in case you make a discovery worth hundreds of millions of dollars!

Record the physical properties of the product from your experiment, the yield in grams, and the percent yield. Analyze your results. When things did not turn out as expected, explain why. When your record of an experiment is complete, another chemist should be able to understand your account and determine what you did, how you did it, and what conclusions you reached. That is, from the information in your notebook, a chemist should be able to repeat your work.

Preparing a Laboratory Record

Use the following steps to prepare your laboratory record. The letters correspond to the completed laboratory records that appear at the end of this chapter. Because your laboratory notebook is so important, two examples, written in alternative forms, are presented.

A. Number each page. Allow space at the front of the notebook for a table of contents. Use a hardbound, lined notebook, and keep all notes in ink.
B. Date each entry.
C. Give a short title to the experiment, and enter it in the table of contents.
D. State the purpose of the experiment.
E. Write balanced equation(s) for the reaction(s).
F. Give a reference to the source of the experimental procedure.
G. Prepare a table of quantities and physical constants. Look up the needed data in the *Handbook of Chemistry and Physics*.
H. Write equation(s) for the principal side reaction(s).
I. Write out the procedure with just enough information so that you can follow it easily. Do not merely copy the procedure from the text. Note any hazards and safety precautions. A highly experienced chemist might write a procedure in a formal report as follows: "Dibenzalacetone was prepared by condensing at room temperature 1 mmol acetone with 2 mmol benzaldehyde in 1.6 mL of 95% ethanol to which was added 2 mL of aqueous 3 M sodium hydroxide solution. After 30 min the product was collected and crystallized from 70% ethanol to give 0.17 g (73%) of flat yellow plates of dibenzalacetone, mp 110.5–111.5°C." Note that in this formal report, no jargon (e.g., EtOH for 95% ethanol, FCHO for benzaldehyde) is used and that the names of reagents are written out (sodium hydroxide, not NaOH). In this report the details of measuring, washing, drying, crystallizing, collecting the product, and so on are assumed to be understood by the reader.
J. Don't forget to note how to dispose of the byproducts from the experiment using the "Cleaning Up" section of the experimental procedure.
K. Record what you do as you do it. These observations are the most important part of the experiment. Note that conclusions do not appear among these observations.
L. Calculate the theoretical yield in grams. The experiment calls for exactly 2 mmol fluorobenzaldehyde and 1 mmol acetone, which will produce 1 mmol

of product. The equation for the experiment also indicates that the product will be formed from exactly a 2:1 ratio of the reactants. Experiments are often designed to have one reactant in great excess. In this experiment, a very slight excess of fluorobenzaldehyde was used inadvertently, so acetone becomes the limiting reagent.

M. Once the product is obtained, dried, and weighed, calculate the ratio of product actually isolated to the amount theoretically possible. Express this ratio as the percent yield.
N. Write out the mechanism of the reaction. If it is not given in the text of the experiment, look it up in your lecture text.
O. Draw conclusions from the observations. Write this part of the report in narrative form in complete English sentences. This part of the report can, of course, be written after leaving the laboratory.
P. Analyze TLC, IR, and NMR spectra if they are a part of the experiments. Rationalize observed versus reported melting points.
Q. Answer assigned questions from the end of the experiment.
R. This page presents an alternative method for entering the experimental procedure and observations in the notebook. Before coming to the laboratory, enter the procedure in outline form on the left side of the page. Then enter observations in brief form on the right side of the page as the experiment is carried out. Draw a single line through any words incorrectly entered. Do not erase or obliterate entries in the notebook, and never remove pages from the notebook.

Two samples of completed laboratory records follow. An alternative method for recording procedure, cleaning up, and observations is given on p. 25. The other parts of the report are the same.

(A) p. 35
(B) Sept. 23, 2010

(C) DIFLUORODIBENZALACETONE

(D) *Purpose:* To observe and carry out an aldol condensation reaction.

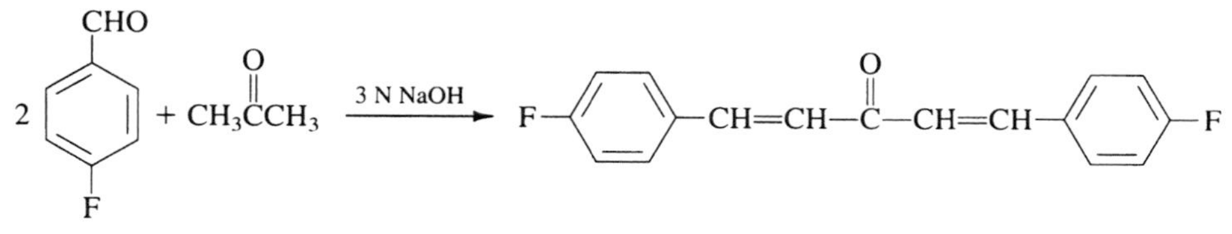

(E) 2 (4-F-C₆H₄-CHO) + CH₃COCH₃ →[3 N NaOH] F—C₆H₄—CH=CH—CO—CH=CH—C₆H₄—F

4-Fluorobenzaldehyde   Acetone   **4,4′-Difluorodibenzalacetone**

(F) *Reference:* Williamson and Masters, *Macroscale and Microscale Organic Experiments*, p. 792.

Table of Quantities and Physical Constants (G)

Substance	Mol wt	G/mol Used or Produced	Mol Needed (from eq.)	Density	mp	bp	Solubility
4-Fluorobenzaldehyde	124.11	0.248/0.002	0.002	1.157		181°C	Slightly soluble in H₂O; soluble in ethanol, diethyl ether, acetone
Acetone	58.08	0.058/0.001	0.001	0.790		56°C	Soluble in H₂O, ethanol, diethyl ether, acetone
4,4′-Difluoro-dibenzalacetone	270.34	0.270/0.001			167°C		Insoluble in H₂O; slightly soluble in ethanol, diethyl ether, acetone
4-Fluorobenzalacetone	164.19				54°C		Insoluble in H₂O; slightly soluble in ethanol, diethyl ether, acetone
Sodium hydroxide	40.01						Soluble in H₂O; insoluble in ethanol, diethyl ether, acetone

(H) *Side Reactions:* No important side reactions, but if excess acetone is used, product will be contaminated with 4-fluorobenzalacetone.

p. 36

(I) *Procedure:* To a reaction tube, add 0.248 g (0.002 mole) of 4-fluorobenzaldehyde and then 1.6 mL of an ethanol solution that contains 58 mg of acetone. Mix the solution and observe it for 1 min. Remove sample for TLC. Then add 2 mL of 3 M NaOH(*aq*). Cap the tube with a septum and shake it at intervals over a 30-min period. Collect product on Hirsch funnel, wash crystals thoroughly with water, press as dry as possible, determine crude weight, and save sample for mp. Recrystallize product from 70% ethanol in a 10-mL Erlenmeyer flask. Cool solution slowly, scratch or add seed crystal, cool in ice, and collect product on Hirsch. Wash crystals with a few drops of ice-cold ethanol. Run TLC on silica gel plates using hexane to elute, develop in iodine chamber. Run NMR spectrum in deuterochloroform and IR spectrum as a mull.

(J) *Cleaning Up:* Neutralize filtrate with HCl and flush solution down the drain. Put recrystallization filtrate in organic solvents container.

(K) *Observations:* Into a reaction tube, 0.250 g of 4-fluorobenzaldehyde was weighed (0.248 called for). Using the 1.00-mL graduated pipette and pipette pump, 1.61 mL of the stock ethanol solution containing 58 mg of acetone was added (1.6 mL called for). The contents of the tube were mixed by flicking the tube, and then a drop of the reaction mixture was removed and diluted with 1 mL of hexane for later TLC. The water-clear solution did not change in appearance during 1 min.

Then about 2 mL of 3 M NaOH(*aq*) was added using graduations on side of reaction tube. The tube was capped with a septum and shaken. The clear solution changed to a light yellow color immediately and got slightly warm. Then after about 50 sec, the entire solution became opaque. Then yellow oily drops collected on sides and bottom of tube. Shaken every 5 min for 1/2 hr. The oily drops crystallized, and more crystals formed. At the end of 30 min, the opaque solution became clear, and yellow crystals had formed. Tube filled with crystals.

Product was collected on 12-mm filter paper on a Hirsch funnel; transfer of material to funnel was completed using the filtrate to wash out the tube. Crystals were washed with about 15 mL of water. The filtrate was slightly cloudy and yellow. It was poured into a beaker for later disposal. The crude product was pressed as dry as possible with a cork. Wt (damp) 270 mg. Sample saved for mp.

Recrystallized from about 50 mL of 70% ethanol on sand bath. Almost forgot to add boiling stick! Clear, yellow solution cooled slowly by wrapping tube in cotton. A seed crystal from Julie added to start crystallization. After about 25 min, flask was placed in ice bath for 15 min, then product was collected on Hirsch funnel (filter paper), washed with a few drops of ice-cold solvent, pressed dry and then spread out on paper to dry. Wt 190 mg. Very nice flat, yellow crystals, mp 179.5–180.5°C. The crude material before recrystallization had mp 177.5–179°C.

TLC on silica gel using hexane to elute gave only one spot, R_f 0.23, for the starting material and only one spot for the product, R_f 0.66. See attached plate.

p. 37

The NMR spectrum that was run on the Bruker in deuterochloroform containing TMS showed a complex group of peaks centered at about 7 ppm with an integrated value of 17.19 and a sharp, clear quartet of peaks centered at 6.5 ppm with an integral of 8.62. The coupling constant of the AB quartet was 7.2 Hz. See attached spectrum.

The IR spectrum, run as a Nujol mull on the Mattson FT IR, showed intense peaks at 1590 and 1655 cm^{-1} as well as small peaks at 3050, 3060, and 3072 cm^{-1}. See attached spectrum.

Cleaning Up: Recrystallization filtrate put in organic solvents container. First filtrate neutralized with HCl and poured down the drain.

(L) *Theoretical Yield Calculations:*

$$0.250 \text{ g 4-fluorobenzaldehyde used } \frac{0.250 \text{ g}}{124.11 \text{ g/mole}} = 0.00201 \text{ mole}$$

$$0.058 \text{ g of acetone used } \frac{0.058 \text{ g}}{58.08 \text{ g/mole}} = 0.001 \text{ mole}$$

The moles needed (from equation) are 2 mmol of the 4-fluorobenzaldehyde and 1 mmol of acetone. Therefore, acetone is the limiting reagent, and 1.00 mmol of product should result. The MW of the product is 270.34 g/mol.

$$0.001 \text{ mol} \times 270.34 \text{ g/mol}$$
$$= 0.270 \text{ g, theoretical yield of difluorodibenzalacetone}$$

(M) *Percent Yield of Product:*

$$\frac{0.19 \text{ g}}{0.270 \text{ g}} \times 100 = 70\%$$

(N) *The Mechanism of the Reaction:*

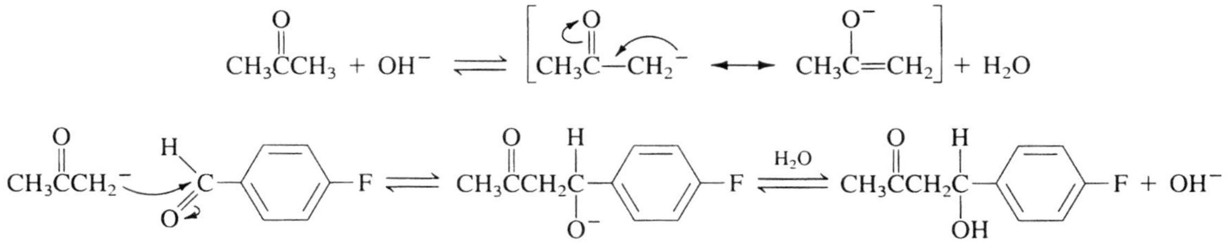

$$\underset{\substack{\nwarrow\\ \text{OH}^-}}{\text{CH}_3\overset{\text{O}}{\overset{\|}{\text{C}}}-\overset{\text{H}}{\underset{\text{H}}{\text{C}}}-\overset{\text{H}}{\underset{\text{OH}}{\text{C}}}-\bigcirc-\text{F}} \longrightarrow \text{CH}_3\overset{\text{O}}{\overset{\|}{\text{C}}}-\overset{\text{H}}{\underset{\text{H}}{\text{C}}}=\overset{\text{H}}{\underset{}{\text{C}}}-\bigcirc-\text{F} + \text{H}_2\text{O} + \text{OH}^-$$

(O) *Discussion and Conclusions:* Mixing the 4-fluorobenzaldehyde, acetone, and ethanol gave no apparent reaction because the solution did not change in appearance, but less than a minute after adding the NaOH, a yellow oil appeared and the reaction mixture got slightly warm, indicating that a reaction was taking place. The mechanism given in the text (above) indicates that hydroxide ion is a catalyst for the reaction, confirmed by this observation. The reaction takes place spontaneously at room temperature. The reaction was judged to be complete when the liquid surrounding the crystals became clear (it was opaque) and the amount of crystals in the tube did not appear to change. This took 30 min. The crude product was collected and washed with water. Forgot to cool it in ice before filtering it off. The crude product weighed more than the theoretical because it must still have been damp. However, this amount of crude material indicates that the reaction probably went pretty well to completion. The crude product was recrystallized from 5 mL of 70% ethanol. This was probably too much, since it dissolved very rapidly in that amount. This and the fact that the reaction mixture was not cooled in ice probably accounts for the relatively low 70% yield. Probably could have obtained second crop of crystals by concentrating filtrate. It turned very cloudy when water was added to it.

Theoretically, it is possible for this reaction to give three different products (*cis, cis-*; *cis, trans-*; and *trans, trans-*isomers). Because the mp of the crude and recrystallized products were close to each other and rather sharp, and because the TLC of the product gave only one spot, it is presumed that the product is just one of these three possible isomers. The NMR spectrum shows a sharp quartet of peaks from the vinyl protons, which also indicates only one isomer. Since the coupling constant observed is 7.2 Hz, the protons must be *cis* to each other. Therefore, the product is the *cis, cis-*isomer.

(P) The infrared spectrum shows two strong peaks at 1590 and 1655 cm^{-1}, indicative of an α-β-unsaturated ketone. The small peaks at 3050, 3060, and 3072 cm^{-1} are consistent with aromatic and vinyl protons. The yellow color indicates that the molecule must have a long conjugated system.

The fact that the TLC of the starting material showed only one peak is probably due to the evaporation of the acetone from the TLC plate. The spot with R_f 0.23 must be from the 4-fluorobenzaldehyde.

(Q) Answers to assigned questions are written at the end of the report.

p. 39

ⓡ	**Sept. 23, 2010 Procedure**	**Sept. 24, 2010 Observations**
	Weigh 0.248 g of 4-fluorobenzaldehyde into reaction tube.	Actually used 0.250 g.
	Add 1.6 mL of ethanol stock soln that contains 0.058 g of acetone.	Actually used 1.61 mL.
	Mix, observe for 1 min.	Nothing seems to happen. Water clear soln.
	Add 2 mL of 3 M NaOH(*aq*).	Clear, then faintly yellow but clear, then after about 50 sec, the entire soln very suddenly turned opaque. Got slightly warm. Light yellow color.
	Cap tube with septum; shake at 5-min intervals for 30 min.	Oily drops separate on sides and bottom of tube. These crystallized; more crystals formed. Tube became filled with crystals. Liquid around crystals became clear.
	Filter on Hirsch funnel.	Filtered (filter paper) on Hirsch.
	Wash crystals with much water.	Washed with about 150 mL of H_2O. Transfer of crystals done using filtrate.
	Press dry.	Crystals pressed dry with cork.
	Weigh crude.	Crude (damp) wt 0.27 g.
	Save sample for mp.	Crude mp 177.5–179°C.
	Recrystallize from 70% ethanol; wash with a few drops of ice-cold solvent.	Used about 50 mL of 70% EtOH. Too much. Dissolved very rapidly.
	Dry product.	Dried on paper for 30 min.
	Weigh.	Wt 0.19 g.
	Take mp.	Mp 179.5–180.5°C.
	Run IR as mull.	Done.
	Run NMR in $CDCl_3$.	Done.
	Neutralize first filtrate with HCl, pour down drain.	Done.
	Org. filtrate in waste organic bottle.	Done.

2 CHAPTER

Laboratory Safety, Courtesy, and Waste Disposal

> **PRELAB EXERCISE:** Read this chapter carefully. Locate the emergency eyewash station, the safety shower, the fire extinguisher, and the emergency exits in your laboratory. Check your safety glasses or goggles for size and transparency. Learn which reactions must be carried out in the hood. Learn to use your laboratory fire extinguisher; learn how to summon help and how to put out a clothing fire. Learn first aid procedures for acid and alkali spills on the skin. Learn how to tell if your laboratory hood is working properly. Learn which operations under reduced pressure require special precautions. Check to see that compressed gas cylinders in your lab are firmly fastened to benches or walls. Learn the procedures for properly disposing of solid and liquid waste in your laboratory.

Small-scale (microscale) organic experiments are much safer to conduct than their macroscale counterparts that are run on a scale up to 100 times larger. However, for either microscale or macroscale experiments, the organic chemistry laboratory is an excellent place to learn and practice safety. The commonsense procedures practiced here also apply to other laboratories as well as to the shop, kitchen, and studio.

General laboratory safety information—particularly applicable to this organic chemistry laboratory course—is presented in this chapter. But it is not comprehensive. Throughout this text you will find specific cautions and safety information presented as margin notes printed in red. For a brief and thorough discussion of the topics in this chapter, you should read *Safety in Academic Chemistry-Laboratories.*[1] There are also some specific admonitions regarding contact lenses (see "Eye Safety").

IMPORTANT GENERAL RULES

- Know the safety rules of your particular laboratory.
- Know the locations of emergency eyewashes, fire extinguishers, safety showers, and emergency exits.
- Never eat, drink, smoke, or apply cosmetics while in the laboratory.

---

[1] American Chemical Society Joint Board-Council Committee on Chemical Safety. *Safety in Academic Chemistry Laboratories, Vol. 1: Accident Prevention for College and University Students*, 7th ed.; American Chemical Society: Washington, DC, 2003 (0-8412-3864-2).

Dress sensibly.

- Wear gloves and aprons when handling corrosive materials.
- Never work alone.
- Perform no unauthorized experiments and do not distract your fellow workers; horseplay has no place in the laboratory.
- Dress properly for lab work. Do not wear open-toed shoes; your feet must be completely covered; wear shoes that have rubber soles and no heels or sneakers.
- Confine long hair and loose clothes. Do not wear shorts.
- Immediately report any accident to your instructor.
- If the fire extinguisher is used, report this to your instructor.
- Never use mouth suction to fill a pipette.
- Always wash your hands before leaving the laboratory.
- Do not use a solvent to remove a chemical from your skin. This will only hasten the absorption of the chemical through the skin.
- Do not use cell phones or tape, CD, MP3, iPods, or similar music players while working in the laboratory.
- Refer to the chemical supplier's hazard warning information or Material Safety Data Sheet (MSDS) when handling a new chemical for the first time.

Eye protection

Information on Eye Protection

Eye protection is extremely important. The requirement that you wear approved eye protection in the laboratory is mandatory and is most effective in preventing the largest number of potential accidents of serious consequence. Safety glasses or goggles of some type must be worn at all times that you are in the laboratory, whether engaged in experimental work or not. Ordinary prescription eyeglasses do not offer adequate protection.

There are three types of eye protection acceptable for use in the organic chemistry laboratory; they are described below. Laboratory safety glasses should be constructed of plastic or tempered glass. If you do not have such glasses, wear goggles that afford protection from splashes and objects coming from the side as well as from the front. Chemical splash goggles are the preferred eye protection. One of the most important features of safety glasses/goggles is the brow bar. It is critical to have proper eye protection above the eyes; a brow bar satisfies this requirement for adequate splash protection. If plastic safety glasses are permitted in your laboratory, they should have side shields (Fig. 2.1). Eye safety cannot be overemphasized in the chemistry laboratory.

1. **Goggles** are pliable, form a complete cup around the eyes, and are held in place by a strap that wraps around the head. Goggles offer the highest level of splash protection compared to other types of protective eyewear. The disadvantages are that they may fog up and limit peripheral vision. Most goggles will fit over prescription eyeglasses.
2. **Safety Glasses** look similar to prescription glasses but have side shields and a browbar to provide extra splash protection. They are generally considered to be more comfortable and to offer better peripheral vision compared to goggles. The disadvantages are that they do not fit well, if at all, over prescription eyeglasses and they offer less protection than goggles since they do not completely cover the eye.

FIG. 2.1
(a) Chemical splash goggles
(b) Safety glasses

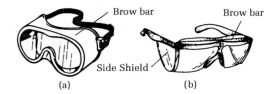

3. **Visor Goggles** are a type of "half goggle"; the upper half is similar to a goggle and the lower half is similar to safety glasses. As such, they have some of the advantages and disadvantages of both safety glasses and goggles. They are reasonably comfortable, afford good splash protection, have better peripheral vision, and show less tendency to fog than goggles. They fit over many types of prescription eyeglasses.

Contact Lens Users

In 1994, the Occupational Safety and Health Administration (OSHA) concluded that "contact lenses do not pose additional hazards to the wearer, and has determined that additional regulation addressing the use of contact lenses is unnecessary."[2] Other reports support this position.[3,4] It has been determined "that contact lenses can be worn in most work environments provided the same approved eye protection is worn as required of other workers in the area. Approved eye protection refers to safety glasses or goggles."[5]

LABORATORY COURTESY

Please show up on time and be prepared for the day's work. Clean up and leave promptly at the end of the lab period. Clean your desktop and sink before you leave the lab. Be certain that no items such as litmus paper, used filter papers, used cotton, or stir bars collect in the sink. Dispose of all trash properly. Please keep the balances clean. Always replace the caps on reagent bottles after use.

WORKING WITH FLAMMABLE SUBSTANCES

Relative flammability of organic solvents.

Flammable substances are the most common hazard in the organic laboratory. Two factors can make today's organic laboratory much safer than its predecessor: (1) making the scale of the experiments as small as possible and (2) not using flames. Diethyl ether (bp 35°C), the most flammable substance you will usually work with in this course, has an ignition temperature of 160°C, which means that a hot plate at that temperature will set it afire. For comparison, n-hexane (bp 69°C), a constituent of gasoline, has an ignition temperature of 225°C. The flash points of these organic liquids—that is, the temperatures at which they will catch fire if exposed to a flame or spark—are below 220°C. These are very flammable liquids; however, if you are

---

[2]OSHA Personal Protective Equipment for General Industry Standard, 29 CFR 1910; final rule, April 6, 1994, p. 16343.
[3]*Chemical Health and Safety* **1995**, 16.
[4]*Chemical Health and Safety* **1997**, 33.
[5]Ramsey, H.; and Breazeale, W. H. J. Jr. *Chem. Eng. News* **1998**, 76 (22), 6.

careful to eliminate all possible sources of ignition, they are not difficult to work with. Except for water, almost all of the liquids you will use in the laboratory are flammable.

Bulk solvents should be stored in and dispensed from appropriate safety containers (Fig. 2.2). These and other liquids will burn in the presence of the proper amount of their flammable vapors, oxygen, and a source of ignition (most commonly a flame or spark). It is usually difficult to remove oxygen from a fire, although it is possible to put out a fire in a beaker or a flask by simply covering the vessel with a flat object, thus cutting off the supply of air. Your lab notebook might do in an emergency. The best prevention is to pay close attention to sources of ignition—open flames, sparks, and hot surfaces. Remember, the vapors of flammable liquids are **always** heavier than air and thus will travel along bench tops, down drain troughs, and remain in sinks. For this reason all flames within the vicinity of a flammable liquid must be extinguished. Adequate ventilation is one of the best ways to prevent flammable vapors from accumulating. Work in an exhaust hood when manipulating large quantities of flammable liquids.

FIG. 2.2
Solvent safety can.

Flammable vapors travel along bench tops.

Keep ignition sources away from flammable liquids.

If a person's clothing catches fire and there is a safety shower close at hand, then shove the person under it and turn the shower on. Otherwise, push the person down and roll him or her over to extinguish the flames (stop, drop, and roll!). It is extremely important to prevent the victim from running or standing because the greatest harm comes from breathing the hot vapors that rise past the mouth. The safety shower might then be used to extinguish glowing cloth that is no longer aflame. A so-called fire blanket should not be used because it tends to funnel flames past the victim's mouth, and clothing continues to char beneath it. It is, however, useful for retaining warmth to ward off shock after the flames are extinguished.

An organic chemistry laboratory should be equipped with a carbon dioxide or dry chemical (monoammonium phosphate) *fire extinguisher* (Fig. 2.3). To use this type of extinguisher, lift it from its support, pull the ring to break the seal, raise the horn, aim it at the base of the fire, and squeeze the handle. Do not hold on to the horn because it will become extremely cold. Do not replace the extinguisher; report the incident so the extinguisher can be refilled.

FIG. 2.3
Carbon dioxide fire extinguisher.

Flammable vapors plus air in a confined space are explosive.

When disposing of certain chemicals, be alert for the possibility of *spontaneous combustion*. This may occur in oily rags; organic materials exposed to strong oxidizing agents such as nitric acid, permanganate ions, and peroxides; alkali metals such as sodium; or very finely divided metals such as zinc dust and platinum catalysts. Fires sometimes start when these chemicals are left in contact with filter paper.

WORKING WITH HAZARDOUS CHEMICALS

If you do not know the properties of a chemical you will be working with, it is wise to regard the chemical as hazardous. The flammability of organic substances poses the most serious hazard in the organic laboratory. There is a possibility that storage containers in the laboratory may contribute to a fire. Large quantities of organic solvents should not be stored in glass bottles; they should be stored in solvent safety cans. Do not store chemicals on the floor.

A flammable liquid can often be vaporized to form, with air, a mixture that is explosive in a confined space. The beginning chemist is sometimes surprised to learn that diethyl ether is more likely to cause a laboratory fire or explosion than a worker's accidental anesthesia. The chances of being confined in a laboratory with a concentration of ether high enough to cause a loss of consciousness are extremely small, but a spark in such a room would probably destroy the building.

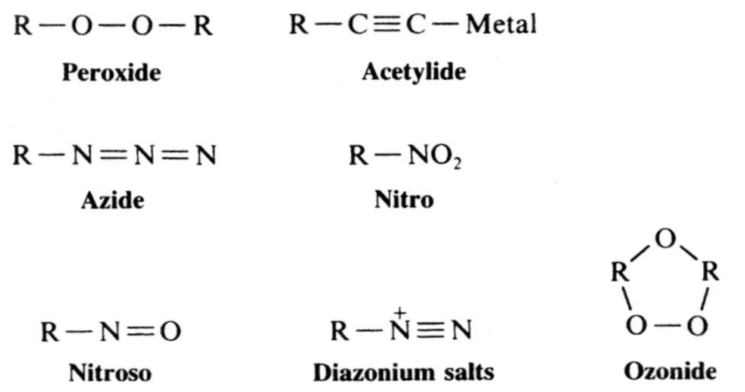

FIG. 2.4
Functional groups that can be explosive in some compounds.

R—O—O—R
Peroxide

R—C≡C—Metal
Acetylide

R—N=N=N
Azide

R—NO$_2$
Nitro

R—N=O
Nitroso

R—N$^+$≡N
Diazonium salts

$$\begin{array}{c} O \\ R' \quad 'R \\ | \quad | \\ O - O \end{array}$$
Ozonide

The probability of forming an explosive mixture of volatile organic liquids with air is far greater than that of producing an explosive solid or liquid. The chief functional groups that render compounds explosive are the *peroxide, acetylide, azide, diazonium, nitroso, nitro, and ozonide groups* (Fig. 2.4). Not all members of these groups are equally sensitive to shock or heat. You would find it difficult to detonate trinitrotoluene (TNT) in the laboratory, but nitroglycerine is treacherously explosive. Peroxides present special problems that are discussed in the next section.

You will need to contend with the corrosiveness of many of the reagents you will handle. The principal danger here is to the eyes. Proper eye protection is mandatory, and even small-scale experiments can be hazardous to the eyes. It takes only a single drop of a corrosive reagent to do permanent damage. Handling concentrated acids and alkalis, dehydrating agents, and oxidizing agents calls for commonsense care to avoid spills and splashes and to avoid breathing the often corrosive vapors.

Certain organic chemicals present acute toxicity problems from short-duration exposure and chronic toxicity problems from long-term or repeated exposure. Exposure can result from ingestion, contact with the skin, or, most commonly, inhalation. Currently, great attention is being focused on chemicals that are teratogens (chemicals that often have no effect on a pregnant woman but cause abnormalities in a fetus), mutagens (chemicals causing changes in the structure of the DNA, which can lead to mutations in offspring), and carcinogens (cancer-causing chemicals). Small-scale experiments significantly reduce these hazards but do not completely eliminate them.

Safety glasses or goggles must be worn at all times.

WORKING WITH EXPLOSIVE HAZARDS

1. Peroxides

Certain functional groups can make an organic molecule become sensitive to heat and shock, such that it will explode. Chemists work with these functional groups only when there are no good alternatives. One of these functional groups, the peroxide group (R—O—O—R), is particularly insidious because it can form spontaneously when oxygen and light are present (Fig. 2.5). Ethers, especially cyclic ethers and those made from primary or secondary alcohols (such as tetrahydrofuran, diethyl ether, and diisopropyl ether), form peroxides. Other compounds that form peroxides are aldehydes, alkenes that have allylic hydrogen atoms (such as

Ethers form explosive peroxides.

FIG. 2.5
Some compounds that form peroxides.

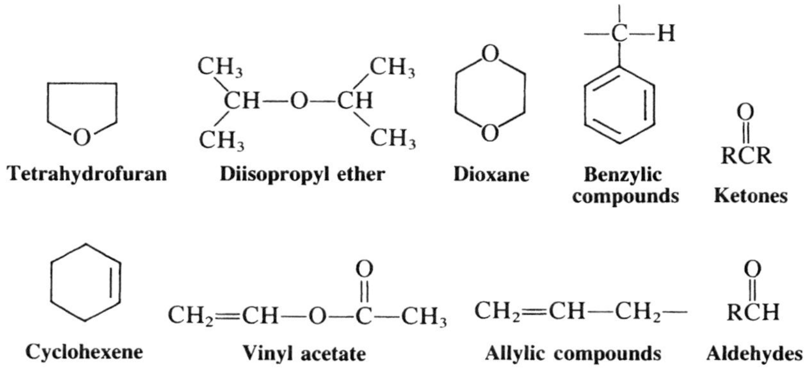

Tetrahydrofuran, **Diisopropyl ether**, **Dioxane**, **Benzylic compounds**, **Ketones** (RCR with =O)

Cyclohexene, **Vinyl acetate** (CH₂=CH—O—C(=O)—CH₃), **Allylic compounds** (CH₂=CH—CH₂—), **Aldehydes** (RCH with =O)

cyclohexene), compounds having benzylic hydrogens on a tertiary carbon atom (such as isopropyl benzene), and vinyl compounds (such as vinyl acetate). Peroxides are low-power explosives but are extremely sensitive to shock, sparks, light, heat, friction, and impact. The greatest danger from peroxide impurities comes when the peroxide-forming compound is distilled. The peroxide has a higher boiling point than the parent compound and remains in the distilling flask as a residue that can become overheated and explode. For this reason, one should never distill a liquid for too long a period of time so that the distilling flask completely dries out, and the distillation of a peroxide-containing liquid should be run in a hood with the sash down to help contain a possible explosion.

⚠ Never distill to dryness.

The Detection of Peroxides	The Removal of Peroxides
To a solution of 0.01 g of sodium iodide in 0.1 mL of glacial acetic acid, add 0.1 mL of the liquid suspected of containing a peroxide. If the mixture turns brown, a high concentration of peroxide is present; if it turns yellow, a low concentration of peroxide is present.	Pouring the solvent through a column of activated alumina will remove peroxides and simultaneously dry the solvent. Do not allow the column to dry out while in use. When the alumina column is no longer effective, wash the column with 5% aqueous ferrous sulfate and discard the column as nonhazardous waste.

Problems with peroxide formation are especially critical for ethers. Ethers (R-O-R) form peroxides readily. Because ethers are frequently used as solvents, they are often used in quantity and then removed to leave reaction products. Cans of diethyl ether should be dated when opened. If opened cans are not used within one month, they should be treated for peroxides and disposed of.

t-Butyl methyl ether, $(CH_3)_3C$-O-CH_3, with a primary carbon on one side of the oxygen and a tertiary carbon on the other, does not form peroxides easily. It is highly desirable to use this in place of diethyl ether for extraction. Refer to the discussion in Chapter 7.

You may have occasion to use 30% *hydrogen peroxide*. This material causes severe burns if it contacts the skin and decomposes violently if contaminated with metals or their salts. Be particularly careful not to contaminate the reagent bottle.

Richard Reid, the Shoe Bomber, was arrested in 2001 for attempting to set off a peroxide explosion on a trans-Atlantic air flight. He had packed 240 g of the peroxide-initiated explosive in the heel of his shoe, a highly hazardous undertaking considering the sensitivity of peroxides to any type of friction. Theoretically, he could have made his peroxide in the airplane lavatory by mixing 30% hydrogen peroxide (not easily obtained) with another common organic liquid and an acid catalyst at 0°C, allowing the mixture to react with thorough stirring and cooling for a number of hours, preferably overnight, and then isolating the crystalline peroxide by filtration. To prevent the possibility of anyone doing this in the future, no liquids can be carried on to passenger airplanes.

2. Closed Systems

A closed system is defined as not being open to the atmosphere. Any sealed system is a closed system. If a closed system is not properly prepared, an explosion may result from the system being under pressure, caused from gas or heat evolution from the reaction or from applied heat to the system. One way to prevent an explosion of a closed system is to use glassware that can withstand the pressure, and to evacuate the system under vacuum before it is closed to the atmosphere.

Most reactions are run in open systems; that is, they are run in apparatus that are open to atmosphere, either directly or through a nitrogen line hooked up to a bubbler, which is open to atmosphere.

WORKING WITH CORROSIVE SUBSTANCES

Handle strong acids, alkalis, dehydrating agents, and oxidizing agents carefully so as to avoid contact with the skin and eyes, and to avoid breathing the corrosive vapors that attack the respiratory tract. All strong, concentrated acids attack the skin and eyes. Concentrated *sulfuric acid* is both a dehydrating agent and a strong acid and will cause very severe burns. *Nitric acid and chromic acid* (used in cleaning solutions) also cause bad burns. *Hydrofluoric* acid is especially harmful and causes deep, painful, and slow-healing wounds. It should be used only after thorough instruction. Do not add water to concentrated sulphuric acid. The heat of solution is so large that the acid may boil and spatter. It can best be diluted by pouring the acid into water, with stirring. You should wear approved safety glasses or goggles, protective gloves, and an apron when handling these materials.

Sodium, potassium, and ammonium *hydroxides* are common bases that you will encounter. Sodium and potassium hydroxides are extremely damaging to the eyes, and ammonium hydroxide is a severe bronchial irritant. Like sulfuric acid, sodium hydroxide, *phosphorous pentoxide*, and *calcium oxide* are powerful dehydrating agents. Their great affinity for water will cause burns to the skin. Because they release a great deal of heat when they react with water, to avoid spattering they should always be added to water rather than water being added to them. That is, the heavier substance should always be added to the lighter one so that layers don't form where one liquid floats on another; the desired result is a rapid mixing that occurs as a consequence of the law of gravity.

⚠ Add H_2SO_4, P_2O_5, CaO, and NaOH to water, not the reverse.

You will receive special instructions when it comes time to handle metallic sodium, lithium aluminum hydride, and sodium hydride, three substances that can react explosively with water.

Wipe up spilled hydroxide pellets rapidly.

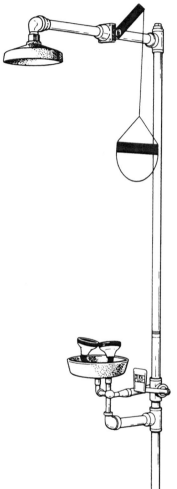

FIG. 2.6

Emergency shower and eyewash station.

Among the strong oxidizing agents, *perchloric acid* ($HClO_4$) is probably the most hazardous. It can form heavy metal and organic *perchlorates* that are *explosive,* and it can react explosively if it comes in contact with organic compounds.

If one of these substances gets on the skin or in the eyes, wash the affected area with very large quantities of water, using the safety shower and/or eyewash station (Fig. 2.6) until medical assistance arrives. Do not attempt to neutralize the reagent chemically. Remove contaminated clothing so that thorough washing can take place. Take care to wash the reagent from under the fingernails.

Take care not to let the reagents, such as sulfuric acid, run down the outside of a bottle or flask and come in contact with your fingers. Wipe up spills immediately with a very damp sponge, especially in the area around the balances. Pellets of sodium and potassium hydroxide are very hygroscopic and will dissolve in the water they pick up from the air; they should therefore be wiped up very quickly. When handling large quantities of corrosive chemicals, wear protective gloves, a face mask, and a neoprene apron. The corrosive vapors can be avoided by carrying out work in a good exhaust hood.

Do not use a plastic syringe with a metal needle to dispense corrosive inorganic reagents, such as concentrated acids or bases.

WORKING WITH TOXIC SUBSTANCES

Many chemicals have very specific toxic effects. They interfere with the body's metabolism in a known way. For example, the cyanide ion combines irreversibly with hemoglobin to form cyanometmyoglobin, which can no longer carry oxygen. Aniline acts in the same way. Carbon tetrachloride and other halogenated compounds can cause liver and kidney failure. Carcinogenic and mutagenic substances deserve special attention because of their long-term insidious effects. The ability of certain carcinogens to cause cancer is very great; for example, special precautions are needed when handling aflatoxin B_1. In other cases, such as with dioxane, the hazard is so low that no special precautions are needed beyond reasonable, normal care in the laboratory.

Women of childbearing age should be careful when handling any substance of unknown properties. Certain substances are highly suspected as teratogens and will cause abnormalities in an embryo or fetus. Among these are benzene, toluene, xylene, aniline, nitrobenzene, phenol, formaldehyde, dimethylformamide (DMF), dimethyl sulfoxide (DMSO), polychlorinated biphenyls (PCBs), estradiol, hydrogen sulfide, carbon disulfide, carbon monoxide, nitrites, nitrous oxide, organolead and mercury compounds, and the notorious sedative thalidomide. Some of these substances will be used in subsequent experiments. Use care when working with these (and all) substances. One of the leading known causes of embryotoxic effects is ethyl alcohol in the form of maternal alcoholism, but the amount of ethanol vapor inhaled in the laboratory or absorbed through the skin is so minute that it is unlikely to have morbid effects.

It is impossible to avoid handling every known or suspected toxic substance, so it is wise to know what measures should be taken. Because the eating of food or the consumption of beverages in the laboratory is strictly forbidden and because one should never taste material in the laboratory, the possibility of poisoning by mouth is remote. Be more careful than your predecessors—the hallucinogenic properties of LSD and **all** artificial sweeteners were discovered by accident. The two most important measures to be taken, then, are (1) avoiding skin contact by wearing the *proper* type of protective gloves (see "Gloves") and (2) avoiding inhalation by working in a good exhaust hood.

Many of the chemicals used in this course will be unfamiliar to you. Their properties can be looked up in reference books, a very useful one being the *Aldrich Handbook of Fine Chemicals*.[6] Note that 1,4-dichlorobenzene is listed as a "toxic irritant" and naphthalene is listed as an "irritant." Both are used as mothballs. Camphor, used in vaporizers, is classified as a "flammable solid irritant." Salicylic acid, which we will use to synthesize aspirin (Chapter 41), is listed as a "moisture-sensitive toxic." Aspirin (acetylsalicylic acid) is classified as an "irritant." Caffeine, which we will isolate from tea or cola syrup (Chapter 7), is classified as "toxic." Substances not so familiar to you, for example, 1-naphthol and benzoic acid, are classified, respectively, as "toxic irritant" and "irritant." To put things in perspective, nicotine is classified as "highly toxic." Pay attention to these health warnings. In laboratory quantities, common chemicals can be hazardous. Wash your hands carefully after coming in contact with laboratory chemicals. Consult the *Hazardous Laboratory Chemicals Disposal Guide*[7] for information on truly hazardous chemicals.

Because you have not had previous experience working with organic chemicals, most of the experiments you will carry out in this course will not involve the use of known carcinogens, although you will work routinely with flammable, corrosive, and toxic substances. A few experiments involve the use of substances that are suspected of being carcinogenic, such as hydrazine. If you pay proper attention to the rules of safety, you should find working with these substances no more hazardous than working with ammonia or nitric acid. The single, short-duration exposure you might receive from a suspected carcinogen, should an accident occur, would probably have no long-term consequences. The reason for taking the precautions noted in each experiment is to learn, from the beginning, good safety habits.

GLOVES

Be aware that protective gloves in the organic laboratory may not offer much protection. Polyethylene and latex rubber gloves are very permeable to many organic liquids. An undetected pinhole may bring with it long-term contact with reagents. Disposable polyvinyl chloride (PVC) gloves offer reasonable protection from contact with aqueous solutions of acids, bases, and dyes, but no one type of glove is useful as a protection against all reagents. It is for this reason that no less than 25 different types of chemically resistant gloves are available from laboratory supply houses. Some gloves are quite expensive and will last for years.

If disposable gloves are available, fresh nitrile gloves can be worn whenever handling a corrosive substance and disposed of once the transfer is complete. When not wearing gloves, it is advised that you wash your hands every 15 minutes to remove any traces of chemicals that might be on them.

USING THE LABORATORY HOOD

Modern practice dictates that in laboratories where workers spend most of their time working with chemicals, there should be one exhaust hood for every two people. However, this precaution is often not possible in the beginning organic chemistry laboratory. In this course you will find that for some experiments, the hood

---

[6]Free copies of this catalog can be obtained from http://www.sigmaaldrich.com/Brands/Aldrich.html.
[7]Armour, M-A. *Hazardous Laboratory Chemicals Disposal Guide*, 3rd ed.; CRC Press LLC: Boca Raton, FL, 2003. This extremely useful book is now available without charge on Google Books.

must be used and for others it is advisable; in these instances, it may be necessary to schedule experimental work around access to the hoods. Many experiments formerly carried out in the hood can now be carried out at the lab desk because the concentration of vapors is significantly minimized when working at a microscale.

Keep the hood sash closed.

The hood offers a number of advantages when working with toxic and flammable substances. Not only does it draw off the toxic and flammable fumes, but it also affords an excellent physical barrier on all four sides of a reacting system when the sash is pulled down. If a chemical spill occurs, it may be contained within the hood.

It is your responsibility each time you use a hood to see that it is working properly. You should find some type of indicating device that will give you this information on the hood itself. A simple propeller on a cork works well. Note that the hood is a backup device. Never use it alone to dispose of chemicals by evaporation; use an aspirator tube or carry out a distillation. Toxic and flammable fumes should be trapped or condensed in some way and disposed of in the prescribed manner. The sash should be pulled down unless you are actually carrying out manipulations with the experimental apparatus. The water, gas, and electrical controls should be on the outside of the hood, so it is not necessary to open the hood to make adjustments. The ability of the hood to remove vapors is greatly enhanced if the apparatus is kept as close to the back of the hood as possible, where the air movement is the strongest. Everything should be at least 15 cm back from the hood sash. Chemicals should not be permanently stored in the hood, but should be removed to ventilated storage areas. If the hood is cluttered with chemicals, you will not achieve a good, smooth airflow or have adequate room for experiments.

WORKING AT REDUCED PRESSURE

Implosion

Whenever a vessel or system is evacuated, an implosion could result from atmospheric pressure on the empty vessel. It makes little difference whether it's a total vacuum or just 10 mm Hg; the pressure difference is almost the same (760 versus 750 mm Hg). An implosion may occur if there is a star crack in the flask or if the flask is scratched or etched. Only with heavy-walled flasks specifically designed for vacuum filtration is the use of a safety shield (Fig. 2.7) ordinarily unnecessary. Although caution should still be observed, the chances of implosion of the apparatus used for microscale experiments are remote.

Dewar flasks (thermos bottles) are often found in the laboratory without shielding. These should be wrapped with friction tape or covered with a plastic net to prevent the glass from flying about in case of an implosion (Fig. 2.8). Similarly, vacuum desiccators should be wrapped with tape before being evacuated.

FIG. 2.7
Safety shield.

FIG. 2.8
Dewar flask with safety net in place.

WORKING WITH COMPRESSED GAS CYLINDERS

Many reactions are carried out under an inert atmosphere so that the reactants and/or products will not react with oxygen or moisture in the air. Nitrogen and argon are the inert gases most frequently used. Oxygen is widely used both as a reactant and to provide a hot flame for glassblowing and welding. It is used in the oxidative coupling of alkynes (Chapter 24). Helium is the carrier gas used in gas chromatography. Other gases commonly used in the laboratory are ammonia, which is often used as a solvent; chlorine, used for chlorination reactions; acetylene, used in combination with oxygen for welding; and hydrogen, used for high- and low-pressure hydrogenation reactions.

Always clamp gas cylinders

The following rule applies to all compressed gases: Compressed gas cylinders should be firmly secured at all times. For temporary use, a clamp that attaches to the laboratory bench top and has a belt for the cylinder will suffice (Fig. 2.9). Eyebolts and chains should be used to secure cylinders in permanent installations. Flammable gases should be stored 20 feet from oxidizing gases.

A variety of outlet threads are used on gas cylinders to prevent incompatible gases from being mixed because of an interchange of connections. Both right- and left-handed external and internal threads are used. Left-handed nuts are notched to differentiate them from right-handed nuts. Right-handed threads are used on nonfuel and oxidizing gases, and left-handed threads are used on fuel gases, such as hydrogen. Never grease the threads on tank or regulator valves because there is the possibility that the grease could ignite under certain conditions.

Cylinders come equipped with caps that should be left in place during storage and transportation. These caps can be removed by hand. Under these caps is a cylinder valve. It can be opened by turning the valve counterclockwise. However, because most compressed gases in full cylinders are under very high pressure (commonly up to 3000 lb/in.^{2}), a pressure regulator must be attached to the cylinder. This pressure regulator is almost always of the diaphragm type and has two gauges—one indicating the pressure in the cylinder, the other the outlet pressure (Fig. 2.10). On the outlet, low-pressure side of the regulator is a small needle valve and then the outlet connector. After connecting the regulator to the cylinder, unscrew the diaphragm valve (turn it counterclockwise) before

Clockwise **movement of the diaphragm valve handle *increases* pressure.**

FIG. 2.9
Gas cylinder clamp.

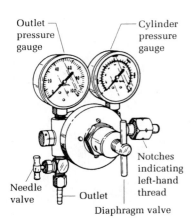

FIG. 2.10
Gas pressure regulator. Turn the two-flanged diaphragm valve clockwise to increase outlet pressure.

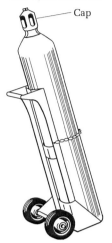

FIG. 2.11
Gas cylinder cart.
Cap

opening the cylinder valve on the top of the cylinder. This valve should be opened only as far as necessary. For most gas flow rates in the laboratory, this will be a very small amount. The gas flow or pressure is increased by turning the two-flanged diaphragm valve **clockwise**. When the gas is not being used, turn off the cylinder valve (clockwise) on the top of the cylinder (Fig. 2.9). Before removing the regulator from the cylinder, reduce the flow or pressure to zero. Cylinders should never be emptied to zero pressure and left with the valve open because the residual contents will become contaminated with air. Empty cylinders should be labeled "empty." Their valves should be closed and capped, and the cylinders should be returned to the storage area and separated from full cylinders. Gas cylinders should never be dragged or rolled from place to place, but should be fastened onto and moved in a cart designed for that purpose (Fig. 2.11). The cap should be in place whenever the cylinder is moved. If you detect even a small leak from any valve or connection, immediately seek the help of an instructor to remedy the problem. If there is a major leak of a corrosive or flammable gas, notify those around you to leave the area and seek help immediately.

ODORIFEROUS CHEMICALS

Never attempt to identify an unknown organic compound by smelling it.

Clean up spills rapidly.

Some organic chemicals just smell bad. Among these are the thiols (organic derivatives of hydrogen sulfide, also called mercaptans), isonitriles, many amines (e.g., cadaverine), and butyric acid. Washing apparatus and, if necessary, hands in a solution of a quaternary ammonium salt may solve the problem. These compounds apparently complex with many odoriferous substances, allowing them to be rinsed away. Commercial products (e.g., Zephiran, Roccal, San-O-Fec, and others) are available at pet and farm supply stores.

WASTE DISPOSAL—CLEANING UP

Spilled solids should simply be swept up and placed in the appropriate solid waste container. This should be done promptly because many solids are hygroscopic and become difficult if not impossible to sweep up in a short time. This is particularly true of sodium hydroxide and potassium hydroxide; these strong bases should be dissolved in water and neutralized with sodium bisulfate before disposal.

The method used to clean up spills depends on the type and amount of chemical spilled. If more than 1 or 2 g or mL of any chemical, particularly a corrosive or volatile one, is spilled, you should consult your instructor for the best way to clean up the spill. If a large amount of volatile or noxious liquid is spilled as might happen if a reagent bottle is dropped and broken, advise those in the area to leave the laboratory and contact your instructor immediately. If a spill involves a large amount of flammable liquid, be aware of any potential ignition sources and try to eliminate them. Large amounts of spilled acid can be neutralized with granular limestone or cement; large amounts of bases can be neutralized with solid sodium bisulfate, $NaHSO_4$. Large amounts of volatile liquids can be absorbed into materials such as vermiculite, diatomaceous earth, dry sand, kitty litter, or paper towels and these materials swept up and placed in a separate disposal container.

Mercury requires special measures—see your instructor

For spills of amounts less than 2 g of chemical, proceed as follows. Acid spills should be neutralized by dropping solid sodium carbonate onto them, testing the pH, wiping up the neutralized material with a sponge, and rinsing the neutral salt solution down the drain. Bases should be neutralized by sprinkling solid sodium bisulfate onto them, checking the pH, and wiping up with a sponge or towel. Do not use paper towels to wipe up spills of strong oxidizers such as dichromates or nitrates; the towels can ignite. Bits of sodium metal will also cause paper towels to ignite. Sodium metal is best destroyed with n-butyl alcohol. Always wear gloves when cleaning up a spill.

Cleaning Up. In the not-too-distant past, it was common practice to wash all unwanted liquids from the organic laboratory down the drain and to place all solid waste in the trash basket. For environmental reasons, this is never a wise practice and is no longer allowed by law.

Organic reactions usually employ a solvent and often involve the use of a strong acid, a strong base, an oxidant, a reductant, or a catalyst. None of these should be washed down the drain or placed in the wastebasket. Place the material, classified as waste, in containers labeled for nonhazardous solid waste, organic solvents, halogenated organic solvents, or hazardous wastes of various types.

Waste containers:
   Nonhazardous solid waste
   Organic solvents
   Halogenated organic solvents
   Hazardous waste (various types)

Nonhazardous waste encompasses such solids as paper, corks, TLC plates, solid chromatographic absorbents such as alumina or silica that are dry and free of residual *organic solvents*, and solid drying agents such as calcium chloride or sodium sulfate that are also dry and free of residual organic solvents. These will ultimately end up in a sanitary landfill (the local dump). Any chemicals that are leached by rainwater from this landfill must not be harmful to the environment. In the organic solvents container are placed the solvents that are used for recrystallization and for running reactions, cleaning apparatus, and so forth. These solvents can contain dissolved, solid, nonhazardous organic solids. This solution will go to an incinerator where it will be burned. If the solvent is halogenated (e.g., dichloromethane) or contains halogenated material, it must go in the *halogenated organic solvents* container. Ultimately, this will go to a special incinerator equipped with a scrubber to remove HCl from the combustion gases. The organic laboratory should also have several other waste disposal containers for special hazardous, reactive, and noncombustible wastes that would be incompatible with waste organic solvents and other materials. For example, it would be dangerous to place oxidants in lysts with many organics. In particular, separate waste containers should be provided for toxic heavy metal wastes containing mercury, chromium, or lead salts, and so forth. The cleaning up sections throughout this text will call your attention to these special hazards.

Hazardous wastes such as sodium hydrosulfite (a reducing agent), platinum catalysts, and Cr^{6+} an oxidizing agent) cannot be burned and must be shipped to a secure landfill. To dispose of small quantities of a hazardous waste (e.g., solid mercury hydroxide), the material must be carefully packed in bottles and placed in a 55-gal ($\approx$-208-L) drum called a lab pack, to which an inert material has been added. The lab pack is carefully documented and hauled off to a site where such waste is disposed of by a bonded, licensed, and heavily regulated waste disposal company. Formerly, many hazardous wastes were disposed of by burial in a secure landfill. The kinds of hazardous waste that can be thus disposed of have become extremely limited in recent years, and much of the waste undergoes various kinds of treatment at the disposal site (e.g., neutralization, incineration, or reduction) to put it in a form that can be safely buried in a secure landfill or flushed to a sewer. There are relatively few places for approved disposal of hazardous waste. For example, there are none in New England, so most hazardous waste from this area is trucked to South Carolina. The charge to small generators of waste is usually based on the volume of waste. So, 1000 mL of a 2% cyanide solution would cost far more to dispose of than 20 g of solid cyanide, even though the total amount of this poisonous substance is the same. It now costs far more to dispose of most hazardous chemicals than it does to purchase them new.

Waste disposal is very expensive.

American law states that a material is not a waste until the laboratory worker declares it a waste. So—for pedagogical and practical reasons—we want you to regard the chemical treatment of the byproducts of each reaction in this text as a part of the experiment.

The law: A waste is not a waste until the laboratory worker declares it a waste.

In the section titled "Cleaning Up" at the end of each experiment, the goal is to reduce the volume of hazardous waste, to convert hazardous waste to less hazardous waste, or to convert it to nonhazardous waste. The simplest example is concentrated sulfuric acid. As a byproduct from a reaction, it is obviously hazardous. But after careful dilution with water and neutralization with sodium carbonate, the sulfuric acid becomes a dilute solution of sodium sulfate, which in almost every locale can be flushed down the drain with a large quantity of water. Anything flushed down the drain must be accompanied by a large excess of water. Similarly, concentrated bases can be neutralized, oxidants such as Cr^{6+} can be reduced, and reductants such as hydrosulfite can be oxidized (by hypochlorite or household bleach). Dilute solutions of heavy metal ions can be precipitated as their insoluble sulfides or hydroxides. The precipitate may still be a hazardous waste, but it will have a much smaller volume.

Cleaning up: reducing the volume of hazardous waste or converting hazardous waste to less hazardous or nonhazardous waste.

One type of hazardous waste is unique: a harmless solid that is damp with an organic solvent. Alumina from a chromatography column and calcium chloride used to dry an ether solution are examples. Being solids, they obviously cannot go in the organic solvents container, and being flammable they cannot go in the nonhazardous waste container. A solution to this problem is to spread the solid out in the hood to let the solvent evaporate. You can then place the solid in the nonhazardous waste container. The savings in waste disposal costs by this operation are enormous. However, be aware of the regulations in your area, as they may not allow evaporation of small amounts of organic solvents in a hood. If this is the case, special containers should be available for disposal of these wet solids.

Disposing of solids wet with organic solvents: alumina and anhydrous calcium chloride pellets.

Our goal in "Cleaning Up" is to make you more aware of all aspects of an experiment. Waste disposal is now an extremely important aspect. Check to be sure the procedure you use is permitted by law in your location. Three sources of information have been used as the basis of the procedures at the end of each experiment: the *Aldrich Catalog Handbook of Fine Chemicals*,[8] which gives brief disposal procedures for every chemical in their catalog; *Prudent Practices in the Laboratory: Handling and Disposal of Chemicals*[9]; and the *Hazardous Laboratory Chemicals Disposal Guide*.[10] The last title listed here should be on the bookshelf of every laboratory. This 464-page book gives detailed information about hundreds of hazardous substances, including their physical properties, hazardous reactions, physiological properties, health hazards, spillage disposal, and waste disposal. Many of the treatment procedures in "Cleaning Up" are adaptations of these procedures. *Destruction of Hazardous Chemicals in the Laboratory*[11] complements this book.

---

[8] See footnote 6 on page 34.
[9] National Research Council. *Prudent Practices in the Laboratory: Handling and Disposal of Chemicals* National Academy Press: Washington, DC, 1995.
[10] See footnote 7 on page 34.
[11] Lunn, G.; Sansone, E. B. *Destruction of Hazardous Chemicals in the Laboratory*; Wiley: New York, 1994.

The area of waste disposal is changing rapidly. Many levels of laws apply—local, state, and federal. What may be permissible to wash down the drain or evaporate in the hood in one jurisdiction may be illegal in another, so before carrying out any waste disposal, check with your college or university waste disposal officer.

BIOHAZARDS

The use of microbial growth bioassays is becoming common in chemistry laboratories. The use of infectious materials presents new hazards that must be recognized and addressed. The first step in reducing hazards when using these materials is to select infectious materials that are known not to cause illness in humans and are of minimal hazard to the environment. A number of procedures should be followed to ensure safety when these materials are used: Individuals need to wash their hands after they handle these materials and before they leave the laboratory; work surfaces need to be decontaminated at the end of each use; and all infectious materials need to be decontaminated before disposal.

QUESTIONS

1. Write a balanced equation for the reaction between the iodide ion, a peroxide, and the hydrogen ion. What causes the orange or brown color?
2. Why does the horn of the carbon dioxide fire extinguisher become cold when the extinguisher is used?
3. Why is water not used to extinguish most fires in an organic laboratory? Consider the following scenario. An organic chemistry student is setting up an experiment on the benchtop, just outside the hood. He is wearing gloves as he dispenses diethyl ether, a volatile, flammable solvent. By accident he spills some ether on his hand but continues on with the experiment. The experiment is complete, and he cleans up. He puts the ether down the drain and washes the glassware. He then takes off his gloves and washes his hands. List all of the safety issues (if any) with this scenario.

CHAPTER 3

Melting Points and Boiling Points

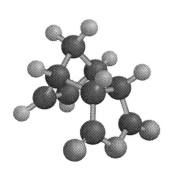

When you see this icon, sign in at this book's premium website at www.cengage.com/login to access videos, Pre-Lab Exercises and other online resources.

PRELAB EXERCISE: Draw the organic compounds, identify the intermolecular attractive forces for each, and list them in order of increasing boiling point as predicted by the relative strength of those intermolecular forces: (a) butane, (b) *i*-butanol, (c) potassium acetate, (d) acetone.

PART 1: FIVE CONCEPTS FOR PREDICTING PHYSICAL PROPERTIES

In organic chemistry, structure is everything. A molecule's structure determines both its physical properties and its reactivity. Since the dawn of modern chemistry 200 years ago, over 20 million substances, most of them organic compounds, have been isolated, and their properties and reactions have been studied. It became apparent from these studies that certain structural features in organic molecules would affect the observed properties in a predictable way and that these millions of organic compounds could be organized into classes based on molecular size, composition, and the pattern of bonds between atoms. Chemists also saw trends in certain properties based on systematic changes in these structural features. This organized knowledge allows us to look at a compound's structure and to predict the physical properties of that compound.

Physical properties, such as melting point, boiling point, and solubility, are largely determined by *intermolecular attractive forces*. You learned about these properties in previous chemistry courses. Because a solid understanding of these concepts is critical to understanding organic chemistry, we will review the different types of forces in the context of structural organic chemistry. Using five simple concepts, you should be able to look at the structures of a group of different organic molecules and predict which might be liquids, gases, or solids and which might be soluble in water. You can often predict the boiling point, melting point, or solubility of one structure relative to other structures. In fact, as your knowledge grows, you may be able to predict a compound's approximate melting or boiling temperature based on its structure. Your understanding of intermolecular attractive forces will be very useful in this chapter's experiments on melting and boiling points, and those in Chapters 5 and 6 that involve distillation and boiling points.

1. LONDON ATTRACTIVE FORCES (OFTEN CALLED VAN DER WAALS FORCES)

Organic molecules that contain only carbon and hydrogen (hydrocarbons) are weakly attracted to each other by London forces. Though weak, these attractive forces increase as molecular size increases. Thus, the larger the molecule, the greater the attractive force for neighboring molecules and the greater the energy required to get two molecules to move apart. This trend can be seen if we compare the melting points and boiling points of three hydrocarbons of different size: methane, hexane, and tetracosane.

CH_4
Methane
mp −182°C
bp −162°C
Gas at room temp.

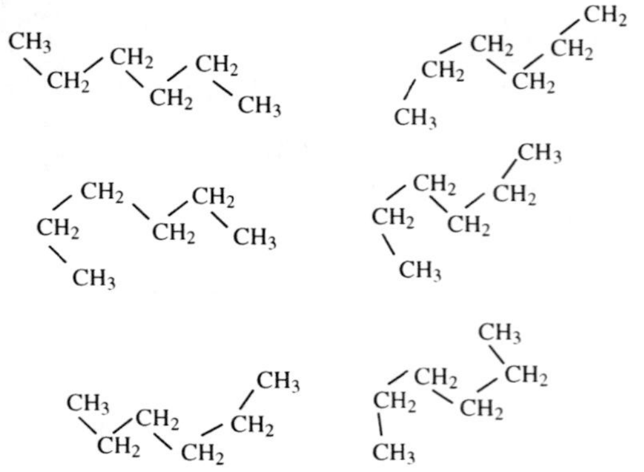

Hexane, C_6H_{14}
mp −95°C
bp +69°C
Liquid at room temp.

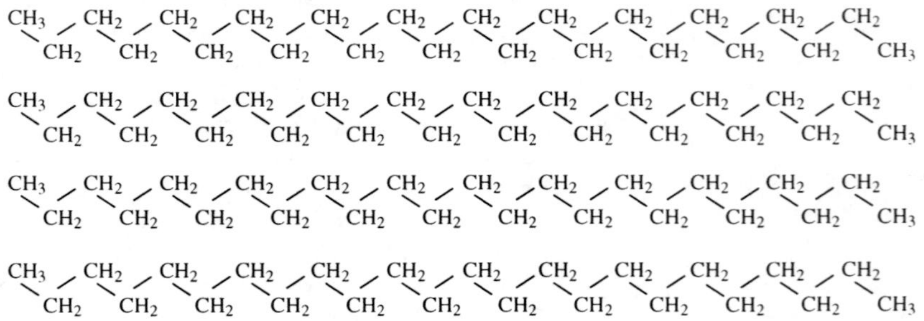

Tetracosane, $C_{24}H_{50}$
mp +51°C
bp +391°C
Solid at room temp.

We know that methane is called natural gas because methane's physical state at room temperature (20°C = 68°F) is a gas. Its London forces are so weak that methane must be cooled to −162°C at 1 atm of pressure before the molecules will stick together enough to form a liquid. Hexane is a very common liquid solvent found on most organic laboratory shelves. The intermolecular forces between its molecules are strong enough to keep them from flying apart, but the molecules are still able to flex and slide by each other to form a fluid. Hexane must be heated above 69°C, which is 231°C hotter than methane, to convert all its molecules to a gas. Tetracosane, a C_{24} solid hydrocarbon, is four times larger than hexane, and its London forces are strong enough to hold the molecules rigidly in place at room temperature. Tetracosane is one of the many long-chain hydrocarbons found in candle wax, which must be heated in order to disrupt the intermolecular forces and melt the wax into a liquid. A lot of energy is required to convert liquid tetracosane to a gas, as evidenced by its extremely high boiling point (391°C).

2. DIPOLE-DIPOLE ATTRACTIVE FORCES

The attractive forces between molecules increases when functional groups containing electronegative atoms such as chlorine, oxygen, and nitrogen are present because these atoms are more electronegative than carbon. These atoms pull electrons toward themselves, making their end of the bond slightly negatively charged (δ^-) and leaving the carbon slightly positively charged (δ^+), as shown for isopropyl chloride and acetone.

Isobutane
mp –137°C
bp 0°C
Gas at room temp.

Isopropyl chloride
mp –117°C
bp +35°C
Liquid at room temp.

Acetone
mp –94°C
bp +56°C
Liquid at room temp.

A bond with a slight charge separation is termed a *polar* bond, and polar atomic bonds often give a molecule a *dipole*: slightly positive and negative ends symbolized by an arrow in the direction of the negative charge (+→). Attraction of the positive end of one molecule's dipole to the negative end of another's dipole occurs between polar molecules, which increases the intermolecular attractive force. Dipole-dipole attractive forces are stronger than London forces, as demonstrated in the previous examples that show an increase in melting point and boiling point when a methyl group of isobutane is replaced by chlorine or oxygen.

3. HYDROGEN BONDING

Hydrogen bonding is an even stronger intermolecular attractive force, as evidenced by the large increase in the melting point and boiling points of the alcohol methanol (MW = 32) relative to those of the hydrocarbon ethane (MW = 30), both of comparable molecular weight. Hydrogen bonding occurs with organic molecules containing O—H groups (for example, alcohols and carboxylic acids) or N—H groups (for example, amines or amides). The hydrogen in these groups is attracted to the unshared pair of electrons on the O or N of another molecule, forming a hydrogen bond, often symbolized by a dashed line, which is shown for methanol.

CH₃—CH₃
Ethane
mp –172°C
bp –88°C
Gas at room temp.

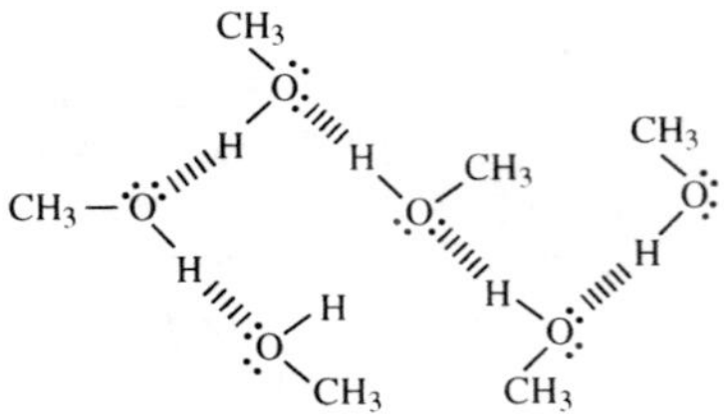

Methanol
mp –97°C
bp +65°C
Liquid at room temp.

As this example indicates, the hydrogen bonds extend throughout the liquid. One can think of these hydrogen bonds as molecular Velcro that can be pulled apart if there is sufficient energy. Hydrogen bonding plays a major role in the special physical behavior of water and is a major determinant of the chemistry of proteins and nucleic acids in living systems.

4. IONIC ATTRACTIVE FORCES

Recall that ionic substances, such as table salt (NaCl), are usually solids with high melting points (>300°C) due to the strong attractive forces between positive and negative ions. Most organic molecules contain only covalent bonds and have no ionic attractive forces between them. However, there are three important exceptions involving acidic or basic functional groups that can form ionic structures as the pH is raised or lowered.

1. The hydrogen on the —OH of the carboxyl group in carboxylic acids, such as acetic acid, is acidic (H⁺ donating) and reacts with bases such as potassium hydroxide (KOH) and sodium bicarbonate ($NaHCO_3$) to form salts. The process is reversed by adding an acid to lower the pH.

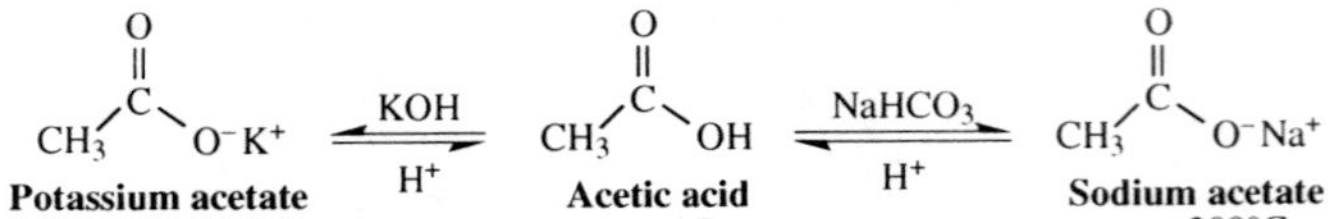

Potassium acetate
mp 306°C
⇌ KOH / H⁺
Acetic acid
mp 17°C
⇌ $NaHCO_3$ / H⁺
Sodium acetate
mp >300°C

The dry salts are ionic and have very high melting points, which is expected for ionic substances. Note that this acidity is *not* observed for alcohols where the —OH group is attached to a singly bonded (sp^3 or saturated) carbon.

2. The hydrogen on an —OH group that is attached to an aromatic ring is weakly acidic and reacts with strong bases such as sodium hydroxide (NaOH) to form high melting ionic salts, as evidenced by the reaction of phenol to sodium phenolate. Again, the reaction is reversed by the addition of an acid.

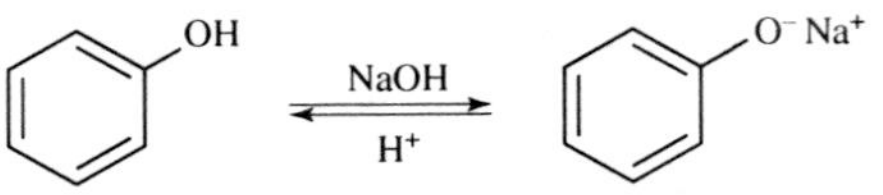

Phenol
mp 41°C
⇌ NaOH / H⁺
Sodium phenolate
mp 382°C

Note again that this acidity is *not observed* for alcohols where the —OH group is attached to a singly bonded (sp³ or saturated) carbon.

3. Amines (but not amides) are basic (H⁺ accepting) and will react with acid to form ionic amine salts with elevated melting points, as shown, for example, for isopropyl amine.

$$\begin{array}{c} CH_3 \\ \diagdown \\ CH-NH_2 \\ \diagup \\ CH_3 \end{array} \underset{\text{base}}{\overset{\text{HCl}}{\rightleftarrows}} \begin{array}{c} CH_3 \\ \diagdown \\ CH-NH_3^+ \; Cl^- \\ \diagup \\ CH_3 \end{array}$$

Isopropyl amine
mp –95°C

Isopropyl amine hydrochloride
mp 162°C

Amine salts can be converted back to amines by adding a base to raise the pH.

5. COMPETING INTERMOLECULAR FORCES AND SOLUBILITY

For pure compounds containing identical molecules, the total attractive force between molecules is the sum of all the attractive forces listed previously, both weak and strong. These forces tend to work together to raise melting and boiling points as the size of the molecule's hydrocarbon skeleton increases and as polar, hydrogen bonded, or ionic functional groups are incorporated into the molecule. However, solubility involves the interaction of two different molecules, which may have different types of attractive forces. When we try to dissolve one substance in another, we have to disrupt the attractive forces in both substances to get the molecules of the two substances to intermingle. For example, to get water (polar with hydrogen bonding) and the hydrocarbons (nonpolar and no hydrogen bonding) in motor oil to mix and dissolve in one another, we would have to disrupt the London attractive forces between the oil molecules and the hydrogen bonds between the water molecules. Because London forces are weak, separating the oil molecules does not require much energy. However, breaking apart the much stronger hydrogen bonds by inserting oil molecules between the water molecules requires considerable energy and is unfavorable. Therefore, oil or even the simplest hydrocarbon, methane, is insoluble in water. This is the molecular basis of the old adage "Oil and water don't mix."

In addition to carbon and hydrogen, the majority of organic molecules contain other elements, such as nitrogen and oxygen, in functional groups that can be polar, exhibit hydrogen bonding, show ionic tendencies, or have any combination thereof. Can we predict the water solubility of these organic substances? Let's look at some examples and see.

Figure 3.1 shows a collection of small organic molecules of about the same size and molecular weight, listed in order of increasing boiling point or melting point, which is consistent with the types of intermolecular forces discussed. With the exception of the hydrocarbon butane, all of these substances are very soluble in water—at least 100 g will dissolve in 1 L of water. It appears that the intermolecular attractive forces between these polar, hydrogen bonded, or ionic molecules and water compensates for the disruption of hydrogen bonding between water molecules so the organic molecules can move into and intermingle with the water molecules—in other words, dissolve.

Figure 3.2 shows a collection of larger organic molecules than those in Figure 3.1, again listed in order of increasing melting and boiling point, which is consistent with our knowledge of the strength of intermolecular attractive forces. The important difference for this group is that the hydrocarbon portion of each molecule is

FIG. 3.1
Some small organic molecules containing 4 atoms (carbon, oxygen, and nitrogen) in order of increasing melting or boiling points.

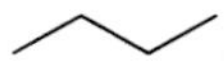

C$_4$ Hydrocarbon, butane
mp −138°C; bp 0°C
Insoluble in H$_2$O

C$_3$ Amine
mp −43°C; bp 48°C
Soluble in H$_2$O; Soluble in organic solvents

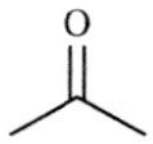

C$_3$ Ketone, acetone
mp −94°C; bp 56°C
Soluble in H$_2$O; Soluble in organic solvents

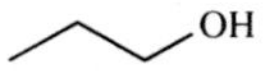

C$_3$ Alcohol
mp −127°C; bp 97°C
Soluble in H$_2$O; Soluble in organic solvents

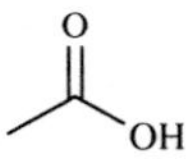

C$_2$ Carboxylic acid, acetic acid
mp 17°C; bp 117°C
Soluble in H$_2$O; Soluble in organic solvents

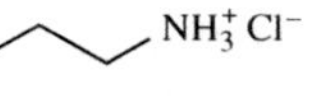

C$_3$ Amine hydrochloride
mp 161°C
Soluble in H$_2$O; Insoluble in organic solvents

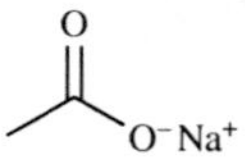

Sodium salt of C$_2$ carboxylic acid
mp 324°C
Soluble in H$_2$O; Insoluble in organic solvents

FIG. 3.2
Some organic molecules containing 8 atoms (carbon, oxygen, and nitrogen) in order of increasing melting or boiling points.

C$_8$ Hydrocarbon, octane
mp −57°C; bp 126°C
Insoluble in H$_2$O

C$_7$ Amine
mp −18°C; bp 154°C
Insoluble in H$_2$O; Soluble in organic solvents

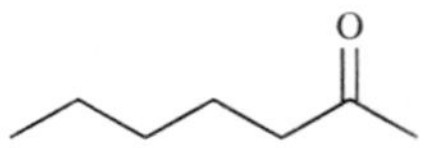

C$_7$ Ketone
mp −35°C; bp 150°C
Insoluble in H$_2$O; Soluble in organic solvents

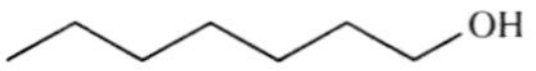

C$_7$ Alcohol
mp −34°C; bp 175°C
Insoluble in H$_2$O; Soluble in organic solvents

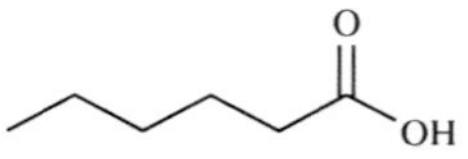

C$_6$ Carboxylic acid
mp −2°C; bp 200°C
Insoluble in H$_2$O; Soluble in organic solvents

C$_7$ Amine hydrochloride
mp 242°C
Soluble in H$_2$O; Insoluble in organic solvents

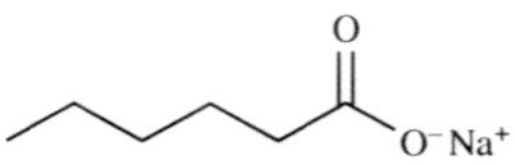

Sodium salt of C$_6$ carboxylic acid
mp 245°C
Soluble in H$_2$O; Insoluble in organic solvents

four carbons larger than for those in Figure 3.1. We might predict that the larger hydrocarbon portion makes them behave more like the water-insoluble hydrocarbon octane. Indeed, the larger hydrocarbon portion of these molecules greatly reduces their solubility in water to less than 5 g/L of water except for the two ionic compounds. These two, the amine hydrochloride and the sodium carboxylate compounds, have higher solubility—tens of grams per liter of water—proof that ionic charges can interact strongly with water molecules.

FIG. 3.3
Some organic molecules containing 18 atoms (carbon, oxygen, and nitrogen) in order of increasing melting or boiling points.

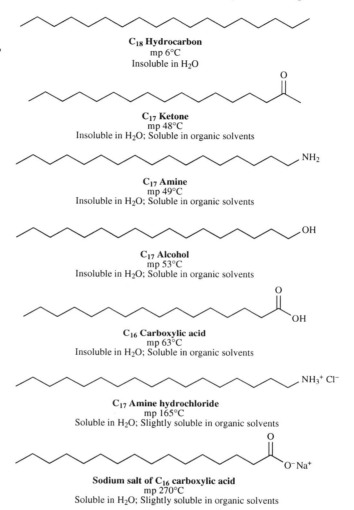

C_{18} **Hydrocarbon**
mp 6°C
Insoluble in H_2O

C_{17} **Ketone**
mp 48°C
Insoluble in H_2O; Soluble in organic solvents

C_{17} **Amine**
mp 49°C
Insoluble in H_2O; Soluble in organic solvents

C_{17} **Alcohol**
mp 53°C
Insoluble in H_2O; Soluble in organic solvents

C_{16} **Carboxylic acid**
mp 63°C
Insoluble in H_2O; Soluble in organic solvents

C_{17} **Amine hydrochloride**
mp 165°C
Soluble in H_2O; Slightly soluble in organic solvents

Sodium salt of C_{16} carboxylic acid
mp 270°C
Soluble in H_2O; Slightly soluble in organic solvents

This trend in water solubility based on the size of the hydrocarbon portion continues for the set of even larger organic molecules shown in Figure 3.3 containing C_{16} to C_{18} hydrocarbon chains. Most are virtually insoluble in water, and even the ionic forms have solubilities of less than 1 g/L of water.

In addition to water, many other liquid solvents are used in the organic laboratory to dissolve substances, including acetone, dichloromethane (CH_2Cl_2), ethanol (CH_3CH_2OH), diethyl ether ($CH_3CH_2OCH_2CH_3$), hexane, methanol, and toluene ($C_6H_5CH_3$). Predicting the solubility of different organic compounds, such as those shown in Figures 3.1, 3.2, and 3.3, in these solvents can be done using the intermolecular attractive force concepts discussed here. For example, because the molecules in Figure 3.3 have long hydrocarbon skeletons, you would predict that these would probably be soluble in the hydrocarbon hexane. This predictive rule can be summed up as "Like dissolves like." You might also predict that the two ionic materials would be the least soluble in hexane because of the strong intermolecular forces in these solids, as evidenced by their high melting points, which are so unlike the weak London forces in hexane. The solubility of organic compounds in organic solvents and water at low, neutral, and high pH will be considered in more detail when you learn about recrystallization in Chapter 4 and extraction in Chapter 7.

PART 2: Melting Points

A. THERMOMETERS

There are a few types of thermometers that can be used to read melting point (and boiling point) temperatures: mercury-in-glass thermometers, non-mercury thermometers, and digital thermometers. Mercury-in-glass thermometers provide highly accurate readings and are ideal for use at high temperatures (260°C–400°C). Care should be taken not to break the thermometer, which will release the toxic mercury. Teflon-coated mercury thermometers are usable up to 260°C and are less likely to spill mercury if broken. If breakage does happen, inform your instructor immediately because special equipment is required to clean up mercury spills. A digital thermometer (Fig. 3.4) has a low heat capacity and a fast response time. It is more robust than a glass thermometer and does not, of course, contain mercury. Non-mercury thermometers may be filled with isoamyl benzoate (a biodegradable liquid) or with a custom organic red-spirit liquid instead of mercury. These thermometers give reasonably accurate readings at temperatures up to 150°C, but above this temperature they need to be carefully calibrated. These thermometers should be stored vertically to prevent thread separation.

Mercury is toxic. Immediately report any broken thermometers to your instructor so that proper clean-up with a mercury disposal kit can occur.

B. MELTING POINTS

The melting point of a pure solid organic compound is one of its characteristic physical properties, along with molecular weight, boiling point, refractive index, and density. A pure solid will melt reproducibly over a narrow range of temperatures, typically less than 1°C. The process of determining this melting point is done on a truly micro scale using less than 1 mg of material. The apparatus is simple, consisting of a thermometer, a capillary tube to hold the sample, and a heating bath.

Melting points—a micro technique

Melting points are determined for three reasons: (1) If the compound is a known one, the melting point will help to characterize the sample; (2) If the compound is new, then the melting point is recorded to allow future characterization by others; (3) The range of the melting point is indicative of the purity of the compound – an impure compound will melt over a wide range of temperatures. Recrystallization of the compound will purify it and decrease the melting point range. In addition, the entire range will be displaced upward. For example, an impure sample might melt from 120°C to 124°C, and after recrystallization will melt at 125°C–125.5°C. A solid is considered pure if the melting point does not rise after recrystallization.

Characterization

An indication of purity

A crystal is an orderly arrangement of molecules in a solid. As heat is added to the solid, the molecules will vibrate and perhaps rotate, but still remain a solid. At a characteristic temperature, the crystal will suddenly acquire the necessary

FIG. 3.4
A digital thermometer.

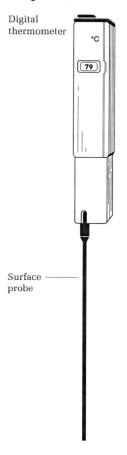

Digital thermometer

Surface probe

Melting point generalizations

A phase diagram

Melting point depression

FIG. 3.5
A melting point–composition diagram for mixtures of the solids X and Y.

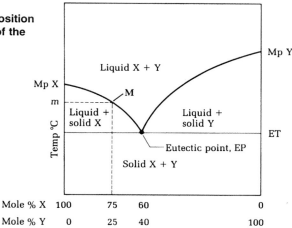

energy to overcome the forces that attract one molecule to another and will undergo translational motion—in other words, it will become a liquid.

The forces by which one molecule is attracted to another include ionic attractions, London forces, hydrogen bonds, and dipole-dipole attractions. Most, but by no means all, organic molecules are covalent in nature and melt at temperatures below 300°C. Typical inorganic compounds are ionic and have much higher melting points (e.g., sodium chloride melts at 800°C). Ionic organic molecules often decompose before melting, as do compounds having many strong hydrogen bonds, such as sucrose.

Other factors being equal, larger molecules melt at higher temperatures than smaller molecules. Among structural isomers, the more symmetrical isomer will have the higher melting point. Among optical isomers, the R and S enantiomers will have the same melting points, but the racemate (a mixture of equal parts of R and S) will usually possess a different melting point. Diastereomers, another type of stereoisomer, will have different melting points. Molecules that can form hydrogen bonds will usually possess higher melting points than their counterparts of similar molecular weight.

The melting point behavior of impure compounds is best understood by considering a simple binary mixture of compounds X and Y (Fig. 3.5). This melting point–composition diagram shows melting point behavior as a function of composition. The melting point of a pure compound is the temperature at which the vapor pressures of the solid and liquid are equal. But in dealing with a mixture, the situation is different. Consider the case of a mixture of 75% X and 25% Y. At a temperature below the eutectic temperature (ET), the mixture is solid Y and solid X. At ET, the solid begins to melt. The melt is a solution of Y dissolved in liquid X. The vapor pressure of the solution of X and Y together is less than that of pure X at the melting point; therefore, the temperature at which X will melt is lower when mixed with Y. This is an application of Raoult's law (*see* Chapter 5). As the temperature is increased, more and more of solid X melts until it is all gone at point **M** (temperature *m*). The melting point range is thus from ET to *m*. In practice, it is very difficult to detect the ET when a melting point is determined in a capillary because it represents the point at which an infinitesimal amount of the solid mixture has begun to melt.

The eutectic point

In this hypothetical example, the liquid solution becomes saturated with Y at the eutectic point (EP). This is the point at which X and Y and their liquid solutions are in equilibrium. A mixture of X and Y containing 60% X will appear to have a sharp melting point at the ET.

The melting point range of a mixture of compounds is generally broad, and the breadth of the range is an indication of purity. The chances of accidentally coming on the eutectic composition are small. Recrystallization will enrich the predominant compound while excluding the impurity and will, therefore, decrease the melting point range.

It should be apparent that the impurity must be soluble in the compound, so an insoluble impurity such as sand or charcoal will not depress the melting point. The impurity does not need to be a solid. It can be a liquid such as water (if it is soluble) or an organic solvent, such as the one used to recrystallize the compound; this advocates the necessity for drying the compound before determining its melting point.

Advantage is taken of the depression of melting points of mixtures to prove whether or not two compounds having the same melting points are identical. If X and Y are identical, then a mixture of the two will have the same melting point; if X and Y are not identical, then a small amount of X in Y or of Y in X will reduce the melting point.

Mixed melting points

Apparatus

Melting Point Capillaries

Before using a melting point apparatus, the sample needs to be prepared for analysis. Most melting point determinations require that the sample be placed in a capillary tube. The experiments in this book require capillary tubes for sample preparation. Capillaries may be obtained commercially or may be produced by drawing out 12-mm soft-glass tubing. The tubing is rotated in the hottest part of a Bunsen burner flame until it is very soft and begins to sag. It should not be drawn out during heating; it is removed from the flame and after a slight hesitation drawn steadily and not too rapidly to arm's length. With some practice it is possible to produce 10–15 good tubes in a single drawing. The long capillary tube can be cut into 100-mm lengths with a glass scorer. Each tube is sealed by rotating one open end in the edge of a small flame until the glass melts and seals the bottom, as seen in Figure 3.6.

FIG. 3.6
Sealing a melting point capillary tube.

Filling Melting Point Capillaries. The dry sample is ground to a fine powder on a watch glass or a piece of weighing paper on a hard surface using the flat portion of a spatula. It is formed into a small pile, and the open end of the melting point capillary is forced down into the pile. The sample is shaken into the closed end of the capillary by rapping sharply on a hard surface or by dropping it down a 2-ft length of glass tubing onto a hard surface. The height of the sample should be no more than 2–3 mm.

Samples that sublime

Sealed Capillaries. Some samples sublime (go from a solid state directly to the vapor phase without appearing to melt), or undergo rapid air oxidation and decompose at the melting point. These samples should be sealed under vacuum. This can be accomplished by forcing a capillary through a hole previously made in a rubber septum and evacuating the capillary using a water aspirator or a mechanical vacuum pump (Fig. 3.7). Using the flame from a small micro burner, the tube is gently heated about 15 mm above the tightly packed sample. This will cause any

FIG. 3.7
Evacuation of a melting point capillary prior to sealing.

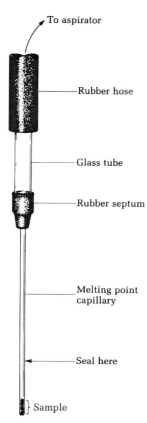

A small rubber band can be made by cutting off a very short piece of 1/4" gum rubber tubing.

Thomas–Hoover Uni-Melt

FIG. 3.8
A simple melting point apparatus.

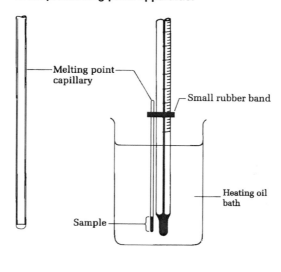

FIG. 3.9
The Thomas–Hoover Uni-Melt melting point apparatus.

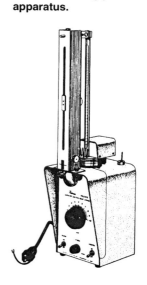

material in this region to sublime away. It is then heated more strongly in the same place to collapse and seal the tube, taking care that the tube is straight when it cools. It is also possible to seal the end of a long Pasteur pipette, add the sample, pack it down, and seal off a sample under vacuum in the same way.

Melting Point Devices. The apparatus required for determining an accurate melting point need not be elaborate. The same results are obtained on the simplest as on the most complex devices. The simplest setup involves attaching the sample-filled, melting point capillary to a thermometer using a rubber band and immersing them into a silicone oil bath (Fig. 3.8). This rubber band must be above the level of the oil bath; otherwise, it will break in the hot oil. The sample should be close to and on a level with the center of the thermometer bulb. This method can analyze compounds whose melting points go up to ~350°C. If determinations are to be done on two or three samples that differ in melting point by as much as 10°C, two or three capillaries can be secured to the thermometer together; the melting points can be observed in succession without removing the thermometer from the bath. As a precaution against the interchange of tubes while they are being attached, use some system of identification, such as one, two, and three dots made with a marking pencil.

More sophisticated melting point devices, some of which can attain temperatures of 500°C, will now be described.

The Thomas–Hoover Uni-Melt apparatus (Fig. 3.9) will accommodate up to seven capillaries in a small, magnified, lighted beaker of high-boiling silicone oil that is stirred and heated electrically. The heating rate is controlled with a variable transformer that is part of the apparatus. The rising mercury column of the thermometer can be observed with an optional traveling periscope device so the eye need not move away from the capillary. For industrial, analytical, and control work, the Mettler apparatus determines the melting point automatically and displays the result in digital form.

FIG. 3.10

The Mel-Temp melting point apparatus.

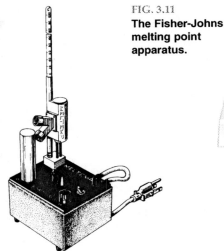

FIG. 3.11

The Fisher-Johns melting point apparatus.

The Mel-Temp apparatus (Fig. 3.10) consists of an electrically heated aluminum block that accommodates three capillaries. The sample is illuminated through the lower port and observed with a six-power lens through the upper port. The heating rate can be controlled and, with a special thermometer, the apparatus can be used up to 500°C—far above the useful limit of silicone oil (which is about 350°C). For this melting point apparatus, it is advisable to use a digital thermometer rather than a mercury-in-glass thermometer.

The Fisher-Johns melting point apparatus (Fig. 3.11) is used to determine the melting point of a single sample. Instead of a capillary tube, the sample is placed between two thin glass disks that are placed on an aluminum heating stage. Heating is controlled by a variable transformer, and melting is observed through a magnifier; the melting temperature is read from a mercury-in-glass thermometer. This apparatus can be used for compounds that melt between 20°C and 300°C.

Determining the Melting Point

The rate of heating is the most important factor in obtaining accurate melting points. Heat no faster than 1°C per minute.

The accuracy of the melting point depends on the accuracy of the thermometer, so the first exercise in the following experiments will be to calibrate the thermometer. Melting points of pure, known compounds will be determined and deviations recorded so a correction can be applied to future melting points. Be forewarned, however, that thermometers are usually fairly accurate.

The most critical factor in determining an accurate melting point is the rate of heating. At the melting point, the temperature increase should not be greater than 1°C per minute. This may seem extraordinarily slow, but it is necessary in order for heat from the oil bath or the heating block to be transferred equally to the sample and to the glass and mercury of the thermometer.

From your own experience, you know the rate at which ice melts. Consider performing a melting point experiment on an ice cube. Because water melts at 0°C, you would need to have a melting point bath a few degrees below zero. To observe the true melting point of the ice cube, you would need to raise the temperature extraordinarily slowly. The ice cube would appear to begin to melt at 0°C and, if you waited for temperature equilibrium to be established, it would all be melted at

0.5°C. If you were impatient and raised the temperature too rapidly, the ice might appear to melt over a range of 0°C to 20°C. Similarly, melting points determined in capillaries will not be accurate if the rate of heating is too fast.

EXPERIMENTAL PROCEDURES

1. CALIBRATION OF THE THERMOMETER

Determine the melting point of standard substances (Table 3.1) over the temperature range of interest. The difference between the values found and those expected constitutes the correction that must be applied to future temperature readings. If the thermometer has been calibrated previously, then determine one or more melting points of known substances to familiarize yourself with the technique. If the determinations do not agree within 1°C, then repeat the process. Both mercury-in-glass and digital thermometers will need to be calibrated; non-mercury thermometers are not typically used for melting point determination.

TABLE 3.1 *Melting Point Standards*

Compound		Structure	Melting Point (°C)
Naphthalene	(a)	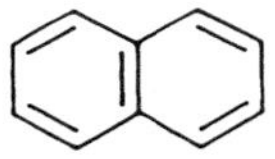	80–82
Urea	(b)	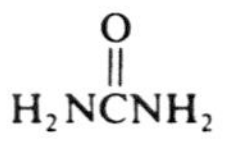 H_2NCNH_2 (O)	132.5–133
Sulfanilamide	(c)	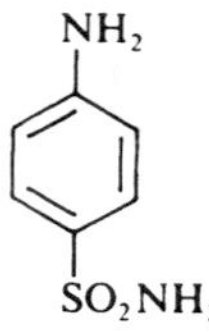 (aniline with SO_2NH_2)	164–165
4-Toluic acid	(d)	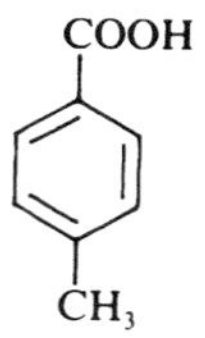 (p-COOH, CH_3 benzene)	180–182
Anthracene	(e)	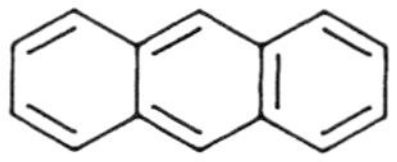	214–217
Caffeine (evacuated capillary)	(f)	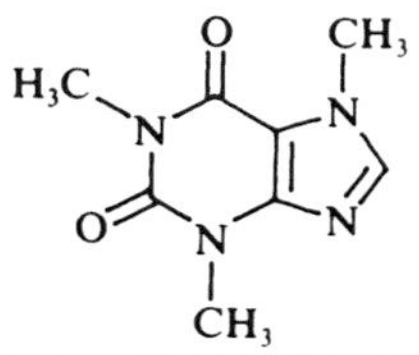	234–236.5

2. MELTING POINTS OF PURE UREA AND CINNAMIC ACID

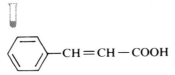

Cinnamic acid

Heat rapidly to within 20°C of the melting point.

Heat slowly (<1°C/min) near the melting temperature.

Using a metal spatula, crush the sample to a fine powder on a hard surface such as a watch glass. Push the open end of a melting point capillary into the powder and force the powder down in the capillary by tapping the capillary or by dropping it through a long glass tube held vertically and resting on a hard surface. The column of solid should be no more than 2–3 mm in height and should be tightly packed.

If the approximate melting temperature is known, the bath can be heated rapidly until the temperature is about 20°C below this point; the heating during the last 15°C–20°C should slow down considerably *so the rate of heating at the melting point is no more than 1°C per minute* while the sample is melting. As the melting point is approached, the sample may shrink because of crystalline structure changes. However, the melting process begins when the first drops of liquid are seen in the capillary and ends when the last trace of solid disappears. For a pure compound this whole process may occur over a range of only 0.5°C; hence, it is necessary to slowly increase the temperature during the determination.

Determine the melting point (mp) of either urea (mp 132.5°C–133°C) or cinnamic acid (mp 132.5°C–133°C). Repeat the determination; if the two determinations do not check within 1°C, repeat a third time.

3. MELTING POINTS OF UREA-CINNAMIC ACID MIXTURES

> IN THIS EXPERIMENT you will see dramatic evidence of the phenomenon of melting point depression, which will allow you to prepare a phase diagram like that shown in Figure 3.5 (on page 49).

Make mixtures of urea and cinnamic acid in the approximate proportions 1:4, 1:1, and 4:1 by putting side by side the correct number of equal-sized small piles of the two substances and then mixing them. Grind the mixture thoroughly for at least a minute on a watch glass using a metal spatula. Note the ranges of melting of the three mixtures and use the temperatures of complete liquefaction to construct a rough diagram of melting point versus composition.

4. UNKNOWNS

Determine the melting point of one or more of the unknowns selected by your instructor and identify the substance based on its melting point (Table 3.2). Prepare two capillaries of each unknown. Run a very fast determination on the first sample to ascertain the approximate melting point. Cool the melting point bath to just below the melting point and make a slow, careful determination using the other capillary.

5. AN INVESTIGATION: DETERMINATION OF MOLECULAR WEIGHT USING MELTING POINT DEPRESSION

Before the mass spectrometer came into common usage, the molal freezing point depression of camphor was used to determine molecular weights. Whereas a 1% solid solution of urea in cinnamic acid will cause a relatively small melting

TABLE 3.2 *Melting Point Unknowns*

Compound	Melting Point (°C)
Benzophenone	49–51
Maleic anhydride	52–54
4-Nitrotoluene	54–56
Naphthalene	80–82
Acetanilide	113.5–114
Benzoic acid	121.5–122
Urea	132.5–133
Salicylic acid	158.5–159
Sulfanilamide	165–166
Succinic acid	184.5–185
3,5-Dinitrobenzoic acid	205–207
p-Terphenyl	210–211

point depression, a 1%-by-weight solid solution in camphor of any organic compound with a molecular weight of 100 will cause a 4.0°C depression in the melting point of the camphor. Quantitative use of this relationship can be used to determine the molecular weight of an unknown. You can learn more about the details of this technique in very old editions of *Organic Experiments*[1] or by searching the Web for "colligative properties, molal freezing point depression." Visit this book's website for more information.

PART 3: BOILING POINTS

The boiling point of a pure organic liquid is one of its characteristic physical properties, just like its density, molecular weight, and refractive index, and the melting point of its solid state. The boiling point is used to characterize a new organic liquid, and knowledge of the boiling point helps to compare one organic liquid to another, as in the process of identifying an unknown organic substance.

A comparison of boiling points with melting points is instructive. The process of determining the boiling point is more complex than that for the melting point: It requires more material, and because it is less affected by impurities, it is not as good an indication of purity. Boiling points can be determined on a few microliters of a liquid, but on a small scale it is difficult to determine the boiling point range. This requires enough material to distill—about 1 to 2 mL. Like the melting point, the boiling point of a liquid is affected by the forces that attract one molecule to another—ionic attraction, London forces, dipole-dipole interactions, and hydrogen bonding, as discussed in Part 1 of this chapter.

STRUCTURE AND BOILING POINT

In a homologous series of molecules, the boiling point increases in a perfectly regular manner. The normal saturated hydrocarbons have boiling points ranging from −161°C for methane to 330°C for n-$C_{19}H_{40}$, an increase of about 27°C for each

[1] Fieser L. F. *Organic Experiments*, 2nd ed.; D.C. Heath: Lexington, MA, 1968; 38–42.

CH₂ group. It is convenient to remember that *n*-heptane with a molecular weight of 100 has a boiling point near 100°C (98.4°C). A spherical molecule such as 2,2-dimethylpropane has a lower boiling point than *n*-pentane because it does not have as many points of attraction to adjacent molecules. For molecules of the same molecular weight, those with dipoles, such as carbonyl groups, will have higher boiling points than those without, and molecules that can form hydrogen bonds will boil even higher. The boiling point of such molecules depends on the number of hydrogen bonds that can be formed. An alcohol with one hydroxyl group will boil at a lower temperature than an alcohol with two hydroxyl groups if they both have the same molecular weight. A number of other generalizations can be made about boiling point behavior as a function of structure; you will learn about these throughout your study of organic chemistry.

BOILING POINT AS A FUNCTION OF PRESSURE

Boiling points decrease about 0.5°C for each 10-mm decrease in atmospheric pressure.

Because the boiling point of a pure liquid is defined as the temperature at which the vapor pressure of the liquid exactly equals the pressure exerted on it, the boiling point will be a function of atmospheric pressure. At an altitude of 14,000 ft, the boiling point of water is 81°C. At pressures near that of the atmosphere at sea level (760 mm), the boiling point of most liquids decreases about 0.5°C for each 10-mm decrease in atmospheric pressure. This generalization does not hold for greatly reduced pressures because the boiling point decreases as a nonlinear function of pressure (see Fig. 5.2 on page 89). Under these conditions, a nomograph relating the observed boiling point, the boiling point at 760 mm, and the pressure in millimeters should be consulted (see Fig. 6.19 on page 124). This nomograph is not highly accurate because the change in boiling point as a function of pressure also depends on the type of compound (polar, nonpolar, hydrogen bonding, etc.). Consult the *CRC Handbook of Chemistry and Physics*[2] for the correction of boiling points to standard pressure.

Mercury is toxic. Immediately report any broken thermometers to your instructor.

Most mercury-in-glass laboratory thermometers have a mark around the stem that is 3 in. (76 mm) from the bottom of the bulb. This is the immersion line; the thermometer will record accurate temperatures if immersed to this line. If you break a mercury thermometer, immediately inform your instructor, who will use special apparatus to clean up the mercury. Mercury vapor is very toxic and can be fatal if the fumes are produced by the heating of liquid mercury.

CALIBRATING THE THERMOMETER

If you have not previously carried out a calibration, test the 0°C point of your thermometer with a well-stirred mixture of crushed ice and distilled water. To check the 100°C point, put 2 mL of water in a test tube with a boiling chip to prevent bumping, and boil the water gently over a hot sand bath with the thermometer in the vapor from the boiling water. Make sure that the thermometer does not touch the side of the test tube. Then immerse the bulb of the thermometer into the liquid and see if you can observe superheating. Check the atmospheric pressure to determine the true boiling point of the water.

---

[2]Lide, D. R., ed. *CRC Handbook of Chemistry and Physics*, 86th ed.; CRC Press: Boca Raton, FL, 2005.

DISTILLATION CONSIDERATIONS

Prevention of Superheating—Boiling Sticks and Boiling Stones

Superheating occurs when a very clean liquid in a very clean vessel is heated to a temperature above its boiling point without ever actually boiling. That is, if a thermometer is placed in the superheated liquid, the thermometer will register a temperature higher than the boiling point of the liquid. If boiling does occur under these conditions, it occurs with explosive violence. To avoid this problem, boiling stones or boiling sticks are always added to liquids before heating them to boiling—whether to determine a boiling point or to carry out a reaction or distillation. These boiling stones or sticks provide the nuclei on which the bubbles of vapor indicative of a boiling liquid can form; be careful not to confuse the bubbling for boiling. Some boiling stones, also called boiling chips, are composed of porous unglazed porcelain. When dry, this material is filled with air in numerous fine capillaries. With heating, this air expands to form the fine bubbles upon which even boiling can take place. Once the liquid cools, it will fill these capillaries and the boiling chip will become ineffectual, so another chip must be added each time the liquid is reheated to boiling. Wooden boiling sticks about 1.5 mm in diameter—often called applicator sticks—also promote even boiling and, unlike stones, are removed easily from the solution. Neither boiling sticks nor stones work well for vacuum distillation (*see* Chapter 6).

Closed Systems

Distillations that are run at atmospheric pressure need to be open to the atmosphere to avoid pressure buildup, which could lead to an explosion. Therefore, always make sure that a distillation setup is not a closed system—unless, of course, you are running a vacuum distillation.

BOILING POINT DETERMINATION: APPARATUS AND TECHNIQUE

Boiling Point Determination by Distillation

When enough material is available (at least 3 mL), the best method for determining the boiling point of a liquid is to distill it (*see* Chapter 5). Distillation allows the boiling range to be determined and thus gives an indication of purity. Bear in mind, however, that a constant boiling point is not a guarantee of homogeneity and thus purity. Constant-boiling azeotropes such as 95% ethanol are common.

All procedures involving volatile and/or flammable solvents should be conducted in a fume hood.

Boiling Point Determination Using a Digital Thermometer and a Reaction Tube

Boiling points can be measured rapidly and accurately using an electronic digital thermometer, as depicted in Figure 3.12. Although digital thermometers are currently too expensive for each student to have, several of these in the laboratory can greatly speed up the determination of boiling points. Also, digital thermometers are much safer to use because there is no danger from toxic mercury vapor if the thermometer is accidentally dropped.

Photo: Boiling Point Determination with a Digital Thermometer

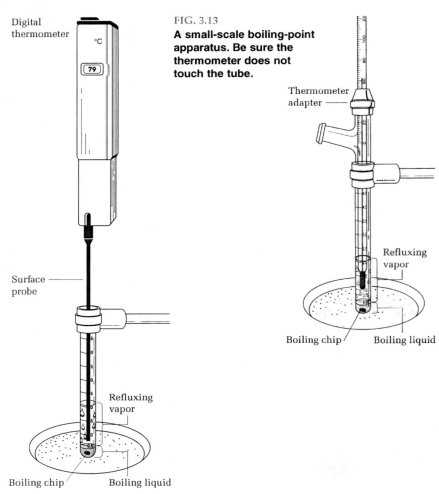

FIG. 3.12
Using a digital thermometer for determining boiling points.

FIG. 3.13
A small-scale boiling-point apparatus. Be sure the thermometer does not touch the tube.

The surface probe of the digital thermometer is the active element. Unlike the bulb of mercury at the end of a thermometer, this element has a very low heat capacity and a very fast response time, so boiling points can be determined very quickly with this apparatus. About 0.2 mL to 0.3 mL of the liquid and a boiling chip are heated on a sand bath until the liquid refluxes about 3 cm up the tube. The thermometer probe should not touch the side of the reaction tube, but should be placed approximately 5 mm above the liquid. The boiling point is the highest temperature recorded by the thermometer and is maintained for about 1 min. The application of heat will drive tiny bubbles of air from the boiling chip; do not mistake these tiny bubbles for true boiling; this mistake can readily happen if the unknown has a very high boiling point.

Boiling Point Determination in a Reaction Tube

If a digital thermometer is not available, use the apparatus shown in Figure 3.13. Using a distilling adapter at the top of a reaction tube allows access to the atmosphere. Place 0.3 mL of the liquid along with a boiling stone in a 10 × 100-mm reaction tube, clamp a thermometer so that the bulb is just above the level of the liquid, and then heat the liquid with a sand bath. It is *very important* that no part of the thermometer touch the reaction tube. Heating is regulated so that the boiling liquid refluxes about 3 cm up the thermometer, but does not boil out of the apparatus. If you cannot see the refluxing liquid, carefully run your finger down the side of the reaction tube until you feel heat. This indicates where the liquid is refluxing. Droplets of liquid must drip from the thermometer bulb to thoroughly heat the mercury. The boiling point is the highest temperature recorded by the thermometer and is maintained over about a 1-minute time interval.

The application of heat will drive out tiny air bubbles from the boiling chip. Do not mistake these tiny bubbles for true boiling. This can occur if the unknown has a very high boiling point. It may take several minutes to heat up the mercury in the thermometer bulb. True boiling is indicated by drops dripping from the thermometer, with a constant temperature recorded on the thermometer. If the temperature is not constant, then you are probably not observing true boiling.

Boiling Point Determination Using a 3-mm to 5-mm Tube

Smaller-scale boiling point apparatus

For smaller quantities, a 3-mm to 5-mm diameter tube is attached to the side of a thermometer by a rubber band (Fig. 3.14) and heated in a liquid bath. The tube, which can be made from tubing 3 mm to 5 mm in diameter, contains a small inverted capillary. This is made by cutting a 6-mm piece from the sealed end of a melting point capillary, inverting it, and sealing the closed end of the capillary. A centimeter ruler is printed on the inside back cover of this book.

When the sample is heated in this device, the air in the inverted capillary will expand, and an occasional bubble will escape. At the true boiling point, a continuous and rapid stream of bubbles will emerge from the inverted capillary. When this occurs, the heating is stopped, and the bath is allowed to cool. A time will come when bubbling ceases and the liquid just begins to rise in the inverted capillary. The temperature at which this happens is recorded. The liquid is allowed to partially fill the small capillary, and then heat is applied carefully until the first bubble emerges from the capillary. The temperature is recorded at that point. The two temperatures approximate the boiling point range for the liquid.

As the liquid was being heated, the air expanded in the inverted capillary and was replaced by vapor of the liquid. The liquid was actually slightly superheated when rapid bubbles emerged from the capillary, but on cooling, a point was

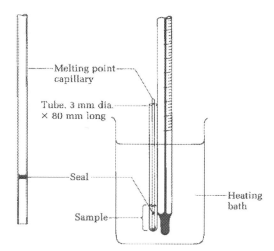

FIG. 3.14
A smaller-scale boiling point apparatus.

reached at which the pressure on the inside of the capillary matched the outside (atmospheric) pressure. This is, by definition, the boiling point.

Cleaning Up. Place the boiling point sample in either the halogenated or nonhalogenated waste container. Do not pour it down the sink.

QUESTIONS

1. What effect would poor circulation of the melting point bath liquid have on the observed melting point?
2. What is the effect of an insoluble impurity, such as sodium sulfate, on the observed melting point of a compound?
3. Three test tubes, labeled A, B, and C, contain substances with approximately the same melting points. How could you prove that the test tubes contain three different chemical compounds?
4. One of the most common causes of inaccurate melting points is too rapid heating of the melting point bath. Under these circumstances, how will the observed melting point compare to the true melting point?
5. Strictly speaking, why is it incorrect to speak of a melting *point?*
6. What effect would the incomplete drying of a sample (for example, the incomplete removal of a recrystallization solvent) have on the melting point?
7. Why should the melting point sample be finely powdered?
8. You suspect that an unknown is acetanilide (mp 113.5°C–114°C). Give a qualitative estimation of the melting point when the acetanilide is mixed with 10% by weight of naphthalene.
9. You have an unknown with an observed melting point of 90°C–93°C. Is your unknown compound A with a reported melting point of 95.5°C–96°C, or compound B with a reported melting point of 90.5°C–91°C? Explain.
10. Why is it important to heat the melting point bath or block slowly and steadily when the temperature gets close to the melting point?
11. Why is it important to pack the sample tightly in the melting point capillary?
12. An unknown compound is suspected to be acetanilide (mp 113.5°C–114°C). What would happen to the melting point if this unknown were mixed with (a) an equal quantity of pure acetanilide? (b) an equal quantity of benzoic acid?
13. Which would be expected to have the higher boiling point, *t*-butyl alcohol (2-methyl-2-propanol) or *n*-butyl alcohol (1-butanol)? Explain.

See Web Links

14. What is the purpose of the side arm of the thermometer adapter in Figure 3.13?

REFERENCES

Weissberger, Arnold, and Bryant W. Rossiter (eds.). *Physical Methods of Chemistry,* Vol. 1, Part V. New York: Wiley-Interscience, 1971.

J. Kofron and J.K. Hardy of the University of Akron have an excellent website with photographs illustrating mixed melting points: http://ull.chemistry.uakron.edu/organic_lab/melting_point.

CHAPTER 4

Recrystallization

> **PRELAB EXERCISE:** Write an expanded outline for the seven-step process of recrystallization.

Recrystallization: the most important purification method for solids, especially for small-scale experiments.

Recrystallization is the most important method for purifying solid organic compounds. It is also a very powerful, convenient, and efficient method of purification, and it is an important industrial technique that is still relevant in the chemical world today. For instance, the commercial purification of sugar is done by recrystallization on an enormous scale.

A pure, crystalline organic substance is composed of a three-dimensional array of molecules held together primarily by London forces. These attractive forces are fairly weak; most organic solids melt in the range between 22°C and 250°C. An impure organic solid will not have a well-defined crystal lattice because impurities do not allow the crystalline structure to form. The goal of recrystallization is to remove impurities from a solid to allow a perfect crystal lattice to grow.

There are four important concepts to consider when discussing the process of recrystallization:

1. Solubility
2. Saturation level
3. Exclusion
4. Nucleation

Recrystallization involves dissolving the material to be purified (the *solute*) in an appropriate hot *solvent* to yield a solution (*solubility*). As the solvent cools, the solution becomes saturated with respect to the solute (*saturation level*), which then recrystallizes. As the perfectly regular array of a crystal is formed, impurities are excluded (*exclusion*), and the crystal is thus a single pure substance. Soluble impurities remain in solution because they are not concentrated enough to saturate the solution. Recrystallization of a solute is initiated at a point of *nucleation*, which can be a seed crystal, a speck of dust, or a scratch on the wall of the test tube.

In this chapter, you will carry out the recrystallization process, one of the most important laboratory operations of organic chemistry, by following its **seven steps**. Then you will perform several actual recrystallization experiments.

The Seven Steps of Recrystallization

The process of recrystallization can be broken into seven discrete steps: (1) choosing the solvent and solvent pairs; (2) dissolving the solute; (3) decolorizing the solution with pelletized Norit; (4) filtering suspended solids; (5) recrystallizing the solute; (6) collecting and washing the crystals; and (7) drying the crystals. A detailed description of each of these steps is given in the following sections.

STEP 1. CHOOSING THE SOLVENT AND SOLVENT PAIRS

Similia similibus solvuntur.

In choosing the solvent, the chemist is guided by the dictum "like dissolves like." Even the nonchemist knows that oil and water do not mix, and that sugar and salt dissolve in water but not in oil. Hydrocarbon solvents such as hexane will dissolve hydrocarbons and other nonpolar compounds, and hydroxylic solvents such as water and ethanol will dissolve polar compounds. It is often difficult to decide, simply by looking at the structure of a molecule, just how polar or nonpolar it is and which solvent would be best. Therefore, the solvent is often chosen by experimentation. If an appropriate single solvent cannot be found for a given substance, a solvent pair system may be used. The requirement for this solvent pair is miscibility; both solvents should dissolve in each other for use as a recrystallization solvent system.

Video: *The Reaction Tube in Use*

The ideal solvent

The best recrystallization solvent (and none is ideal) will dissolve the solute when the solution is hot but not when the solution is cold; it will either not dissolve the impurities at all or it will dissolve them very well (so they do not recrystallize out along with the solute); it will not react with the solute; and it will be nonflammable, nontoxic, inexpensive, and very volatile (so it can be removed from the crystals).

Some common solvents and their properties are presented in Table 4.1 in order of decreasing polarity of the solvent. Solvents adjacent to each other in the list will dissolve in each other; that is, they are miscible with each other, and each solvent will, in general, dissolve substances that are similar to it in chemical structure. These solvents are used both for recrystallization and as solvents in which reactions are carried out.

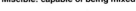
Miscible: capable of being mixed

Procedure

Choosing a Solvent

Video: *Picking a Solvent*

To choose a solvent for recrystallization, place a few crystals of the impure solute in a small test tube or centrifuge tube and add a very small drop of the solvent. Allow the drop to flow down the side of the tube and onto the crystals. If the crystals dissolve instantly at about 22°C, that solvent cannot be used for recrystallization because too much of the solute will remain in solution at low temperatures. If the crystals do not dissolve at about 22°C, warm the tube on a hot sand bath and observe the crystals. If they do not go into solution, add 1 more drop of the solvent. If the crystals go into solution at the boiling point of the solvent and then recrystallize when the tube is cooled, you have found a good recrystallization solvent. If not, remove the solvent by evaporation and try a different solvent. In this trial-and-error process, it is easiest to try low-boiling solvents first because they can be easily removed. Occasionally, no single satisfactory solvent can be found, so mixed solvents, or *solvent pairs*, are used.

TABLE 4.1 *Recrystallization Solvents*

Solvent	Boiling Point (°C)	Density (g/mL)	Remarks
Water (H_2O)	100	1.000	It is the solvent of choice because it is cheap, nonflammable, nontoxic, and will dissolve a large variety of polar organic molecules. Its high boiling point and high heat of vaporization make it difficult to remove from crystals.
Acetic acid (CH_3COOH)	118	1.049	It will react with alcohols and amines, and it is difficult to remove from crystals. It is not a common solvent for recrystallizations, although it is used as a solvent when carrying out oxidation reactions.
Dimethyl sulfoxide (DMSO; CH_3SOCH_3)	189	1.100	It is not a commonly used solvent for recrystallization, but it is used for reactions.
Methanol (CH_3OH)	64	0.791	It is a very good solvent that is often used for recrystallization. It will dissolve molecules of higher polarity than other alcohols.
95% Ethanol (CH_3CH_2OH)	78	0.789	It is one of the most commonly used recrystallization solvents. Its high boiling point makes it a better solvent for less polar molecules than methanol. It evaporates readily from crystals. Esters may undergo an interchange of alcohol groups on recrystallization.
Acetone (CH_3COCH_3)	56	0.791	It is an excellent solvent, but its low boiling point means there is not much difference in the solubility of a compound at its boiling point compared to about 22°C.
2-Butanone; also known as methyl ethyl ketone (MEK; $CH_3COCH_2CH_3$)	80	0.805	It is an excellent solvent that has many of the most desirable properties of a good recrystallization solvent.
Ethyl acetate ($CH_3COOC_2H_5$)	78	0.902	It is an excellent solvent that has about the right combination of moderately high boiling point and the volatility needed to remove it from crystals.
Dichloromethane; also known as methylene chloride (CH_2Cl_2)	40	1.325	Although a common extraction solvent, dichloromethane has too low a boiling point to make it a good recrystallization solvent. It is useful in a solvent pair with ligroin.
Diethyl ether; also known as ether ($CH_3CH_2OCH_2CH_3$)	35	0.706	Its boiling point is too low for recrystallization, although it is an extremely good solvent and fairly inert. It is used in a solvent pair with hexanes.
Methyl *t*-butyl ether ($CH_3OC(CH_3)_3$)	52	0.741	It is a relatively new and inexpensive solvent. It does not easily form peroxides; it is less volatile than diethyl ether, but it has the same solvent characteristics. (*See also* Chapter 7.)
1,4-Dioxane ($C_4H_8O_2$)	101	1.034	It is a very good solvent that is not too difficult to remove from crystals; it is a mild carcinogen, and it forms peroxides.
Toluene ($C_6H_5CH_3$)	111	0.865	It is an excellent solvent that has replaced the formerly widely used benzene (a weak carcinogen) for the recrystallization of aryl compounds. Because of its boiling point, it is not easily removed from crystals.

(continued)

TABLE 4.1 (continued)

Solvent	Boiling Point (°C)	Density (g/mL)	Remarks
Pentane (C_5H_{12})	36	0.626	It is a widely used solvent for nonpolar substances. It is not often used alone for recrystallization, but it is good in combination with several other solvents as part of a solvent pair.
Hexane (C_6H_{14})	69	0.659	It is frequently used to recrystallize nonpolar substances. It is inert and has the correct balance between boiling point and volatility. It is often used as part of a solvent pair.
Cyclohexane (C_6H_{12})	81	0.779	It is similar in all respects to hexane.

Note: The solvents in this table are listed in decreasing order of polarity. Adjacent solvents in the list are, in general, miscible with each other.

TABLE 4.2 Solvent Pairs

Acetic acid–water	Ethyl acetate–cyclohexane
Ethanol–water	Acetone–hexanes
Acetone–water	Ethyl acetate–hexanes
Dioxane–water	t-Butyl methyl ether–hexanes
Acetone–ethanol	Dichloromethane–hexanes
Ethanol–t-butyl methyl ether	Toluene–hexanes

Solvent Pairs

To use a mixed solvent pair, dissolve the crystals in the better solvent (more solubilizing) and add the poorer solvent (less solubilizing) to the hot solution until it becomes cloudy, and the solution is saturated with the solute. The two solvents must, of course, be miscible with each other. Some useful solvent pairs are given in Table 4.2.

STEP 2. DISSOLVING THE SOLUTE

Microscale Procedure

Do not use too much solvent.

Video: Recrystallization

Once a recrystallization solvent has been found, the impure crystals are placed in a reaction tube, the solvent is added dropwise, the crystals are stirred with a microspatula or a small glass rod, and the tube is warmed on a steam bath or a sand bath until the crystals dissolve. Care must be exercised to use the minimum amount of solvent at or near the boiling point. Observe the mixture carefully as solvent is being added. Allow sufficient time for the boiling solvent to dissolve the solute and note the rate at which most of the material dissolves. To hasten the solution process, crush large crystals with a stirring rod, taking care not to break the reaction tube. When you believe most of the material has been dissolved, stop adding solvent. There is a possibility that your sample is contaminated with a small quantity of an insoluble impurity that never will dissolve.

If the solution contains no undissolved impurities and is not colored from impurities, you can simply let it cool, allowing the solute to recrystallize (step 5), and then collect the crystals (step 6). On the other hand, if the solution is colored, it must be treated with activated (decolorizing) charcoal and then filtered before recrystallization (step 3). If it contains solid impurities, it must be filtered before recrystallization takes place (step 4).

On a microscale, there is a tendency to use too much solvent so that on cooling the hot solution, little or no material recrystallizes. This is not a hopeless situation. The remedy is to evaporate some of the solvent (by careful boiling) and repeat the cooling process. Inspect the hot solution to see if crystals form, and if not, continue to evaporate solvent.

A solution (solute dissolved in solvent) can become *superheated*; that is, heated above its boiling point without actually boiling. When boiling does suddenly occur, it can happen with almost explosive violence, a process called *bumping*. To prevent this from happening, a *wood applicator stick* can be added to the solution (Fig. 4.1). Air trapped in the wood comes out of the stick and forms the nuclei on which even boiling can occur. Porous porcelain *boiling chips* work in the same way; only a single chip is required to prevent bumping. Never add a boiling chip or a boiling stick to a hot solution because the hot solution may be superheated and boil over or bump.

Prevention of bumping

Do not use wood applicator sticks (boiling sticks) in place of boiling chips in a reaction. Use them only for recrystallization.

Macroscale Procedure

Place the substance to be recrystallized in an Erlenmeyer flask (never use a beaker), add enough solvent to cover the crystals, and then heat the flask on a steam bath (if the solvent boils below 90°C) or a hot plate until the solvent boils. (Note: Adding a boiling stick or a boiling chip to the solution will promote even boiling. It is easy to superheat the solution; that is, heat it above the boiling point with no boiling taking place. Once the solution does boil, it does so with explosive violence; it bumps.) Never add a boiling chip or boiling stick to a hot solution because this might cause the solution to boil over.

Video: *Macroscale Crystallization*

Stir the mixture or, better, swirl it (Fig. 4.2) to promote dissolution. Add solvent gradually, keeping it at the boiling point, until all of the solute dissolves. A glass rod with a flattened end can sometimes be useful in crushing large particles of solute to speed up the dissolving process. Be sure no flames are nearby when working with flammable solvents.

All procedures involving volatile and/or flammable solvents should be conducted in a fume hood.

Be careful not to add too much solvent. Note how rapidly most of the material dissolves; stop adding solvent when you suspect that almost all of the desired material has dissolved. It is best to err on the side of too little solvent rather than too much. Undissolved material noted at this point could be an insoluble impurity

FIG. 4.1
A reaction tube being used for recrystallization. The wood applicator stick ("boiling stick") promotes even boiling and is easier to remove than a boiling chip. The Thermowell sand is cool on top and hotter deeper down, so it provides a range of temperatures. The reaction tube is long and narrow; it can be held in the hand while the solvent refluxes. Do not use a boiling stick in place of a boiling chip in a reaction.

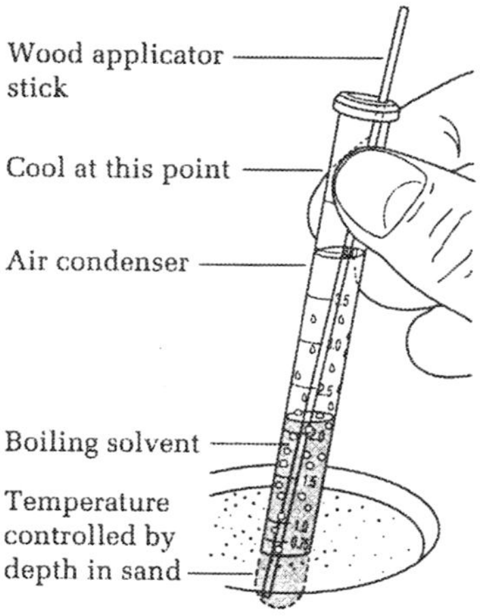

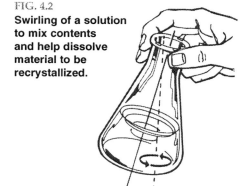

FIG. 4.2
Swirling of a solution to mix contents and help dissolve material to be recrystallized.

that never will dissolve. Allow the solvent to boil, and if no further material dissolves, proceed to step 4 to remove suspended solids from the solution by filtration, or if the solution is colored, go to step 3 to carry out the decolorization process. If the solution is clear, proceed to step 5.

STEP 3. DECOLORIZING THE SOLUTION WITH PELLETIZED NORIT

Video: Decolorization of a Solution with Norit

Activated charcoal = decolorizing carbon = Norit

The vast majority of pure organic chemicals are colorless or a light shade of yellow; consequently, this step is not usually required. Occasionally, a chemical reaction will produce high molecular weight byproducts that are highly colored. The impurities can be adsorbed onto the surface of activated charcoal by simply boiling the solution with charcoal. Activated charcoal is made by the pyrolysis of carbonaceous material such as coconut shells, wood and lignite and activated with steam. It has an extremely large surface area per gram (several hundred square meters) and can bind a large number of molecules to this surface. On a commercial scale, the impurities in brown sugar are adsorbed onto charcoal in the process of refining sugar.

Add a small amount (0.1% of the solute weight is sufficient) of pelletized Norit to the colored solution and then boil the solution for a few minutes. Be careful not to add the charcoal pieces to a superheated solution; the charcoal will function like hundreds of boiling chips and will cause the solution to boil over. Remove the Norit by filtration as described in step 4.

STEP 4. FILTERING SUSPENDED SOLIDS

The filtration of a hot, saturated solution to remove solid impurities or charcoal can be performed in a number of ways. These processes include gravity filtration, pressure filtration, decantation, or removing the solvent with a Pasteur pipette. Vacuum filtration is not used because the hot solvent will cool during the process, and the product will recrystallize in the filter. Filtration can be one of the most vexing operations in the laboratory if the desired compound recrystallizes during filtration. Test the solution or a small portion of it before filtration to ensure that no crystals form at about 22°C. Like decolorization with charcoal, the removal of solid impurities by filtration is usually not necessary.

 Microscale Procedure

(A) Removal of Solution with a Pasteur Pipette

If the solid impurities are large in size, they can be removed by filtration of the liquid through the small space between the flat end of a Pasteur pipette and the bottom of a reaction tube (Fig. 4.3). Expel air from the pipette by squeezing the pipette bulb as the pipette is being pushed to the bottom of the tube. Use a small additional quantity of solvent to rinse the tube and pipette. Anhydrous calcium chloride, a drying agent, is easily removed in this way. The removal of very fine material, such as traces of charcoal, is facilitated by filtration of the solution through a small piece of filter paper (3 mm²) placed in a reaction tube. This process is even easier if the filter paper is the thick variety, such as that from which Soxhlet extraction thimbles are made.[1]

Video: Filtration of Crystals Using the Pasteur Pipette

---

[1]Belletire, J. L.; Mahmoodi, N. O. *J. Chem. Educ.* **1989**, *66*, 964.

Chapter 4 ■ Recrystallization

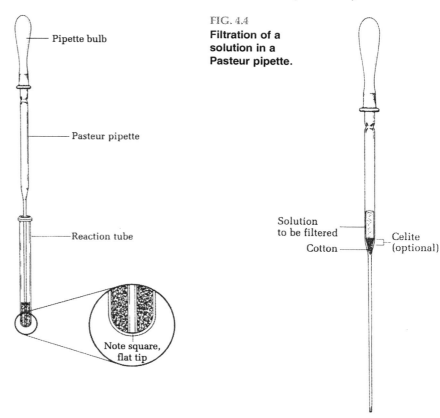

FIG. 4.3
Filtration using a Pasteur pipette and a reaction tube.

FIG. 4.4
Filtration of a solution in a Pasteur pipette.

(B) Filtration in a Pasteur Pipette

To filter 0.1 mL to 2 mL of a solution, dilute the solution with enough solvent so that the solute will not recrystallize at about 22°C. Prepare a filter pipette by pushing a tiny bit of cotton into a Pasteur pipette, put the solution to be filtered into this filter pipette using another Pasteur pipette, and then force the liquid through the filter using air pressure from a pipette bulb (Fig. 4.4). Fresh solvent should be added to rinse the pipette and cotton. The filtered solution is then concentrated by evaporation. One problem encountered with this method is using too much cotton packed too tightly in the pipette so that the solution cannot be forced through it. To remove very fine impurities such as traces of decolorizing charcoal, a 3-mm to 4-mm layer of Celite filter aid can be added to the top of the cotton.

 Photo: *Preparation of a Filter Pipette*

(C) Removal of Fine Impurities by Centrifugation

To remove fine solid impurities from up to 4 mL of solution, dilute the solution with enough solvent so that the solute will not recrystallize at about 22°C. Counterbalance the reaction tube and centrifuge for about 2 minutes at high speed in a laboratory centrifuge. The clear supernatant can be decanted (poured off) from the solid on the bottom of the tube. Alternatively, with care, the solution can be removed with a Pasteur pipette, leaving the solid behind.

 Video: Microscale Crystallization

Use filter paper on top of the frit.

Using the chromatography column for pressure filtration.

Decant: to pour off. A fast, easy separation procedure.

FIG. 4.5
A pressure filtration apparatus. The solution to be filtered is added through the aperture, which is closed by a finger as pressure is applied.

(D) Pressure Filtration with a Micro Büchner Funnel

The technique applicable to volumes from 0.1 mL to 5 mL is to use a micro Büchner funnel. It is made of polyethylene and is fitted with a porous polyethylene frit that is 6 mm in diameter. This funnel fits in the bottom of an inexpensive disposable polyethylene pipette in which a hole is cut (Fig. 4.5). The solution to be filtered is placed in the polyethylene pipette using a Pasteur pipette. The thumb covers the hole in the plastic pipette and pressure is applied to filter the solution. It is good practice to place a 6-mm-diameter piece of filter paper over the frit, which would otherwise become clogged with insoluble material.

The glass chromatography column from the kit can be used in the same way. A piece of filter paper is placed over the frit. The solution to be filtered is placed in the chromatography column, and pressure is applied to the solution using a pipette bulb. In both procedures, dilute the solution to be filtered so that it does not recrystallize in the apparatus, and use a small amount of clean solvent to rinse the apparatus. The filtered solution is then concentrated by evaporation.

 Macroscale Procedure

(A) Decantation

On a large scale, it is often possible to pour off (decant) the hot solution, leaving the insoluble material behind. This is especially easy if the solid is granular like sodium sulfate. The solid remaining in the flask and the inside of the flask should be rinsed with a few milliliters of the solvent in order to recover as much of the product as possible.

(B) Gravity Filtration

The most common method for the removal of insoluble solid material is gravity filtration through fluted filter paper (Fig. 4.6). This is the method of choice for removing

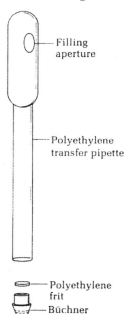

FIG. 4.6
Gravity filtration of hot solution through fluted filter paper.

FIG. 4.7
Fluting a piece of filter paper.

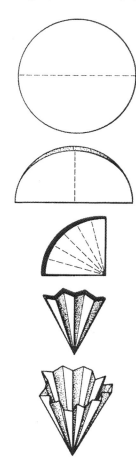

⚠ Be aware that the vapors of low-boiling solvents can ignite on an electric hot plate.

FIG. 4.8
Assemblies for gravity filtration. Stemless funnels have diameters of 2.5, 4.2, 5.0, and 6.0 cm.

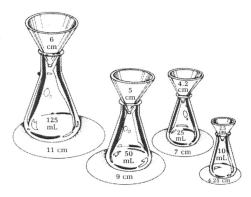

finely divided charcoal, dust, lint, and so on. The following equipment is needed for this process: three labeled Erlenmeyer flasks on a steam bath or a hot plate (flask A contains the solution to be filtered, flask B contains a few milliliters of solvent and a stemless funnel, and flask C contains several milliliters of the crystallizing solvent to be used for rinsing purposes), a fluted piece of filter paper, a towel for holding the hot flask and drying out the stemless funnel, and boiling chips for all solutions.

A piece of filter paper is fluted as shown in Figure 4.7 and is then placed in a stemless funnel. Appropriate sizes of Erlenmeyer flasks, stemless funnels, and filter paper are shown in Figure 4.8. The funnel is stemless so that the saturated solution being filtered will not have a chance to cool and clog the stem with crystals. The filter paper should fit entirely inside the rim of the funnel; it is fluted to allow rapid filtration. Test to see that the funnel is stable in the neck of flask B. If not, support it with a ring attached to a ring stand. A few milliliters of solvent and a boiling chip should be placed in flask B into which the solution is to be filtered. This solvent is brought to a boil on a steam bath or hot plate, along with the solution to be filtered.

The solution to be filtered (in flask A) should be saturated with the solute at the boiling point. Note the volume and then add 10% more solvent (from flask C). The resulting slightly dilute solution is not as likely to recrystallize in the funnel during filtration. Bring the solution to be filtered to a boil, grasp flask A with a towel, and pour the solution into the filter paper in the stemless funnel equipped in flask B (Fig. 4.6). The funnel should be warm to prevent recrystallization from occurring in the funnel. This can be accomplished in two ways: (1) Invert the funnel over a steam bath for a few seconds, pick up the funnel with a towel, wipe it perfectly dry, place it on top of flask B, and then add the fluted filter paper; or (2) place the stemless funnel in the neck of flask B and allow the solvent to reflux into the funnel, thereby warming it.

Pour the solution to be filtered (in flask A) at a steady rate onto the fluted filter paper (equipped in flask B). Check to see whether recrystallization is occurring in the filter. If it does, add boiling solvent (from flask C heated on a steam bath or a hot plate) until the crystals dissolve, dilute the solution being filtered, and carry on. Rinse flask A with a few milliliters of boiling solvent (from flask C) and rinse the fluted filter paper with this same solvent.

Because the filtrate has been diluted to prevent it from recrystallizing during the filtration process, the excess solvent must now be removed by boiling the solution. This process can be sped up somewhat by blowing a slow current of air into the flask in the hood or using an aspirator tube to pull vapors into the aspirator (Fig. 4.9 and Fig. 4.10). However, the fastest method is to heat the solvent in the

70   *Macroscale and Microscale Organic Experiments*

FIG. 4.9

An aspirator tube in use. A boiling stick may be necessary to promote boiling.

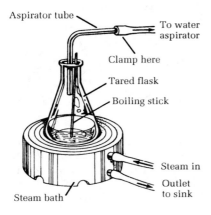

FIG. 4.10

A tube being used to remove solvent vapors.

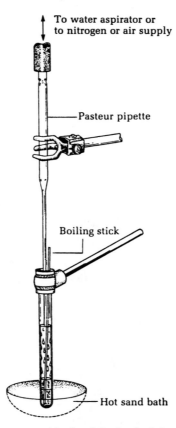

FIG. 4.11

Evaporation of a solvent under a vacuum.

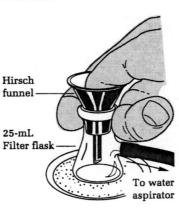

Be sure that you are wearing gloves when doing this step!

filter flask on a sand bath while the flask is connected to a water aspirator. The vacuum is controlled with the thumb (Fig. 4.11).[2] Be sure that you are wearing gloves when doing this step! If your thumb is not large enough, put a one-holed rubber stopper into the Hirsch funnel or the filter flask and again control the vacuum with your thumb. If the vacuum is not controlled, the solution may boil over and go out the vacuum hose.

STEP 5. RECRYSTALLIZING THE SOLUTE

On both a macroscale and a microscale, the recrystallization process should normally start from a solution that is saturated with the solute at the boiling point. If it has been necessary to remove impurities or charcoal by filtration, the solution has been diluted. To concentrate the solution, simply boil off the solvent under an aspirator tube as shown in Figure 4.9 (macroscale) or blow off solvent using a gentle stream of air or, better, nitrogen in the hood as shown in Figure 4.10 (microscale). Be sure to have a boiling chip (macroscale) or a boiling stick (microscale) in the

[2] See also Mayo, D. W.; Pike, R. M.; Butcher, S. M. *Microscale Organic Laboratory;* Wiley: New York, 1986; 97.

solution during this process, but make sure you remove it before initiating recrystallization.

A saturated solution.

Slow cooling is important.

Once it has been ascertained that the hot solution is saturated with the compound just below the boiling point of the solvent, allow it to cool slowly to about 22°C. Slow cooling is a critical step in recrystallization. If the solution is not allowed to cool slowly, precipitation will occur, resulting in impurities "crashing out" of solution along with the desired solute; thus, no exclusion will occur. On a microscale, it is best to allow the reaction tube to cool in a beaker filled with cotton or paper towels which act as insulation, so cooling takes place slowly. Even insulated in this manner, the small reaction tube will cool to about 22°C within a few minutes. Slow cooling will guarantee the formation of large crystals, which are easily separated by filtration and easily washed free of adhering impure solvent. On a small scale, it is difficult to obtain crystals that are too large and occlude impurities. Once the tube has cooled to about 22°C without disturbance, it can be cooled in ice to maximize the amount of product that comes out of solution. On a macroscale, the Erlenmeyer flask is set atop a cork ring or other insulator and allowed to cool gradually to about 22°C. If the flask is moved during recrystallization, many nuclei will form, and the crystals will be small and have a large surface area. They will not be easy to filter and wash clean of the mother liquor. Once recrystallization ceases at about 22°C, the flask should be placed in ice to cool further. Make sure to clamp the flask in the ice bath so that it does not tip over.

 Videos: *Recrystallization, Microscale Crystallization*

Add a seed crystal or scratch the tube.

With slow cooling, recrystallization should begin immediately. If not, add a seed crystal or scratch the inside of the tube with a glass rod at the liquid-air interface. Recrystallization must start on some nucleation center. A minute crystal of the desired compound saved from the crude material will suffice. If a seed crystal is not available, recrystallization can be started on the rough surface of a fresh scratch on the inside of the container.

STEP 6. COLLECTING AND WASHING THE CRYSTALS

Once recrystallization is complete, the crystals must be separated from the ice-cold mother liquor, washed with ice-cold solvent, and dried.

 Microscale Procedure

(A) Filtration Using a Pasteur Pipette

 Videos: *Filtration of Crystals Using the Pasteur Pipette*

The most important filtration technique used in microscale organic experiments employs a Pasteur pipette (Fig. 4.12). About 70% of the crystalline products from the experiments in this text can be isolated in this way. The others will be isolated by filtration on a Hirsch funnel.

The ice-cold crystalline mixture is stirred with a Pasteur pipette and, while air is being expelled from the pipette, forced to the bottom of the reaction tube. The bulb is released, and the solvent is drawn into the pipette through the very small space between the flat tip of the pipette and the curved bottom of the reaction tube. When all the solvent has been withdrawn, it is expelled into another reaction tube. It is sometimes useful to rap the tube containing the wet crystals against a hard surface to pack them so that more solvent can be removed. The tube is returned to the ice bath, and a few drops of cold solvent are added to the crystals. The mixture is stirred to wash the crystals, and the solvent is again removed. This process can be repeated as many times as necessary. Volatile solvents can be removed from the

FIG. 4.12
Filtration using a Pasteur pipette and a reaction tube.

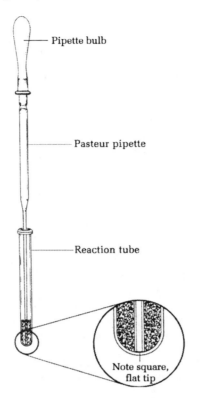

FIG. 4.13
Drying crystals under reduced pressure in a reaction tube.

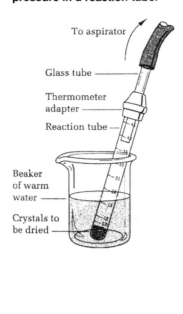

damp crystals under vacuum (Fig. 4.13). Alternatively, the last traces of solvent can be removed by centrifugation using a Wilfilter (Fig. 4.14).

(B) Filtration Using a Hirsch Funnel

When the volume of material to be filtered is greater than 1.5 mL, collect the material on a Hirsch funnel. The Hirsch funnel in the Williamson/Kontes kit[3] is unique. It is composed of polypropylene and has an integral molded stopper that fits a 25-mL filter flask. It comes fitted with a 20-μm polyethylene fritted disk, which is not meant to be disposable, although it costs only about twice as much as an 11-cm piece of filter paper (Fig. 4.15). Although products can be collected directly on this disk, it is good practice to place an 11- or 12-mm diameter piece of No. 1 filter paper on the disk. In this way, the frit will not become clogged with insoluble impurities. The disk of filter paper can be cut to size with a cork borer or leather punch. A piece of filter paper *must* be used on the old-style porcelain Hirsch funnels.

Video: Microscale Filtration on the Hirsch Funnel

Clamp the clean, dry 25-mL filter flask in an ice bath to prevent it from falling over and place the Hirsch funnel with filter paper in the flask. The reason for cooling the filter flask is to keep the mother liquor cold so it will not dissolve the crystals on the Hirsch funnel when fresh cold solvent is used to wash crystals from the container onto the funnel. In a separate flask, cool ~10 mL of solvent in an ice bath; this solvent is used for washing the recrystallization flask and for washing the crystals. Wet the filter paper with the solvent used in the recrystallization, turn on the water aspirator (see "The Water Aspirator and the Trap"), and ascertain that

---

[3]The microscale kit is available through Kontes (www.kontes.com).

FIG. 4.14

The Wilfilter filtration apparatus. Filtration occurs between the flat face of the polypropylene Wilfilter and the top of the reaction tube.

FIG. 4.15

The Hirsch funnel being used for vacuum filtration. This unique design has a removable and replaceable 20-μm polyethylene frit. No adapter is needed because there is a vacuum-tight fit to the filter flask. Always use a piece of filter paper with this funnel.

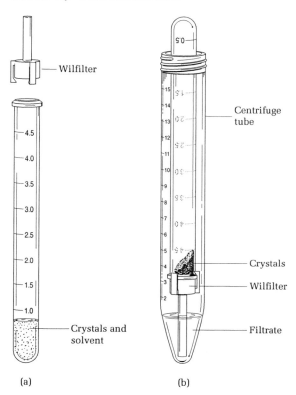
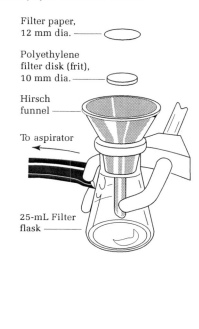

Break the vacuum, add a very small quantity of ice-cold wash solvent, and reapply the vacuum.

the filter paper is pulled down onto the frit. Pour and scrape the crystals and mother liquor onto the Hirsch funnel and, as soon as the liquid is gone from the crystals, break the vacuum at the filter flask by removing the rubber hose.

Rinse the recrystallization flask with ice-cold fresh solvent. Pour this rinse through the Hirsch funnel and reapply vacuum to the filter flask. As soon as all the liquid has disappeared from the crystals, wash the crystals with a few drops of ice-cold solvent. Repeat this washing process as many times as necessary to remove colored material or other impurities from the crystals. In some cases, only one very small wash will be needed. After the crystals have been washed with ice-cold solvent, the vacuum can be left on to dry the crystals. A cork can be used to press the solvent from the crystals, if necessary.

(C) Filtration with a Wilfilter (Replacing a Craig Tube)

The isolation of less than 100 mg of recrystallized material from a reaction tube (or any other container) is not easy. If the amount of solvent is large enough (1 mL or more), the material can be recovered by filtration with a Hirsch funnel. But when the volume of liquid is less than 1 mL, much product is left in the tube during the transfer to a Hirsch funnel. The solvent can be removed with a Pasteur pipette pressed against the bottom of the tube, a very effective filtration technique, but scraping the damp crystals from the reaction tube results in major losses. If the solvent has a relatively low boiling point, it can be evaporated by connecting the tube to a water aspirator (*see* Fig. 4.13 on page 72). Once the crystals are dry, they are easily scraped from the tube with little or no loss. Some solvents—and water is the principal culprit—are not easily removed by evaporation. And even though removal of the solvent under vacuum is not terribly difficult, it takes time.

We have invented a filtration device that circumvents these problems: the Wilfilter. After recrystallization has ceased, most of the solvent is removed from the crystals using a Pasteur pipette in the usual way (*see* Fig. 4.12 on page 72). Then the polypropylene Wilfilter is placed on the top of the reaction tube, followed by a 15-mL polypropylene centrifuge tube (*see* Fig. 4.14 on page 73). The assembly is inverted and placed in a centrifuge such as the International Clinical Centrifuge that holds twelve 15-mL tubes. The assembly, properly counterbalanced, is centrifuged for about 1 minute at top speed. The centrifuge tube is removed from the centrifuge, and the reaction tube is then removed from the centrifuge tube. The three fingers on the Wilfilter keep it attached to the reaction tube. The filtrate is left in the centrifuge tube.

Filtration with the Wilfilter occurs between the top surface of the reaction tube and the flat surface of the Wilfilter. Liquid will pass through that space during centrifugation, but crystals will not. The crystals will be found on top of the Wilfilter and inside the reaction tube. The very large centrifugal forces remove all the liquid, so the crystals will be virtually dry and thus easily removed from the reaction tube by shaking or scraping with the metal spatula.

The Wilfilter replaces an older device known as a Craig tube (Fig. 4.16), which consists of an outer tube of 1-, 2-, or 3-mL capacity with an inner plunger made of Teflon (expensive) or glass (fragile). The material to be recrystallized is transferred to the outer tube and recrystallized in the usual way. The inner plunger is added, and a wire hanger is fashioned so that the assembly can be removed from the centrifuge tube without the plunger falling off. Filtration in this device occurs through the rough surface that has been ground into the shoulder of the outer tube.

The Wilfilter possesses several advantages: a special recrystallization device is not needed, no transfers of material are needed, it is not as limited in capacity (which is 4.5 mL), and its cost is one-fifth that of a Craig tube assembly.

FIG. 4.16
The Craig tube filtration apparatus. Filtration occurs between the rough ground glass surfaces when the apparatus is centrifuged.

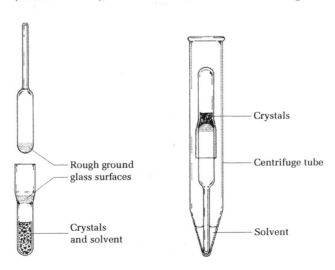

Chapter 4 ■ *Recrystallization* 75

(D) Filtration into a Reaction Tube on a Hirsch Funnel

If it is desired to have the filtrate in a reaction tube instead of spread on the bottom of a 25-mL filter flask, then the process described in the previous section can be carried out in the apparatus shown in Figure 4.17. The vacuum hose is connected to the side arm of the flask using a thermometer adapter and a short length of glass tubing. Evaporate the filtrate in the reaction tube to collect a second crop of crystals.

Photo: *Vacuum Filtration into a Reaction Tube through a Hirsch Funnel*

(E) Filtration into a Reaction Tube with a Micro Büchner Funnel

If the quantity of material being collected is very small, use the bottom of the chromatography column as a micro Büchner funnel, which can be fitted into the top of the thermometer adapter, as shown in Figure 4.18. Again, it is good practice to cover the frit with a piece of 6-mm filter paper (cut with a cork borer).

Video: *Microscale Crystallization*

(F) The Micro Büchner Funnel in an Enclosed Filtration Apparatus

In the apparatus shown in Figure 4.19, recrystallization is carried out in the upper reaction tube in the normal way. The apparatus is then turned upside down, the crystals are shaken down onto a micro Büchner funnel, and a vacuum is applied through the side arm. In this apparatus, crystals can be collected in an oxygen-free atmosphere (Schlenk conditions).

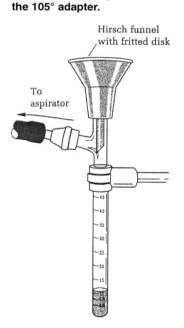

FIG. 4.17

A microscale Hirsch filtration assembly. The Hirsch funnel gives a vacuum-tight seal to the 105° adapter.

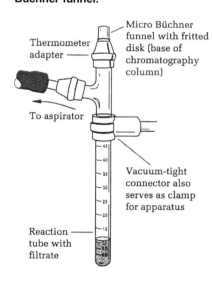

FIG. 4.18

Filtration using a microscale Büchner funnel.

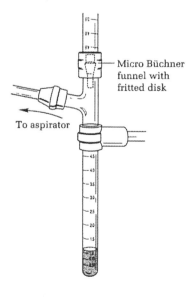

FIG. 4.19

The Schlenk-type filtration apparatus. The apparatus is inverted to carry out the filtration.

FIG. 4.20
Matching filter assemblies. The 6.0-cm polypropylene Büchner funnel (right) resists breakage and can be disassembled for cleaning.

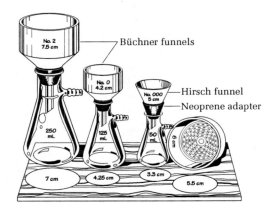

 Macroscale Apparatus

Filtration on a Hirsch Funnel and a Büchner Funnel

If the quantity of material is small (<2 g), a Hirsch funnel can be used in exactly the way that was described in a previous section. For larger quantities, a Büchner funnel is used. Properly matched Büchner funnels, filter paper, and flasks are shown in Figure 4.20. The Hirsch funnel shown in the figure has a 5-cm bottom plate to accept 3.3-cm round filter paper.

Place a piece of filter paper in the bottom of the Büchner funnel. Wet it with solvent and be sure it lies flat so that crystals cannot escape around the edge and go under the filter paper. Then, with the vacuum off, pour the cold slurry of crystals into the center of the filter paper. Apply the vacuum; as soon as the liquid disappears from the crystals, break the vacuum to the flask by disconnecting the hose. Rinse the Erlenmeyer flask with cold solvent. Add this to the crystals and reapply the vacuum just until the liquid disappears from the crystals. Repeat this process as many times as necessary to recover all the crystals from the Erlenmeyer flask, and then leave the vacuum on to dry the crystals.

The Water Aspirator and Trap

The most common way to produce a vacuum for filtration in the organic laboratory is by employing a *water aspirator*. Air is entrained efficiently in the water rushing through the aspirator to produce a vacuum roughly equal to the vapor pressure of the water going through it (17 torr at 20°C, 5 torr at 4°C). A check valve is built into the aspirator, but when the water is turned off, it will often back up into the evacuated system. For this reason, a trap is always installed in the line (Fig. 4.21). *The water passing through the aspirator should always be turned on full force.*

Opening the screw clamp on the trap can open the system to the atmosphere, as well as removing the hose from the small filter flask. Open the system and then turn off the water to avoid having water sucked back into the filter trap. Thin rubber tubing on the top of the trap will collapse and bend over when a good vacuum is established. You will, in time, learn to hear the differences in the sound of an aspirator when it is pulling a vacuum and when it is working on an open system.

Collecting a Second Crop of Crystals

Regardless of the method used to collect crystals on either a macroscale or a microscale, the filtrate and washings can be combined and evaporated to the point of saturation to obtain a second crop of crystals—hence this advocates having a

Video: Microscale Filtration with the Hirsch Funnel

Clamp the filter flask

FIG. 4.21

An aspirator, a filter trap, and a funnel. Clamp the small filter flask to prevent it from tipping over.

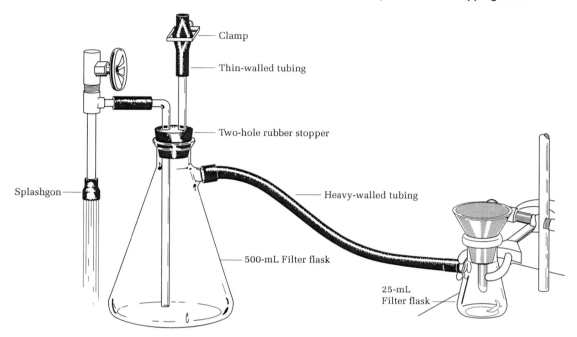

clean receptacle for the filtrate. This second crop will increase the overall yield, but the crystals will not usually be as pure as the first crop.

STEP 7. DRYING THE PRODUCT

Microscale Procedure

 Video: Recrystallization

FIG. 4.22
Drying a solid by reduced air pressure.

If possible, dry the product in the reaction tube after removing the solvent with a Pasteur pipette. Simply connecting the tube to the water aspirator can do this. If the tube is clamped in a beaker of hot water, the solvent will evaporate more rapidly under vacuum, but make sure not to melt the product (see Fig. 4.13 on page 72). Water, which has a high heat of vaporization, is difficult to remove in this way. Scrape the product onto a watch glass and allow it to dry to constant weight, which will indicate that all the solvent has been removed. If the product is collected on a Hirsch funnel or a Wilfilter, the last traces of solvent can be removed by squeezing the crystals between sheets of filter paper before drying them on the watch glass.

Macroscale Procedure

Once the crystals have been washed on a Hirsch funnel or a Büchner funnel, press them down with a clean cork or other flat object and allow air to pass through them until they are substantially dry. Final drying can be done under reduced pressure (Fig. 4.22). The crystals can then be turned out of the funnel and squeezed between sheets of filter paper to remove the last traces of solvent before the final drying on a watch glass.

EXPERIMENTS

1. SOLUBILITY TESTS

Video: *Picking a Solvent*

Test Compounds:

Resorcinol

Anthracene

Benzoic acid

4-Amino-1-naphthalenesulfonic acid, sodium salt

To test the solubility of a solid, transfer an amount roughly estimated to be about 10 mg (the amount that forms a symmetrical mound on the end of a stainless steel spatula) into a reaction tube and add about 0.25 mL of solvent from a calibrated dropper or pipette. Stir with a fire-polished stirring rod that is 4 mm in diameter, break up any lumps, and determine if the solid is readily soluble at room temperature (about 22°C). If the substance is readily soluble at about 22°C in methanol, ethanol, acetone, or acetic acid, add a few drops of water to the solution from a wash bottle to see if a solid precipitates. If so, heat the mixture, adjust the composition of the solvent pair to produce a hot solution saturated at the boiling point, let the solution stand undisturbed, and note the character of the crystals that form.

If the substance fails to dissolve in a given solvent at about 22°C, heat the suspension and observe if a solution occurs. If the solvent is flammable, heat the test tube on a steam bath or in a small beaker of water kept warm on a steam bath or a hot plate. If the solid dissolves completely, it can be declared readily soluble in the hot solvent; if some but not all dissolves, it is said to be moderately soluble, and further small amounts of solvent should then be added until dissolution is complete.

When a substance has been dissolved in hot solvent, cool the solution by holding the flask under the tap and, if necessary, induce recrystallization by rubbing the walls of the tube with a stirring rod to make sure that the concentration permits recrystallization. Then reheat to dissolve the solid, let the solution stand undisturbed, and inspect the character of the ultimate crystals.

Perform solubility tests on the test compounds that are shown in the margin, in each of the following solvents: water (hydroxylic and ionic), toluene (an aromatic hydrocarbon), and hexanes (a mixture of isomers of hexane). Note the degree of solubility in the solvents—cold and hot—and suggest suitable solvents, solvent pairs, or other expedients for the recrystallization of each substance. Record the crystal form, at least to the extent of distinguishing between needles (pointed crystals), plates (flat and thin), and prisms. How do your observations conform to the generalization that like dissolves like?

Cleaning Up. Place organic solvents and solutions of the compounds in the organic solvents waste container. Dilute the aqueous solutions with water and flush down the drain. (For this and all other "Cleaning Up" sections, refer to the complete discussion of waste disposal procedures in Chapter 2.)

2. RECRYSTALLIZATION OF PURE PHTHALIC ACID, NAPHTHALENE, AND ANTHRACENE

Phthalic acid

Naphthalene

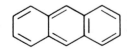
Anthracene

The process of recrystallization can be readily observed using phthalic acid. In the *CRC Handbook of Chemistry and Physics*, in the table "Physical Constants of Organic Compounds," the entry for phthalic acid gives the following solubility data (in grams of solute per 100 mL of solvent). The superscripts refer to temperature in degrees Celsius:

Water	Alcohol	Ether, etc.
0.54^{14}	11.71^{18}	0.69^{15} eth., i. chl.
18^{99}		

The large difference in solubility in water as a function of temperature suggests that water is the solvent of choice. The solubility in alcohol is high at about 22°C. Ether is difficult to use because it is so volatile; the compound is insoluble in chloroform (i. chl.).

Microscale Procedure for Phthalic Acid

Recrystallize 60 mg (0.060 g) of phthalic acid from the minimum volume of water, using the previous data to calculate the required volume. First, turn on an electrically heated sand bath. Add the solid to a 10 × 100 mm reaction tube and add water dropwise with a Pasteur pipette. Use the calibration marks found in Chapter 1 (*see* Fig. 1.18 on page 19) to measure the volume of water in the pipette and the reaction tube. Add a boiling stick (a wood applicator stick) to facilitate even boiling and prevent bumping. After a portion of the water has been added, gently heat the solution to boiling on a hot sand bath in the electric heater. The deeper the tube is placed in the sand, the hotter it will be. As soon as boiling begins, continue to add water dropwise until the entire solid just dissolves. Remove the boiling stick. Cork the tube, clamp it as it cools, and observe the phenomenon of recrystallization.

After the tube reaches about 22°C, cool it in ice, stir the crystals that have formed with a Pasteur pipette, and expel the air from the pipette as the tip is pushed to the bottom of the tube. When the tip is firmly and squarely seated in the bottom of the tube, release the bulb and withdraw the water. Rap the tube sharply on a wood surface to compress the crystals and remove as much of the water as possible with the pipette. Then cool the tube in ice and add a few drops of ice-cold ethanol to the tube to remove water from the crystals. Connect the tube to a water aspirator and warm it in a beaker of hot water (*see* Fig. 4.13 on page 72). Once all the solvent is removed, use a stainless steel spatula to scrape the crystals onto a piece of filter paper, fold the paper over the crystals, and squeeze out excess water before allowing the crystals to dry to constant weight. Weigh the dry crystals and calculate the percent recovery of—the purified compound.

Microscale Procedure for Naphthalene and Anthracene

Following the previous procedure, recrystallize 40 mg of naphthalene from 80% aqueous methanol or 10 mg of anthracene from ethanol. You may find it more convenient to use a hot water bath to heat these low-boiling alcohols. These are more typical of compounds to be recrystallized in later experiments because they are soluble in organic solvents. It will be much easier to remove these solvents from the crystals under vacuum than it is to remove water from phthalic acid. You will seldom encounter the need to recrystallize less than 30 mg of a solid in these experiments.

Cleaning Up. Dilute the aqueous filtrate with water and flush the solution down the drain. Phthalic acid is not considered toxic to the environment and if desired, can be recycled for future recrystallization experiments. Methanol and ethanol filtrates go into the organic solvents waste container.

 Macroscale Procedure

Using the solubility data for phthalic acid to calculate the required volume, recrystallize 1.0 g of phthalic acid from the minimum volume of water. Add the solid to the smallest practical Erlenmeyer flask and then, using a Pasteur pipette, add water dropwise from a full 10-mL graduated cylinder. A boiling stick (a wood applicator stick) facilitates even boiling and will prevent bumping. After a portion of the water has been added, gently heat the solution to boiling on a hot plate. As soon as boiling begins, continue to add water dropwise until the entire solid dissolves. Remove the boiling stick. Place the flask on a cork ring or other insulator and allow it to cool undisturbed to about 22°C, during which time the recrystallization process can be observed. Slow cooling favors large crystals. Then cool the flask in an ice bath, decant (pour off) the mother liquor (the liquid remaining with the crystals), and remove the last traces of liquid with a Pasteur pipette. Scrape the crystals onto a filter paper using a stainless steel spatula, squeeze the crystals between sheets of filter paper to remove traces of moisture, and allow the crystals to dry. Alternatively, the crystals can be collected on a Hirsch funnel. Compare the calculated volume of water to the volume of water actually used to dissolve the acid. Calculate the percent recovery of dry, recrystallized phthalic acid.

Video: Macroscale Crystallization

Cleaning Up. Dilute the filtrate with water and flush the solution down the drain. Phthalic acid is not considered toxic to the environment and if desired, can be recycled for future recrystallization experiments.

3. DECOLORIZING A SOLUTION WITH DECOLORIZING CHARCOAL

Into a reaction tube, place 1.0 mL of a solution of methylene blue dye that has a concentration of 10 mg per 100 mL of water. Add to the tube about 10 or 12 pieces of decolorizing charcoal, shake, and observe the color over a period of 1–2 minutes. Heat the contents of the tube to boiling (reflux) and observe the color by holding the tube in front of a piece of white paper from time to time. How rapidly is the color removed? If the color is not removed in a minute or so, add more charcoal pellets.

Video: Decolorization of a Solution with Norit. Decolorizing using pelletized Norit

Cleaning Up. Place the Norit charcoal pellets in the nonhazardous solid waste container.

4. DECOLORIZATION OF BROWN SUGAR (SUCROSE, $C_{12}H_{22}O_{11}$)

Raw sugar is refined commercially with the aid of decolorizing charcoal. The clarified solution is seeded generously with small sugar crystals, and excess water is removed under vacuum to facilitate recrystallization. The pure white crystalline product is collected by centrifugation. Brown sugar is partially refined sugar and can be easily decolorized using charcoal.

In a 50-mL Erlenmeyer flask, dissolve 15 g of dark brown sugar in 30 mL of water by heating and stirring. Pour half the solution into another 50-mL flask. Heat one of the solutions nearly to its boiling point, allow it to cool slightly, and add 250 mg (0.25 g) of decolorizing charcoal (Norit pellets) to it. Bring the solution back to near the boiling point for 2 minutes; then filter the hot solution into an Erlenmeyer flask through a fluted filter paper held in a previously heated funnel. Treat the other half of the sugar solution in exactly the same way but use only 50 mg of decolorizing charcoal. Collaborate with a fellow student who will heat the solution for only 15 seconds after adding the charcoal. Compare your results.

Cleaning Up. Decant (pour off) the aqueous layer. Place the Norit in the nonhazardous solid waste container. The sugar solution can be flushed down the drain.

5. RECRYSTALLIZATION OF BENZOIC ACID FROM WATER AND A SOLVENT PAIR

Benzoic acid

Recrystallize 50 mg of benzoic acid from water in the same way that phthalic acid was recrystallized. Then in a dry reaction tube, dissolve another 50-mg sample of benzoic acid in the minimum volume of hot methanol and add water to the hot solution dropwise. When the hot solution becomes cloudy and recrystallization has begun, allow the tube to cool slowly to about 22°C; then cool it in ice and collect the crystals. Compare recrystallization in water to that in the solvent pair.

Cleaning Up. The methanol-water solution can be disposed in the organic solvents waste container or, if regulations permit, diluted with water and flushed down the drain.

6. RECRYSTALLIZATION OF NAPHTHALENE FROM A MIXED SOLVENT

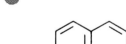

Naphthalene

Add 2.0 g of impure naphthalene (a mixture of 100 g of naphthalene, 0.3 g of a dye such as Congo Red, and another substance such as magnesium sulfate, or dust) to a 50-mL Erlenmeyer flask along with 3 mL of methanol and a boiling stick to promote even boiling. Heat the mixture to boiling over a steam bath or a hot plate, and then add methanol dropwise until the naphthalene just dissolves when the solvent is boiling. The total volume of methanol should be 4 mL. Remove the flask from the heat and cool it rapidly in an ice bath. Note that the contents of the flask set to a solid mass, which would be impossible to handle. Add enough methanol to bring the total volume to 25 mL, heat the solution to its boiling point, remove the flask from the heat, allow it to cool slightly, and add 30 mg of decolorizing charcoal pellets to remove the colored impurity in the solution. Heat the solution to its boiling point for 2 minutes; if the color is not gone, add more Norit and boil again, and then filter through a fluted filter paper in a previously warmed stemless funnel into a 50-mL Erlenmeyer flask. Sometimes filtration is slow because the funnel fits so snugly into the mouth of the flask that a back pressure develops. If you note that raising the funnel increases the flow of filtrate, fold a small strip of paper two or three times and insert it between the funnel and flask. Wash the used flask with 2 mL of hot methanol and use this liquid to wash the filter paper, transferring the

Do not try to grasp Erlenmeyer flasks with a test tube holder.

Support the funnel in a ring stand.

solvent with a Pasteur pipette in a succession of drops around the upper rim of the filter paper. When the filtration is complete, the volume of methanol should be 15 mL. If it is not, evaporate the excess methanol.

Because the filtrate is far from being saturated with naphthalene at this point, it will not yield crystals on cooling. However, the solubility of naphthalene in methanol can be greatly reduced by the addition of water. Heat the solution to its boiling point and add water dropwise from a 10-mL graduated cylinder using a Pasteur pipette (or a precalibrated pipette). After each drop of water, the solution will become cloudy for an instant. Swirl the contents of the flask and heat to redissolve any precipitated naphthalene. After the addition of 3.5 mL of water, the solution will almost be saturated with naphthalene at the boiling point of the solvent. Remove the flask from the heat and place it on a cork ring or other insulating surface to cool, without being disturbed, to about 22°C.

Immerse the flask in an ice bath along with another flask containing a 30:7 mixture of methanol and water. This cold solvent will be used for washing the crystals. The crystals will be separated from the cold recrystallization mixture using vacuum filtration on a small 50-mm Büchner funnel (Fig. 4.23). The water flowing through the aspirator should always be turned on full force. In collecting the product by suction filtration, use a spatula to dislodge crystals and ease them out of the flask. If crystals still remain in the flask, some filtrate can be poured back into the recrystallization flask as a rinse for washing as often as desired because it is saturated with solute. To further purify the crystals, break the suction, pour a few milliliters of the fresh cold solvent mixture into the Büchner funnel, and immediately reapply suction. Repeat this process until the crystals and the filtrate are free of color. Press the crystals with a clean cork to eliminate excess solvent, pull air through the filter cake for a few minutes, and then put the large, flat, platelike crystals out on a filter paper to dry. The yield of pure white crystalline naphthalene should be about 1.6 g. The mother liquor contains about 0.25 g, and about 0.15 g is retained in the charcoal and on the filter paper.

Cleaning Up. Place the Norit in the nonhazardous solid waste container. The methanol filtrate and washings are placed in the organic solvents waste container.

**FIG. 4.23
A suction filter assembly clamped to provide firm support. The funnel must be pressed down on the Filtervac to establish reduced pressure in the flask.**

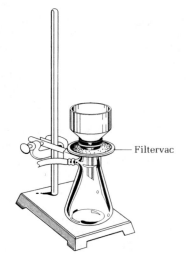

7. PURIFICATION OF AN UNKNOWN

Recall the seven-step recrystallization procedure:

1. Choose the solvent.
2. Dissolve the solute.
3. Decolorize the solution (if necessary).
4. Filter suspended solids (if necessary).
5. Recrystallize the solute.
6. Collect and wash the crystals.
7. Dry the product.

You will purify an unknown provided by your instructor, 2.0 g if working on a macroscale and 100 mg if working on a microscale. Conduct tests for solubility and the ability to recrystallize in several organic solvents, solvent pairs, and water. Conserve your unknown by using very small quantities for the solubility tests. If only a drop or two of solvent is used, heating the test tube on a steam bath or a sand bath can evaporate the solvent, and the residue can be used for another test. If decolorization is necessary, dilute the solution before filtration. It is very difficult to filter a hot, saturated solution from decolorizing carbon without recrystallization occurring in the filtration apparatus. Evaporate the decolorized solution to the point of saturation and proceed with the recrystallization. Submit as much pure product as possible with evidence of its purity (i.e., the melting point). From the posted list, identify the unknown. If an authentic sample is available, your identification can be verified by a mixed melting point determination (see Chapter 3).

Cleaning Up. Place decolorizing charcoal, if used, and filter paper in the nonhazardous solid waste container. Put organic solvents in the organic solvents waste container and flush aqueous solutions down the drain.

RECRYSTALLIZATION PROBLEMS AND THEIR SOLUTIONS

INDUCTION OF CRYSTALLIZATION

Seeding

Occasionally, a sample will not crystallize from solution on cooling, even though the solution is saturated with the solute at an elevated temperature. The easiest method for inducing crystallization is to add to the supersaturated solution a seed crystal that has been saved from the crude material (if it was crystalline before crystallization was attempted). In a probably apocryphal tale, the great sugar chemist Emil Fischer merely had to wave his beard over a recalcitrant solution, and the appropriate seed crystals would drop out, causing recrystallization to occur. In the absence of seed crystals, scratching the inside of the flask with a stirring rod at the liquid-air interface can often induce recrystallization. One theory holds that part of the freshly scratched glass surface has angles and planes corresponding to the crystal structure, and crystals start growing on these spots. Recrystallization is often very slow to begin. Placing the sample in a refrigerator overnight will bring success. Other expedients are to change the solvent (usually to a less soluble one) and to place the sample in an open container where slow evaporation and dust from the air may help induce recrystallization.

Scratching

OILS AND "OILING OUT"

Video: Formation of an Oil Instead of Crystals

Crystallize at a lower temperature

When cooled, some saturated solutions—especially those containing water—deposit not crystals but small droplets referred to as oils. "Oiling out" occurs when the temperature of the solution is above the melting point of the crystals. If these droplets solidify and are collected, they will be found to be quite impure. Similarly, the melting point of the desired compound may be depressed to a point such that a low-melting eutectic mixture of the solute and the solvent comes out of solution. The simplest remedy for this problem is to lower the temperature at which the solution becomes saturated with the solute by simply adding more room-temperature solvent. In extreme cases, it may be necessary to lower this temperature well below 22°C by cooling the solution with dry ice. If oiling out persists, use another solvent.

RECRYSTALLIZATION SUMMARY

Video: Picking a Solvent

Video: Recrystallization

Video: Decolorization of a Solution with Norit

Photo: Preparation of a Filter Pipette, Video: Microscale Crystallization

Photo: Recrystallization; Video: Recrystallization

Photos: Use of the Wilfilter, Filtration Using a Pasteur Pipette; Videos: Microscale Filtration on the Hirsch Funnel, Filtration of Crystals Using the Pasteur Pipette

Photo: Drying Crystals Under Vacuum, Video: Recrystallization

1. **Choosing the solvent.** "Like dissolves like." Some common solvents are water, methanol, ethanol, hexanes, and toluene. When you use a solvent pair, dissolve the solute in the better solvent and add the poorer solvent to the hot solution until saturation occurs. Some common solvent pairs are ethanol-water, ether-hexanes, and toluene—hexanes.

2. **Dissolving the solute.** In an Erlenmeyer flask or reaction tube, add solvent to the crushed or ground solute and heat the mixture to boiling. Add more solvent as necessary to obtain a hot, saturated solution.

3. **Decolorizing the solution.** If it is necessary to remove colored impurities, cool the solution to about 22°C and add more solvent to prevent recrystallization from occurring. Add decolorizing charcoal in the form of pelletized Norit to the cooled solution and then heat it to boiling for a few minutes, making sure to swirl the solution to prevent bumping. Remove the Norit by filtration and then concentrate the filtrate.

4. **Filtering suspended solids.** If it is necessary to remove suspended solids, dilute the hot solution slightly to prevent recrystallization from occurring during filtration. Filter the hot solution. Add solvent if recrystallization begins in the funnel. Concentrate the filtrate to obtain a saturated solution.

5. **Recrystallizing the solute.** Let the hot saturated solution cool to about 22°C spontaneously. Do not disturb the solution. Then cool it in ice. If recrystallization does not occur, scratch the inside of the container or add seed crystals.

6. **Collecting and washing the crystals.** Collect the crystals using the Pasteur pipette method, the Wilfilter, or by vacuum filtration on a Hirsch funnel or a Büchner funnel. If the latter technique is employed, wet the filter paper with solvent, apply vacuum to secure the paper, break vacuum, add crystals and liquid, apply vacuum until solvent just disappears, break vacuum, add cold wash solvent, apply vacuum, and repeat until crystals are clean and filtrate comes through clear.

7. **Drying the product.** Press the wet product on the filter to remove solvent. Then remove it from the filter, squeeze it between sheets of filter paper to remove more solvent, and spread it on a watch glass to dry.

QUESTIONS

1. A sample of naphthalene, which should be pure white, was found to have a grayish color after the usual purification procedure. The melting point was correct, and the melting point range was small. Explain the gray color.
2. How many milliliters of boiling water are required to dissolve 25 g of phthalic acid? If the solution were cooled to 14°C, how many grams of phthalic acid would recrystallize out?
3. Why should activated carbon be used during a recrystallization?
4. If a little activated charcoal does a good job removing impurities in a recrystallization, why not use a larger quantity?
5. Under which circumstances is it wise to use a mixture of solvents to carry out a recrystallization?
6. Why is gravity filtration rather than suction filtration used to remove suspended impurities and charcoal from a hot solution?
7. Why is a fluted filter paper used in gravity filtration?
8. Why are stemless funnels used instead of long-stem funnels to filter hot solutions through fluted filter paper?
9. Why is the final product from the recrystallization process isolated by vacuum filtration rather than gravity filtration?
10. Why should wood applicator sticks not be used when carrying out a chemical reaction?
11. Why should you never use a beaker for recrystallization?

CHAPTER 5

Distillation

PRELAB EXERCISE: Predict what a plot of temperature versus the volume of distillate will look like for the simple distillation and the fractional distillation of (a) a cyclohexane-toluene mixture and (b) an ethanol-water mixture.

Distillation is a common method for purifying liquids and can be used to determine their boiling points.

The origins of distillation are lost in antiquity, when humans in their thirst for more potent beverages found that dilute solutions of fermented alcohol could be separated into alcohol-rich and water-rich portions by heating the solution to boiling and condensing the vapors above the boiling liquid—the process of distillation.

Because ethyl alcohol (ethanol) boils at 78°C and water boils at 100°C, one might naïvely assume that heating a 50:50 mixture of ethanol and water to 78°C would cause the ethanol molecules to leave the solution as a vapor that could be condensed as pure ethanol. However, in such a mixture of ethanol and water, the water boils at about 87°C, and the vapor above the mixture is not 100% ethanol.

A liquid contains closely packed but mobile molecules of varying energy. When a molecule of the liquid approaches the vapor-liquid boundary and possesses sufficient energy, it may pass from the liquid phase into the gas phase. Some of the molecules present in the vapor phase above the liquid may, as they approach the surface of the liquid, reenter the liquid phase and thus become part of the condensed phase. In so doing, the molecules relinquish some of their kinetic energy (i.e., their motion is slowed). Heating the liquid causes more molecules to enter the vapor phase; cooling the vapor reverses this process.

When a closed system is at equilibrium, many molecules are escaping into the vapor phase from the liquid phase, and an equal number are returning from the vapor phase to the liquid phase. The extent of this equilibrium is measured as the vapor pressure. Even when energy is increased and more molecules in the liquid phase have sufficient energy to escape into the vapor phase, equilibrium is maintained because the number moving from the vapor phase into the liquid phase also increases. However, the number of molecules in the vapor phase increases, which increases the vapor pressure. The number of molecules in the vapor phase depends primarily on the volume of the system, the temperature, the combined pressure of all the gaseous components, and the strength of the intermolecular forces exerted in the liquid phase. Review the introduction to Chapter 3 about the types of intermolecular forces.

SIMPLE DISTILLATION

Simple distillation involves boiling a liquid in a vessel (a distilling flask) and directing the resulting vapors through a condenser, in which the vapors are cooled and converted to a liquid that flows down into a collection vessel (a receiving flask). (*See* Fig. 5.5 on page 93.) Simple distillation is used to purify liquid mixtures by separating one liquid component either from nonvolatile substances or from another liquid that differs in boiling point by at least 75°C. The initial condensate will have essentially the same mole ratio of liquids as the vapor just above the boiling liquid. The closer the boiling points of the components of a liquid mixture, the more difficult they are to completely separate by simple distillation.

FRACTIONAL DISTILLATION

Fractional distillation differs from simple distillation in that a fractionating column is placed between the distilling flask and the condenser. This fractionating column allows for successive condensations and distillations and produces a much better separation between liquids with boiling points closer than 75°C. The column is packed with material that provides a large surface area for heat exchange between the ascending vapor and the descending liquid. As a result, multiple condensations and vaporizations occur as the vapors ascend the column. Condensing of the higher-boiling vapor releases heat, which causes vaporization of the lower-boiling liquid on the packing so that the lower-boiling component moves up while the higher-boiling component moves down. Some of the lower-boiling component will run back into the distilling flask. Each successive condensation-vaporization cycle, also called a *theoretical plate*, produces a vapor that is richer in the more volatile fraction. As the temperature of the liquid mixture is increased, the lower-boiling fractions become enriched in the vapor.

Heat exchange between ascending vapor and descending liquid

Column packing

Holdup (noun): unrecoverable distillate that wets the column packing

A large surface area for the packing material is desirable, but the packing cannot be so dense that pressure changes take place within the column to cause non-equilibrium conditions. Also, if the column packing has a very large surface area, it will absorb (hold up) much of the material being distilled. Several different packings for distilling columns have been tried, including glass beads, glass helices, and carborundum chips. One of the best packings in our experience is a copper or steel sponge (brand name—Chore Boy). It is easy to insert into the column; it does not come out of the column as beads do; and it has a large surface area, good heat transfer characteristics, and low holdup. It can be used in both microscale and macroscale apparatus.

The ability of different column packings to separate two materials of differing boiling points is evaluated by calculating the number of theoretical plates. Each theoretical plate corresponds to one condensation-vaporization cycle. Other things being equal, the number of theoretical plates is proportional to the height of the column, so various packings are evaluated according to the *height equivalent to a theoretical plate (HETP)*; the smaller the HETP, the more plates the column will have and the more efficient it will be. The calculation is made by analyzing the proportion of lower- to higher-boiling material at the top of the column and in the distillation pot.[1]

[1]Weissberger, A. ed. *Techniques of Organic Chemistry*, Vol. IV; Wiley-Interscience: New York, 1951.

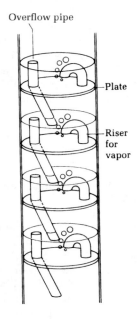

FIG. 5.1

A bubble plate distilling column.

Height equivalent to a theoretical plate (HETP)

Equilibration is slow.

Good fractional distillation takes a long time.

Although not obvious, the most important variable that contributes to good fractional distillation is the rate at which the distillation is carried out. When a series of simple distillations take place within a fractionating column, it is important that complete equilibrium be attained between the ascending vapors and the descending liquid. This process is not instantaneous. It should be an adiabatic process; that is, heat should be transferred from the ascending vapor to the descending liquid with no gain of heat or net heat loss to the surroundings. Advanced distillation systems use thermally insulated, vacuum-jacketed fractionating columns. They also allow the adjustment of the ratio between the amount of condensate that is directed to the receiving flask and the amount returned to the distillation column. A reflux ratio of 30:1 or 50:1 is not uncommon for a 40-plate column. Although a distillation of this type takes several hours, this is far less time than if one to had to do 40 distillations, one after the other, and yields much better separated compounds.

Perhaps it is easiest to understand the series of redistillations that occur in fractional distillation by examining the bubble plate column used to fractionally distill crude oil (Fig. 5.1). These columns dominate the skyline at oil refineries, with some being 150 ft high and capable of distilling 200,000 barrels of crude oil per day. The crude oil enters the column as a hot vapor. Some of this vapor with high-boiling components condenses on one of the plates. The more volatile substances travel through the bubble cap to the next higher plate, where some of the less-volatile components condense. As high-boiling liquid material accumulates on a plate, it descends through the overflow pipe to the next lower plate, and vapor rises through the bubble cap to the next higher plate. The temperature of the vapor that is rising through a cap is above the boiling point of the liquid on that plate. As bubbling takes place, heat is exchanged, and the less volatile components on that plate vaporize and go on to the next plate. The composition of the liquid on a plate is the same as that of the vapor coming from the plate below. So, on each plate, a simple distillation takes place. At equilibrium, vapor containing low-boiling material is ascending through the column, and high-boiling liquid is descending.

As a purification method, distillation, particularly fractional distillation, requires larger amounts of material than recrystallization, liquid/liquid extraction, or chromatography. Performing a fractional distillation on less than 1 g of material is virtually impossible. Fractional distillation can be carried out on a scale of about 3–4 g. As will be seen in Chapters 8, 9, and 10, various types of chromatography are employed for separations of milligram quantities of liquids.

LIQUID MIXTURES

If two different liquid compounds are mixed, the vapor above the mixture will contain some molecules of each component. Let us consider a mixture of cyclohexane and toluene. The vapor pressures, as a function of temperature, are plotted in Figure 5.2. When the vapor pressure of the liquid equals the applied pressure, the liquid boils. Figure 5.2 shows that at 760 mm Hg (standard atmospheric pressure), these pure liquids boil at about 81°C and 111°C, respectively. If one of these pure liquids were to be distilled, we would find that the boiling point of the liquid would equal the temperature of the vapor, and that the temperature of the vapor would remain constant throughout the distillation.

FIG. 5.2
Vapor pressure versus temperature plots for cyclohexane and toluene.

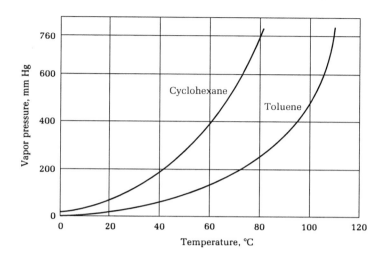

Figure 5.3 is a boiling point composition diagram for the cyclohexane-toluene system. If a mixture of 75 mole percent toluene and 25 mole percent cyclohexane is heated, we find that it boils at 100°C (point A). Above a binary mixture of cyclohexane and toluene, the vapor pressure has contributions from each component. Raoult's law states that the vapor pressure of the cyclohexane is equal to the product of the vapor pressure of pure cyclohexane and the mole fraction of cyclohexane in the liquid mixture:

Raoult's law of partial pressures

$$P_c = P_c^° N_c$$

The mole fraction of cyclohexane is equal to the moles of cyclohexane in the mixture divided by the total number of moles (cyclohexane plus toluene) in the mixture.

where P_c is the partial pressure of cyclohexane, $P_c^°$ is the vapor pressure of pure cyclohexane at the given temperature, and N_c is the mole fraction of cyclohexane in the mixture. Similarly, for toluene,

$$P_t = P_t^° N_t$$

The total vapor pressure above the solution (P_{Tot}) is given by the sum of the partial pressures due to cyclohexane and toluene:

$$P_{Tot} = P_c + P_t$$

Dalton's law states that the mole fraction of cyclohexane (X_c) in the vapor at a given temperature is equal to the partial pressure of the cyclohexane at that temperature divided by the total pressure:

$$X_c = \frac{P_c}{\text{total vapor pressure}}$$

At 100°C, cyclohexane has a partial pressure of 433 mm Hg, and toluene has a partial pressure of 327 mm Hg; the sum of the partial pressures is 760 mm Hg, so the liquid boils. If some of the liquid in equilibrium with this boiling mixture were condensed and analyzed, it would be found to be 433/760, or 57 mole percent cyclohexane (point B, Fig. 5.3). This is the best separation that can be achieved with a simple distillation of this mixture. As the simple distillation proceeds, the boiling point of the mixture moves toward 111°C along the line from point A, and the vapor

FIG. 5.3
Boiling point-composition curves for a mixture of cyclohexane and toluene.

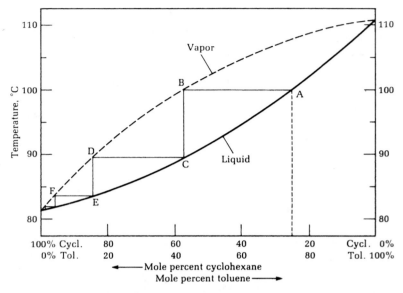

composition becomes richer in toluene as it moves from point B to 110°C. To obtain pure cyclohexane, it would be necessary to condense the liquid at point B and redistill it. When this is done, it is found that the liquid boils at 90°C (point C), and the vapor equilibrium with this liquid is about 85 mole percent cyclohexane (point D). Therefore, to separate a mixture of cyclohexane and toluene, a series of fractions would be collected, and each of these would be partially redistilled. If this fractional distillation were done enough times, the two components could be completely separated.

AZEOTROPES

The ethanol-water azeotrope.

Not all liquids form ideal solutions and conform to Raoult's law. Ethanol and water are two such liquids. Because of molecular interaction, a mixture of 95.5% (by weight) of ethanol and 4.5% of water boils *below* the boiling point of pure ethanol (78.15°C versus 78.3°C). Thus, no matter how efficient the distilling apparatus, 100% ethanol cannot be obtained by distillation of a mixture of, say, 75% water and 25% ethanol. A mixture of liquids of a certain definite composition that distills at a constant temperature without a change in composition is called an *azeotrope*; 95% ethanol is such an azeotrope. The boiling point-composition curve for the ethanol-water mixture is seen in Figure 5.4. To prepare 100% ethanol, the water can be removed chemically (by reaction with calcium oxide) or it can be removed as an azeotrope with still another liquid. An azeotropic mixture of 32.4% ethanol and 67.6% benzene (bp 80.1°C) boils at 68.2°C. A ternary azeotrope containing 74.1% benzene, 18.5% ethanol, and 7.4% water boils at 64.9°C. Absolute alcohol (100% ethanol) is made by adding benzene to 95% ethanol. The water is removed by distilling the ternary azeotrope of benzene, ethanol, and water, bp 64.9°C, followed by the distillation of 100% ethanol, bp 78.4°C

Ethanol and water form a minimum boiling azeotrope. Substances such as formic acid (bp 100.7°C) and water (bp 100°C) form maximum boiling azeotropes. The boiling point of a formic acid-water azeotrope is 107.3°C.

FIG. 5.4

Boiling point–composition curves for a mixture of ethanol and water.

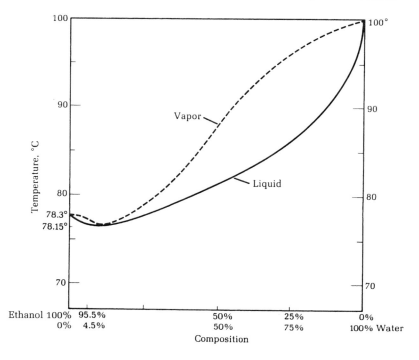

BOILING POINTS AND DISTILLATION

A pure liquid has a constant boiling point. A change in boiling point during distillation is an indication of impurity. The converse proposition, however, is not always true; that is, constancy of a boiling point does not necessarily mean that the liquid consists of only one compound. For instance, two miscible liquids of similar chemical structure that boil at the same temperature individually will have nearly the same boiling point as a mixture. And, as noted previously, azeotropes have constant boiling points that can be either above or below the boiling points of the individual components.

A constant boiling point on distillation does not guarantee that the distillate is a single pure compound.

Distilling a mixture of sugar and water.

When a solution of sugar in water is distilled, the boiling point recorded on a thermometer located in the vapor phase is 100°C (at 760 torr) throughout the distillation, whereas the temperature of the boiling sugar solution itself is initially somewhat above 100°C and continues to rise as the concentration of sugar in the remaining solution increases. The vapor pressure of the solution is dependent on the number of water molecules present in a given volume; hence, with increasing concentration of nonvolatile sugar molecules and decreasing concentration of water, the vapor pressure at a given temperature decreases, and a higher temperature is required for boiling. However, sugar molecules do not leave the solution, and the drop clinging to the thermometer is pure water in equilibrium with pure water vapor.

Boiling point changes with pressure.

When a distillation is carried out in a system open to the air (the boiling point is thus dependent on existing air pressure), the prevailing barometric pressure should be noted and allowance made for appreciable deviations from the accepted boiling point temperature (Table 5.1). Distillation can also be done at the lower pressures that can be achieved using an oil pump or an aspirator, which produces a substantial reduction in the boiling point.

TABLE 5.1 *Variation in Boiling Point with Pressure*

Pressure (mm Hg)	Boiling Point	
	Water (°C)	Benzene (°C)
780	100.7	81.2
770	100.4	80.8
760	100.0	80.3
750	99.6	79.9
740	99.2	79.5
584*	92.8	71.2

*Instituto de Quimica, Mexico City, altitude 7700 ft (2310 m).

EXPERIMENTS

Before beginning any distillation, calibrate the thermometer to ensure accurate readings are made; refer to Part 3 of Chapter 3 for calibration instructions.

1. SIMPLE DISTILLATION

Apparatus for simple distillation.

IN THIS EXPERIMENT, the two liquids to be separated are placed in a 5-mL round-bottomed, long-necked flask that is fitted to a distilling head (Fig. 5.5). The flask has a larger surface area exposed to heat than does the reaction tube, so the necessary thermal energy can be put into the system to cause the materials to distill. The hot vapor rises and completely envelops the bulb of the thermometer before passing over it and down toward the receiver. The downward-sloping portion of the distilling head functions as an air condenser. The objective is to observe how the boiling point of the mixture changes during the course of its distillation. (Another simple distillation apparatus is shown in Figure 5.6. Here, the long air condenser will condense even low-boiling liquids, and the receiver is far from the heat.)

Photo: *Simple Distillation Apparatus*

The rate of distillation is determined by the heat input to the apparatus. This is most easily and effectively controlled by using a spatula to pile up or scrape away hot sand from around the flask.

(A) Simple Distillation of a Cyclohexane-Toluene Mixture

To a 5-mL long-necked, round-bottomed flask, add 2.0 mL of dry cyclohexane, 2.0 mL of dry toluene, and a boiling chip (*see* Fig. 5.5). This flask is joined by means of a Viton (black) connector to a distilling head fitted with a thermometer using a rubber connector. The thermometer bulb should be completely below the side arm of the Claisen head so that the mercury reaches the same temperature as the vapor that distills. The end of the distilling head dips well down into a receiving vial, which rests on the bottom of a 30-mL beaker filled with ice. The distillation is started by piling up hot sand to heat the flask. As soon as boiling begins, the vapors can be seen to rise up the

FIG. 5.5

A small-scale simple distillation apparatus. This apparatus can be adapted for fractional distillation by packing the long neck with a copper sponge. The temperature is regulated by either scraping sand away from or piling sand up around the flask.

FIG. 5.6

A simple distillation apparatus.

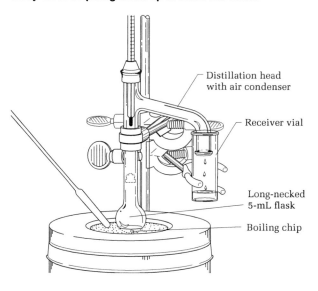

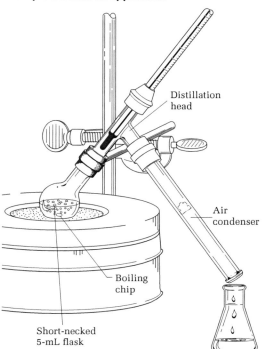

Viton is resistant to hot aromatic vapors.

The thermometer bulb must be *completely below* **the side arm.**

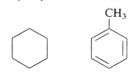

Cyclohexane
bp 81°C
MW 84.16
n_D^{20} 1.4260

Toluene
bp 111°C
MW 92.14
n_D^{20} 1.4960

Throughout this text, information regarding the physical properties of substances has been placed with each structure in the margin. MW is molecular weight, bp is boiling point, den. is density in g/mL, and $nD\ 20$ is the refractive index.

There are 21 ± 3 drops per milliliter.

neck of the flask. Adjust the rate of heating by piling up or scraping away sand from the flask so that it takes *several minutes* for the vapor to rise to the thermometer. **The rate of distillation should be no faster than 2 drops per minute.**

Record the temperature versus the number of drops during the entire distillation process. If the rate of distillation is as slow as it should be, there will be sufficient time between drops to read and record the temperature. Continue the distillation until only about 0.4 mL remains in the distilling flask. **Never distill to dryness.** On a larger scale, explosive peroxides can sometimes accumulate. At the end of the distillation, measure as accurately as possible, perhaps with a syringe, the volume of the distillate and, after it cools, the volume left in the pot (round-bottomed flask); the difference is the holdup of the column if none has been lost by evaporation. Note the barometric pressure, make any thermometer corrections necessary, and make a plot of milliliters (drop number) versus temperature for the distillation.

Cleaning Up. The pot residue should be placed in the organic solvents container. The distillate can also be placed there or recycled.

(B) Simple Distillation of an Ethanol-Water Mixture

In a 5-mL round-bottomed, long-necked flask, place 4 mL of a 10% to 20% ethanol-water mixture. Assemble the apparatus as described previously and carry out the distillation until you believe a representative sample of ethanol has collected in the receiver. In the hood, place 3 drops of this sample on a Pyrex watch glass and try to ignite it with the blue cone of a microburner flame. Does it burn? Is any unburned residue observed? There was a time when alcohol-water mixtures were mixed with gunpowder and ignited to give proof that the alcohol had not been diluted. One hundred proof alcohol is 50% ethanol by volume.

Cleaning Up. The distillate and pot residue can be disposed in the organic solvents waste container or, if regulations permit, diluted with water and flushed down the drain.

2. FRACTIONAL DISTILLATION

Photos: *Column Packing with Chore Boy for Fractional Distillation, Fractional Distillation Apparatus*

> **IN THIS EXPERIMENT**, just as in the last one, you will distill a mixture of two liquids and again record the boiling points as a function of the volume of distillate (drops). The necessity for a very slow rate of distillation cannot be overemphasized.

Apparatus

Assemble the apparatus shown in Figure 5.7. The 10-cm column is packed with 1.5 g of copper sponge and connected to the 5-mL short-necked flask using a black (Viton) connector. The column should be vertical and care should be taken to ensure that the bulb of the thermometer does not touch the side of the distilling head. The column, but not the distilling head, will be insulated with glass wool or cotton at the appropriate time to ensure that the process is adiabatic. Alternatively,

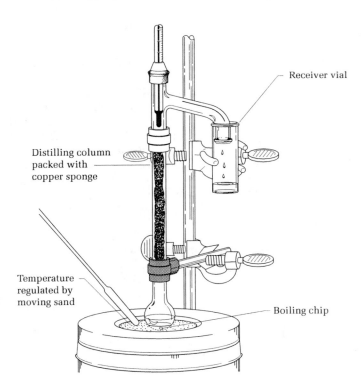

FIG. 5.7

A small-scale fractional distillation apparatus. The 10-cm column is packed with 1.5 g of copper sponge (Chore Boy).

(A) Fractional Distillation of a Cyclohexane-Toluene Mixture

To a short-necked flask, add 2.0 mL of cyclohexane, 2.0 mL of toluene, and a boiling chip. The distilling column is packed with 1.5 g of copper sponge (Fig. 5.7). The mixture is brought to a boil over a hot sand bath. Observe the ring of condensate that should rise slowly through the column; if you cannot at first see this ring, locate it by touching the column with your fingers. It will be cool above the ring and hot below. Reduce the heat by scraping sand away from the flask and wrap the column, but not the distilling head, with glass wool or cotton if it is not already insulated.

> Never distill in an airtight system.
>
> Adjust the heat input to the flask by piling up or scraping away sand around the flask.

The distilling head and the thermometer function as a small reflux condenser. Again, apply the heat, and as soon as the vapor reaches the thermometer bulb, reduce the heat by scraping away sand. **Distill the mixture at a rate no faster than 2 drops per minute** and record the temperature as a function of the number of drops. If the heat input has been *very* carefully adjusted, the distillation will cease, and the temperature reading will drop after the cyclohexane has distilled. Increase the heat input by piling up the sand around the flask to cause the toluene to distill. Stop the distillation when only about 0.4 mL remains in the flask, and measure the volume of distillate and the pot residue as before. Make a plot of the boiling point versus the milliliters of distillate (drops) and compare it to the simple distillation carried out in the same apparatus. Compare your results with those in Figure 5.8.

> Insulate the distilling column but not the Claisen head.
>
> 21 ± 3 drops = 1 mL

Cleaning Up. The pot residue should be placed in the organic solvents container. The distillate can also be placed there or recycled.

(B) Fractional Distillation of an Ethanol-Water Mixture

Distill 4 mL of the same ethanol-water mixture used in the simple distillation experiment, following the procedure used for the cyclohexane-toluene mixture with either the short or the long distilling column. Remove what you regard to be the ethanol fraction and repeat the ignition test. Is any difference noted?

Cleaning Up. The pot residue and distillate can be disposed in the organic solvents waste container or, if regulations permit, diluted with water and flushed down the drain.

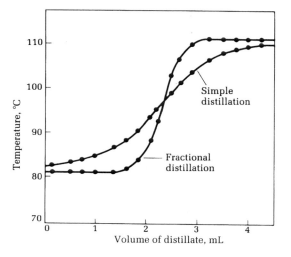

FIG. 5.8
Simple and fractional distillation curves for cyclohexane and toluene.

3. INSTANT MICROSCALE DISTILLATION

Video: *Instant Microscale Distillation*

Frequently, a very small quantity of freshly distilled material is needed in an experiment. For example, two compounds that need to be distilled freshly are aniline, which turns black because of the formation of oxidation products, and benzaldehyde, a liquid that easily oxides to solid benzoic acid. The impurities that arise in both of these compounds have much higher boiling points than the pure compounds, so a very simple distillation suffices to separate them. This can be accomplished as follows.

Place a few drops of the impure liquid in a reaction tube along with a boiling chip. Clamp the tube in a hot sand bath and adjust the heat so that the liquid refluxes about halfway up the tube. Expel the air from a Pasteur pipette, thrust it down into the hot vapor, and then pull the hot vapor into the cold upper portion of the pipette. The vapor will immediately condense and can then be expelled into another reaction tube that is held adjacent to the hot one (Fig. 5.9). In this way enough pure material can be distilled to determine a boiling point, run a spectrum, make a derivative, or carry out a reaction. Sometimes the first drop or two will be cloudy, which indicates the presence of water. This fraction should be discarded in order to obtain pure dry material.

4. SIMPLE DISTILLATION APPARATUS

In any distillation, the flask should be no more than two-thirds full at the start. Great care should be taken not to distill to dryness because, in some cases, high-boiling explosive peroxides can become concentrated.

Assemble the apparatus for macroscale simple distillation, as shown in Figure 5.10, starting with the support ring followed by the electric flask heater and then the flask. One or two boiling stones are put in the flask to promote even boiling. Each ground joint is greased by putting three or four stripes of grease lengthwise around the male joint and pressing the joint firmly into the other without twisting. The air is thus eliminated, and the joint will appear almost transparent. (Do not use excess grease because it will contaminate the product.) Water enters the condenser at the tubulature nearest the receiver. Because of the large heat capacity of water, only a very small stream (3 mm diameter) is needed; too much water pressure will cause the tubing to pop off. A heavy rubber band, or better, a Keck clamp, can be used to hold the condenser to the distillation head. Note that the bulb of the thermometer is below the opening into the side arm of the distillation head.

(A) Simple Distillation of a Cyclohexane-Toluene Mixture

Place a mixture of 30 mL cyclohexane and 30 mL toluene, and a boiling chip in a dry 100-mL round-bottomed flask and assemble the apparatus for simple distillation. After assuring that all connections are tight, heat the flask strongly until boiling begins. Then adjust the heat until the distillate drops at a regular rate of about 1 drop per second. Record both the temperature and the volume of distillate at regular intervals. After 50 mL of distillate is collected, discontinue the distillation. Record the barometric pressure, make any thermometer correction necessary, and plot the boiling point versus the volume of distillate. Save the distillate for fractional distillation.

Chapter 5 ■ *Distillation* 97

FIG. 5.9
An apparatus for instant microscale distillation.

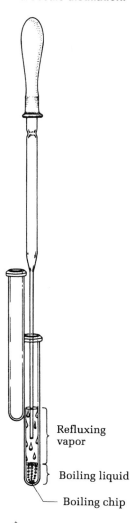

FIG. 5.10
An apparatus for macroscale simple distillation.

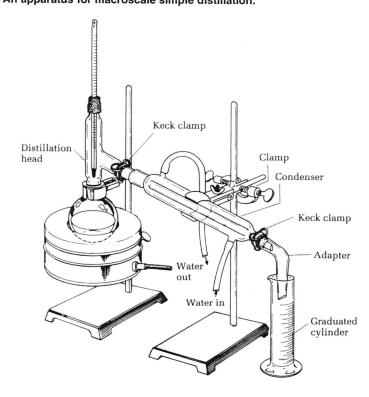

⚠️
Cyclohexane and toluene are flammable; make sure the distilling apparatus is tight.

Do not add a boiling chip to a hot liquid. It may boil over.

Dispose of cyclohexane and toluene in the waste container provided. Do not pour them down the drain.

Cleaning Up. The pot residue should be placed in the organic solvents container. The distillate can also be placed there or recycled.

(B) Simple Distillation of an Ethanol-Water Mixture

In a 500-mL round-bottomed flask, place 200 mL of a 20% aqueous solution of ethanol. Follow the previous procedure for the distillation of a cyclohexane-toluene mixture. Discontinue the distillation after 50 mL of distillate has been collected. Working in the hood, place 3 drops of distillate on a Pyrex watch glass and try to ignite it with the blue cone of a microburner flame. Does it burn? Is any unburned residue observed?

Cleaning Up. The pot residue and distillate can be disposed in the organic solvents waste container or, if regulations permit, diluted with water and flushed down the drain.

5. FRACTIONAL DISTILLATION

Apparatus

Assemble the apparatus shown in Figures 5.11 and 5.12. The fractionating column is packed with one-fourth to one-third of a metal sponge. The column should be perfectly vertical and be insulated with glass wool covered with aluminum foil (shiny side in). However, insulation is omitted for this experiment so that you can observe what is taking place in the column.

FIG. 5.11
An apparatus for macroscale fractional distillation. The position of the thermometer bulb is critical.

FIG. 5.12
A fractionating column and its packing. Use one-third of a copper sponge (Chore Boy).

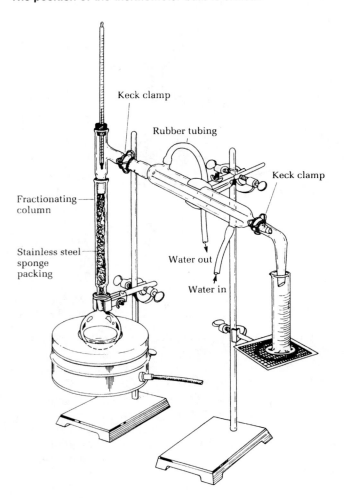

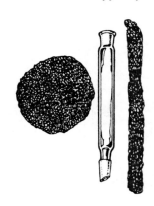

(A) Fractional Distillation of a Cyclohexane-Toluene Mixture

After the flask from the simple macroscale distillation experiment has cooled, pour the 50 mL of distillate back into the distilling flask, add one or two new boiling chips, and assemble the apparatus for fractional distillation. The stillhead delivers into a short condenser fitted with a bent adapter leading into a 10-mL graduated cylinder. Gradually turn up the heat to the electric flask heater until the mixture of cyclohexane and toluene just begins to boil. As soon as boiling starts, turn down the power. Heat slowly at first. A ring of condensate will rise slowly through the column; if you cannot at first see this ring, locate it by cautiously touching the column with your fingers. The rise in temperature should be very gradual so that the column can acquire a uniform temperature gradient. Do not apply more heat until you are sure that the ring of condensate has stopped rising; then increase the heat gradually. In a properly conducted operation, the vapor-condensate mixture reaches the top of the column only after several minutes. Once distillation has commenced, it should continue steadily without any drop in temperature at a rate no greater than 1 mL in 1.5–2 min. Observe the flow and keep it steady by slight increases in heat as required. Protect the column from drafts by wrapping it with aluminum foil, glass wool, or even a towel. This insulation will help prevent flooding of the column, as will slow and steady distillation.

Record the temperature as each milliliter of distillate collects and take more frequent readings when the temperature starts to rise abruptly. Each time the graduated cylinder fills, quickly empty it into a series of labeled 25-mL Erlenmeyer flasks. Stop the distillation when a second constant temperature is reached. Plot a distillation curve and record what you observed inside the column in the course of the fractionation. Combine the fractions that you think are pure, and turn in the product in a bottle labeled with your name, desk number, the name of the product, the boiling point range, and the weight.

Cleaning Up. The pot residue should be placed in the organic solvents waste container. The cyclohexane and toluene fractions can also be placed there or recycled.

(B) Fractional Distillation of Ethanol-Water Mixture

Place the 50 mL of distillate from the simple distillation experiment into a 100-mL round-bottomed flask, add one or two boiling chips, and assemble the apparatus for fractional distillation. Follow the previous procedure for the fractional distillation of a cyclohexane-toluene mixture. Repeat the ignition test. Is any difference noted? Alternatively, distill 60 mL of the 10%–20% ethanol-water mixture that results from the fermentation of sucrose (*see* Chapter 64).

Cleaning Up. The pot residue and distillate can be disposed in the organic solvents waste container or, if regulations permit, diluted with water and flushed down the drain.

6. FRACTIONAL DISTILLATION OF UNKNOWNS

You will be supplied with an unknown, prepared by your instructor, that is a mixture of two solvents listed in Table 5.2, only two of which form azeotropes. The solvents in the mixture will be mutually soluble and differ in boiling point by more than 20°C. The composition of the mixture (in percentages of the two components) will be either 20:80, 30:70, 40:60, 50:50, 60:40, 70:30, or 80:20. Identify the two compounds and determine the percent composition of each. Perform a fractional distillation on

TABLE 5.2 *Some Properties of Common Solvents*

Solvent	Boiling Point (°C)
Acetone	56.5
Methanol	64.7
Hexane	68.8
1-Butanol	117.2
2-Methyl-2-propanol	82.2
Water	100.0
Toluene*	110.6

*Methanol and toluene form an azeotrope with a boiling point of 63.8°C (69% methanol).

4 mL of the unknown for microscale; use at least 50 mL of unknown for macroscale. Fractionate the unknown and identify the components from the boiling points. Prepare a distillation curve. You may be directed to analyze your distillate by gas chromatography (*see* Chapter 10) or refractive index (*see* Chapter 14).

Cleaning Up. Organic material goes in the organic solvents waste container. Water and aqueous solutions can be flushed down the drain.

QUESTIONS

1. In either of the simple distillation experiments, can you account for the boiling point of your product in terms of the known boiling points of the pure components of your mixture? If so, how? If not, why not?
2. From the plots of the boiling point versus the volume of distillate in the simple distillation experiments, what can you conclude about the purity of your product?
3. From the plots of the boiling point versus the volume of distillate in either of the fractional distillations of the cyclohexane-toluene mixture, what conclusion can you draw about the homogeneity of the distillate?
4. From the plots of the boiling point versus the volume of distillate in either of the fractional distillations of the ethanol-water mixture, what conclusion can you draw about the homogeneity of the distillate? Does it have a constant boiling point? If constant, is it a pure substance?
5. What is the effect on the boiling point of a solution (e.g., water) produced by a soluble nonvolatile substance (e.g., sodium chloride)? What is the effect of an insoluble substance such as sand or charcoal? What is the temperature of the vapor above these two boiling solutions?
6. In the distillation of a pure substance (e.g., water), why does all of the water not vaporize at once when the boiling point is reached?
7. In fractional distillation, liquid can be seen running from the bottom of the distillation column back into the distilling flask. What effect does this returning condensate have on the fractional distillation?
8. Why is it extremely dangerous to attempt to carry out a distillation in a completely closed apparatus (one with no vent to the atmosphere)?

Chapter 5 ■ Distillation   101

9. Why is better separation of two liquids achieved by slow rather than fast distillation?
10. Explain why a packed fractionating column is more efficient than an unpacked one.
11. In the distillation of the cyclohexane-toluene mixture, the first few drops of distillate may be cloudy. Explain this occurrence.
12. What effect does the reduction of atmospheric pressure have on the boiling point? Can cyclohexane and toluene be separated if the external pressure is 350 mm Hg instead of 760 mm Hg?
13. When water-cooled condensers are used for distillation or for refluxing a liquid, the water enters the condenser at the lowest point and leaves at the highest. Why?

See Web Links

REFERENCE

An excellent review of distillation at the level of this textbook is *Distillation, an Introduction* by M.T. Tham, http://lorien.ncl.ac.uk/ming/distil/distilop.htm

CHAPTER 7

Extraction

> **PRELAB EXERCISE:** Describe how to separate a mixture of 3-toluic acid and 4'-aminoacetophenone using acid-base liquid/liquid extraction. What species will end up in the aqueous layer if you mix a solution of benzoic acid and aniline in ether with a solution of $NaHCO_3$ (aq)? Draw the structure of this species.

Extraction is one of the oldest chemical operations known to humankind. The preparation of a cup of coffee or tea involves the extraction of flavor and odor components from dried vegetable matter with hot water. Aqueous extracts of bay leaves, stick cinnamon, peppercorns, and cloves, along with alcoholic extracts of vanilla and almond, are used as food flavorings. For the past 150 years or so, organic chemists have extracted, isolated, purified, and characterized the myriad compounds produced by plants that for centuries have been used as drugs and perfumes—substances such as quinine from cinchona bark, morphine from the opium poppy, cocaine from coca leaves, and menthol from peppermint oil. The extraction of compounds from these natural products is an example of solid/liquid extraction—the solid being the natural product and the liquid being the solvent into which the compounds are extracted. In research, a Soxhlet extractor (Fig. 7.1) is often used for solid/liquid extraction.

Although solid/liquid extraction is the most common technique for brewing beverages and isolating compounds from natural products, liquid/liquid extraction is a very common method used in the organic laboratory, specifically when isolating reaction products. Reactions are typically homogeneous liquid mixtures and can therefore be extracted with either an organic or aqueous solvent. Organic reactions often yield a number of byproducts—some inorganic and some organic. Also, because some organic reactions do not go to 100% completion, a small amount of starting material is present at the end of the reaction. When a reaction is complete, it is necessary to do a *workup*, that is, separate and purify the desired product from the mixture of byproducts and residual starting material. Liquid/liquid extraction is a common separation step in this workup, which is then followed by purification of the product. There are two types of liquid/liquid extraction: neutral and acid/base. The experiments in this chapter demonstrate solid/liquid extraction and the two types of liquid/liquid extraction.

FIG. 7.1

The Soxhlet extractor for the extraction of solids such as dried leaves or seeds. The solid is put in a filter paper thimble. Solvent vapor rises in the tube on the right; condensate drops onto the solid in the thimble, leaches out soluble material, and, after initiating an automatic siphon, carries it to the flask where nonvolatile extracted material accumulates. Substances of low solubility can be extracted by prolonged operation.

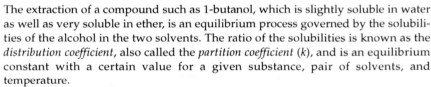

Organic products are often separated from inorganic substances in a reaction mixture by liquid/liquid extraction with an organic solvent. For example, in the synthesis of 1-bromobutane (*see* Chapter 16), 1-butanol, also a liquid, is heated with an aqueous solution of sodium bromide and sulfuric acid to produce the product and sodium sulfate.

$$2\ CH_3CH_2CH_2CH_2OH + 2\ NaBr + H_2SO_4 \rightarrow$$
$$2\ CH_3CH_2CH_2CH_2Br + 2\ H_2O + Na_2SO_4$$

The 1-bromobutane is isolated from the reaction mixture by extraction with *t*-butyl methyl ether, an organic solvent in which 1-bromobutane is soluble and in which water and sodium sulfate are insoluble. The extraction is accomplished by simply adding *t*-butyl methyl ether to the aqueous mixture and shaking it. Two layers will result: an organic layer and an aqueous layer. The *t*-butyl methyl ether is less dense than water and floats on top; it is easily removed/drained away from the water layer and evaporated to leave the bromo product free of inorganic substances, which reside in the aqueous layer.

PARTITION COEFFICIENT

The extraction of a compound such as 1-butanol, which is slightly soluble in water as well as very soluble in ether, is an equilibrium process governed by the solubilities of the alcohol in the two solvents. The ratio of the solubilities is known as the *distribution coefficient*, also called the *partition coefficient* (k), and is an equilibrium constant with a certain value for a given substance, pair of solvents, and temperature.

The *concentration* of the solute in each solvent can be well correlated with the *solubility* of the solute in the pure solvent, a figure that is readily found in solubility tables in reference books. For substance C:

$$k = \frac{\text{concentration of C in } t\text{-butyl methyl ether}}{\text{concentration of C in water}}$$

$$\approx \frac{\text{solubility of C in } t\text{-butyl methyl ether (g/100 mL)}}{\text{concentration of C in water (g/100 mL)}}$$

Consider compound A that dissolves in *t*-butyl methyl ether to the extent of 12 g/100 mL and dissolves in water to the extent of 6 g/100 mL.

$$k = \frac{12\text{g}/100\text{mL } t\text{-butyl methyl ether (g/100 mL)}}{6\text{g}/100\text{mL water}} = 2$$

If a solution of 6 g of A in 100 mL of water is shaken with 100 mL of *t*-butyl methyl ether, then:

$$k = \frac{x\text{g of A}/100\text{mL } t\text{-butyl methyl ether}}{6 - x\text{g of A}/100\text{mL water}}$$

from which

$$x = 4.0 \text{ g of A in the ether layer}$$
$$6 - x = 2.0 \text{ g of A left in the water layer.}$$

It is, however, more efficient to extract the 100 mL of aqueous solution twice with 50-mL portions of *t*-butyl methyl ether rather than once with a 100-mL portion.

$$k = \frac{x \text{g of A}/50\text{mL}}{6 - x \text{g of A}/100\text{mL}} = 2$$

from which

$$x = 3.0 \text{ g of A in the } t\text{-butyl methyl ether layer}$$
$$6 - x = 3.0 \text{ g of A in the water layer.}$$

If this 3.0 g/100 mL of water is extracted again with 50 mL of *t*-butyl methyl ether, we can calculate that 1.5 g of A will be in the ether layer, leaving 1.5 g in the water layer. So two extractions with 50-mL portions of ether will extract 3.0 g + 1.5 g = 4.5 g of A, whereas one extraction with a 100-mL portion of *t*-butyl methyl ether removes only 4.0 g of A. Three extractions with 33-mL portions of *t*-butyl methyl ether would extract 4.7 g. Obviously, there is a point at which the increased amount of A extracted does not repay the effort of multiple extractions, but remember that several small-scale extractions are more effective than one large-scale extraction.

PROPERTIES OF EXTRACTION SOLVENTS

Liquid/liquid extraction involves two layers: the organic layer and the aqueous layer. The solvent used for extraction should possess many properties, including the following:

- It should readily dissolve the substance to be extracted at room temperature.
- It should have a low boiling point so that it can be removed readily.
- It should not react with the solute or the other solvent.
- It should not be highly flammable or toxic.
- It should be relatively inexpensive.

In addition, it should not be miscible with water (the usual second phase). No solvent meets every criterion, but several come close. Some common liquid/liquid extraction solvent pairs are water-ether, water-dichloromethane, and water-hexane. Notice that each combination includes water because most organic compounds are immiscible in water, and therefore can be separated from inorganic compounds. Organic solvents such as methanol and ethanol are not good extraction solvents because they are soluble in water.

Identifying the Layers

One common mistake when performing an extraction is to misidentify the layers and discard the wrong one. It is good practice to save all layers until the desired product is in hand. The densities of the solvents will predict the identities of the top and bottom layers. In general, the densities of nonhalogenated organic solvents are less than 1.0 g/mL and those of halogenated solvents are greater than 1.0 g/mL. Table 7.1 lists the densities of common solvents used in extraction.

Although density is the physical property that determines which layer is on top or on bottom, a very concentrated amount of solute dissolved in either layer can reverse the order. The best method to avoid a misidentification is to perform a drop test. Add a few drops of water to the layer in question and watch the drop

TABLE 7.1 *Common Solvents Listed by Density*

Solvent	Density (g/mL)
Hexane	0.695
Diethyl ether	0.708
t-Butyl methyl ether	0.740
Toluene	0.867
Water	1.000
Dichloromethane	1.325
Chloroform	1.492

Identify layers by a drop test.

$CH_3CH_2—O—CH_2CH_3$

**Diethyl ether
"Ether"**
MW 74.12, den. 0.708
bp 34.6°C, n_D^{20} 1.3530

$$H_3C—\underset{\underset{CH_3}{|}}{\overset{\overset{CH_3}{|}}{C}}—O—CH_3$$

***tert*-Butyl methyl ether**
MW 88.14, den. 0.741
bp 55.2°C, n_D^{20} 1.369

carefully. If the layer is water, then the drop will mix with the solution. If the solvent is the organic layer, then the water drop will create a second layer.

Ethereal Extraction Solvents

In the past, diethyl ether was the most common solvent for extraction in the laboratory. It has high solvent power for hydrocarbons and oxygen-containing compounds. It is highly volatile (bp 34.6°C) and is therefore easily removed from an extract. However, diethyl ether has two big disadvantages: it is highly flammable and poses a great fire threat, and it easily forms peroxides. The reaction of diethyl ether with air is catalyzed by light. The resulting peroxides are higher boiling than the ether and are left as a residue when the ether evaporates. If the residue is heated, it will explode because ether peroxides are volatile and treacherously explosive. In recent years, a new solvent has come on the scene—*tert*-butyl methyl ether.

tert-Butyl methyl ether, called methyl *tert*-butyl ether (MTBE) in industry, has many advantages over diethyl ether as an extraction solvent. Most important, it does not easily form peroxides, so it can be stored for much longer periods than diethyl ether. And, in the United States, it is less than half the price of diethyl ether. It is slightly less volatile (bp 55°C), so it does not pose the same fire threat as diethyl ether, although one must be as careful in handling this solvent as in handling any other highly volatile, flammable substance. The explosion limits for *t*-butyl methyl ether mixed with air are much narrower than for diethyl ether, the toxicity is less, the solvent power is the same, and the ignition temperature is higher (224°C versus 180°C).

The weight percent solubility of diethyl ether dissolved in water is 7.2%, whereas that of *t*-butyl methyl ether is 4.8%. The solubility of water in diethyl ether is 1.2%, while in *t*-butyl methyl ether it is 1.5%. Unlike diethyl ether, *t*-butyl methyl ether forms an azeotrope with water (4% water) that boils at 52.6°C. This means that evaporation of any *t*-butyl methyl ether solution that is saturated with water should leave no water residue, unlike diethyl ether.

The low price and ready availability of *t*-butyl methyl ether came about because it replaced tetraethyl lead as the antiknock additive for high-octane gasoline and as a fuel oxygenate, which helps reduce air pollution, but its water solubility has allowed it to contaminate drinking water supplies in states where leaking underground fuel storage tanks are not well regulated. Consequently, it has been replaced with the much more expensive ethanol. In this text, *t*-butyl methyl ether is *strongly* suggested wherever diethyl ether formerly would have been used in an

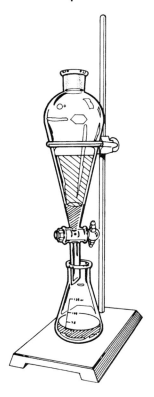

FIG. 7.2

A separatory funnel with a Teflon stopcock.

FIG. 7.3

The correct position for holding a separatory funnel when shaking. Point outlet away from yourself and your neighbors.

extraction. It will not, however, work as the only solvent in the Grignard reaction, probably because of steric hindrance. So whenever the word *ether* appears in this text as an extraction solvent, it is suggested that *t*-butyl methyl ether be used and not diethyl ether.

MIXING AND SEPARATING THE LAYERS

For microscale separations, mixing and separating the layers with a pipette normally incurs very little product loss. Because the two solvents are typically in a reaction tube for microscale extraction, the two layers can be mixed by drawing up and rapidly expelling them with a pipette. Then the layers are allowed to separate, and the bottom layer is separated by drawing it up into a pipette and transferring it to a different container.

For macroscale separations, a separatory funnel (Fig. 7.2) is used to mix and separate the organic and aqueous layers. In macroscale experiments, a frequently used method of working up a reaction mixture is to dilute the mixture with water and extract it with an organic solvent, such as ether, in a separatory funnel. When the stoppered funnel is shaken to distribute the components between the immiscible solvents *t*-butyl methyl ether and water, pressure always develops through volatilization of ether from the heat of the hands, and liberation of a gas (CO_2) (in acid/base extractions) can increase the pressure. Consequently, the funnel is grasped so that the stopper is held in place by one hand and the stopcock by the other, as illustrated in Figure 7.3. After a brief shake or two, the funnel is held in the inverted position shown, and the stopcock is opened cautiously (with the funnel stem pointed away from nearby persons) to release pressure. The mixture can then be shaken more vigorously, with pressure released as necessary. When equilibration is judged to be complete, the slight, constant terminal pressure due to ether is released, the stopper is rinsed with a few drops of ether delivered by a Pasteur pipette, and the layers are allowed to separate. The organic reaction product is distributed wholly or largely into the upper ether layer, whereas inorganic salts, acids or bases pass into the water layer, which can be drawn off. If the reaction was conducted in alcohol or some other water-soluble solvent, the bulk of the solvent is removed in the water layer, and the remainder can be eliminated in two or three washings with 1–2 volumes of water conducted with the techniques used in the first equilibration. The separatory funnel should be supported in a ring stand, as shown in Figure 7.2.

Before adding a liquid to the separatory funnel, check the stopcock. If it is glass, see that it is properly greased, bearing in mind that too much grease will clog the hole in the stopcock and also contaminate the extract. If the stopcock is Teflon, see that it is adjusted to a tight fit in the bore. Store the separatory funnel with the Teflon stopcock loosened to prevent sticking. Because Teflon has a much larger temperature coefficient of expansion than glass, a stuck stopcock can be loosened by cooling the stopcock in ice or dry ice. Do not store liquids in the separatory funnel; they often leak or cause the stopper or stopcock to freeze. To have sufficient room for mixing the layers, fill the separatory funnel no more than three-fourths full. Withdraw the lower layer from the separatory funnel through the stopcock, and pour the upper layer out through the neck.

All too often, the inexperienced chemist discards the wrong layer when using a separatory funnel. Through incomplete neutralization, a desired component may still remain in the aqueous layer, or the densities of the layers may change. Cautious workers save all layers until the desired product has been isolated.

The organic layer is not always the top layer. If in doubt, perform a drop test by adding a few drops of each to water in a test tube.

Practical Considerations When Mixing Layers

Pressure Buildup

The heat of one's hand or heat from acid/base reactions will cause pressure buildup in an extraction mixture that contains a very volatile solvent such as dichloromethane. The extraction container—whether a test tube or a separatory funnel—must be opened carefully to vent this pressure.

Sodium bicarbonate solution is often used to neutralize acids when carrying out acid/base extractions. The result is the formation of carbon dioxide, which can cause foaming and high pressure buildup. Whenever bicarbonate is used, add it very gradually with thorough mixing and frequent venting of the extraction device. If a large amount of acid is to be neutralized with bicarbonate, the process should be carried out in a beaker.

Emulsions

Imagine trying to extract a soap solution (e.g., a nonfoaming dishwashing detergent) into an organic solvent. After a few shakes with an organic solvent, you would have an absolutely intractable emulsion. An emulsion is a suspension of one liquid as droplets in another. Detergents stabilize emulsions, and so any time a detergent-like molecule is in the material being extracted, there is the danger that emulsions will form. Substances of this type are commonly found in nature, so one must be particularly wary of emulsion formation when creating organic extracts of aqueous plant material, such as caffeine from tea. Emulsions, once formed, can be quite stable. You would be quite surprised to open your refrigerator one morning and see a layer of clarified butter floating on the top of a perfectly clear aqueous solution that had once been milk, but milk is the classic example of an emulsion.

Prevention is the best cure for emulsions. This means shaking the solution to be extracted *very gently* until you see that the two layers will separate readily. If a bit of emulsion forms, it may break simply on standing for a sufficient length of time. Making the aqueous layer highly ionic will help. Add as much sodium chloride as will dissolve and shake the mixture gently. Vacuum filtration sometimes works and, when the organic layer is the lower layer, filtration through silicone-impregnated filter paper is helpful. Centrifugation works very well for breaking emulsions. This is easy on a small scale, but often the equipment is not available for large-scale centrifugation of organic liquids.

Shake gently to avoid emulsions.

DRYING AGENTS

The organic solvents used for extraction dissolve not only the compound being extracted but also water. Evaporation of the solvent then leaves the desired compound contaminated with water. At room temperature, water dissolves 4.8% of t-butyl methyl ether by weight, and the ether dissolves 1.5% of water. But ether is virtually insoluble in water saturated with sodium chloride (36.7 g/100 mL). If ether that contains dissolved water is shaken with a saturated aqueous solution of sodium chloride, water will be transferred from the t-butyl methyl ether to the aqueous layer. So, strange as it may seem, ethereal extracts are routinely dried by shaking them with an aqueous saturated sodium chloride solution.

Solvents such as dichloromethane do not dissolve nearly as much water and are therefore dried over a chemical drying agent. Many choices of chemical drying agents are available for this purpose, and the choice of which one to use is governed by four factors: (1) the possibility of reaction with the substance being extracted, (2) the speed with which it removes water from the solvent, (3) the efficiency of the process, and (4) the ease of recovery from the drying agent.

Some very good but specialized and reactive drying agents are potassium hydroxide, anhydrous potassium carbonate, sodium metal, calcium hydride, lithium aluminum hydride, and phosphorus pentoxide. Substances that are essentially neutral and unreactive and are widely used as drying agents include anhydrous calcium sulfate (Drierite), magnesium sulfate, molecular sieves, calcium chloride, and sodium sulfate.

Drierite, $CaSO_4$

Drierite, a specially prepared form of calcium sulfate, is a fast and effective drying agent. However, it is difficult to ascertain whether enough has been used. An indicating type of Drierite is impregnated with cobalt chloride, which turns from blue to red when it is saturated with water. This works well when gases are being dried, but it should not be used for liquid extractions because the cobalt chloride dissolves in many protic solvents.

Magnesium sulfate, $MgSO_4$

Magnesium sulfate is also a fast and fairly effective drying agent, but it is so finely powdered that it always requires careful filtration for removal.

Molecular sieves, zeolites

Molecular sieves are sodium alumino-silicates (zeolites) that have well-defined pore sizes. The 4 Å size adsorbs water to the exclusion of almost all organic substances, making them a fast and effective drying agent. Like Drierite, however, it is impossible to ascertain by appearance whether enough has been used. Molecular sieves in the form of 1/16-in. pellets are often used to dry solvents by simply adding them to the container.

Calcium chloride ($CaCl_2$) pellets are the drying agent of choice for small-scale experiments.

Calcium chloride, recently available in the preferred form of pellets (4 to 80 mesh[1]), is a very fast and effective drying agent. It has the advantage that it clumps together when excess water is present, which makes it possible to know how much to add by observing its behavior. Unlike the older granular form, the pellets do not disintegrate into a fine powder. These pellets are admirably suited to microscale experiments where the solvent is removed from the drying agent with a Pasteur pipette. Calcium chloride is much faster and far more effective than anhydrous sodium sulfate; after much experimentation, we have decided that this is the agent of choice, particularly for microscale experiments. These pellets are used for most of the drying operations in this text. Note, however, that calcium chloride reacts with some alcohols, phenols, amides, and some carbonyl-containing compounds. Advantage is sometimes taken of this property to remove not only water from a solvent but also, for example, a contaminating alcohol (*see* Chapter 16—the synthesis of 1-bromobutane from 1-butanol). Because *t*-butyl methyl ether forms an azeotrope with water, its solutions should, theoretically, not need to be dried because evaporation carries away the water. Drying these ether solutions with calcium chloride pellets removes water droplets that get carried into the ether solution.

Sodium sulfate, Na_2SO_4

Sodium sulfate is a very poor drying agent. It has a very high capacity for water but is slow and not very efficient in the removal of water. Like calcium chloride pellets, it clumps together when wet, and solutions are easily removed from it using a Pasteur pipette. Sodium sulfate has been used extensively in the past and should still be used for compounds that react with calcium chloride.

---

[1]These drying pellets are available from Fisher Scientific, Cat. No. C614–3.

PART 1: THE TECHNIQUE OF NEUTRAL LIQUID/LIQUID EXTRACTION

The workup technique of liquid/liquid extraction has four steps: (1) mixing the layers, (2) separating the layers, (3) drying the organic layer, and (4) removing the solvent. The microscale neutral liquid/liquid extraction technique is described in the following sections.

STEP 1. MIXING THE LAYERS

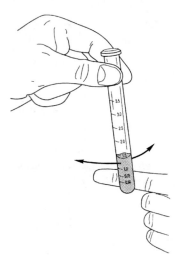

FIG. 7.4
Mixing the contents of a reaction tube by flicking it. Grasp the tube firmly at the very top and flick it vigorously at the bottom. The contents will mix without coming out of the tube.

Always draw out the lower layer and place it in another container.

Once the organic and aqueous layers are in contact with one another, mixing is required to ensure that the desired compound(s) get extracted into the desired layer. First, place 1–2 mL of an aqueous solution of the compound to be extracted in a reaction tube. Add about 1 mL of extraction solvent, for example, dichloromethane. Note, as you add the dichloromethane, whether it is the top or the bottom layer. (Since dichloromethane is more dense than water, predict what layer it will be.) An effective way to mix the two layers is to flick the tube with a finger. Grasp the tube firmly at the very top between the thumb and forefinger and flick it vigorously at the bottom (Fig. 7.4). You will find that this violent motion mixes the two layers well, but nothing will spill out the top. Another good mixing technique is to pull the contents of the reaction tube into a Pasteur pipette and then expel the mixture back into the tube with force. Doing this several times will effect good mixing of the two layers. A stopper can be placed in the top of the tube, and the contents can be mixed by shaking the tube, but the problem with this technique is that the high vapor pressure of the solvent will often force liquid out around the cork or stopper.

STEP 2. SEPARATING THE LAYERS

After thoroughly mixing the two layers, allow them to separate. Tap the tube if droplets of one layer are in the other layer, or on the side of the tube. After the layers separate completely, draw up the lower dichloromethane layer into a Pasteur pipette. Leave behind any middle emulsion layer. The easiest way to do this is to attach the pipette to a pipette pump (Fig. 7.5). This allows very precise control of the liquid being removed. It takes more skill and practice to remove the lower layer cleanly with a 2-mL rubber bulb attached to a pipette because the high vapor pressure of the solvent tends to make it dribble out. To avoid losing any of the solution, it is best to hold a clean, dry, empty tube in the same hand as the full tube to receive the organic layer (Fig. 7.6).

From the discussion of the partition coefficient, you know that several small extractions are better than one large one, so repeat the extraction process with two further 1-mL portions of dichloromethane. An experienced chemist might summarize all the preceding with the following notebook entry, "Aqueous layer extracted 3×1-mL portions CH_2Cl_2," and in a formal report would write, "The aqueous layer was extracted three times with 1-mL portions of dichloromethane."

If you are working on a larger microscale, a microscale separatory funnel (Fig. 7.7) should be used. A separatory funnel, regardless of size, should be filled to only about two-thirds of its capacity so the layers can be mixed by shaking. The microscale separatory funnel has a capacity of 8.5 mL when full, so it is useful for an extraction with a total volume of about 6 mL.

Chapter 7 ■ Extraction   139

FIG. 7.5
The removal of a solvent from a reaction tube with a pipette and pipette pump.

FIG. 7.6
Grasp both reaction tubes in one hand when transferring material from one tube to another with a Pasteur pipette.

FIG. 7.7
A microscale separatory funnel. Remove the polyethylene frit from the micro Büchner funnel before using.

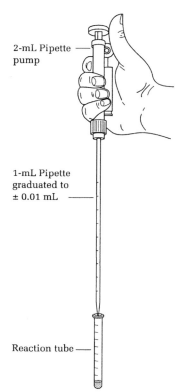

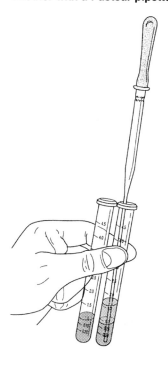

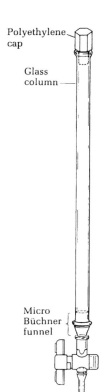

Use a wood boiling stick to poke out the polyethylene frit from the bottom part of the separatory funnel. Store it for later replacement. Close the valve, add up to 5 mL of the solution to be extracted to the separatory funnel, then add the extraction solvent so that the total volume does not exceed 6 mL.

Cap the separatory funnel and mix the contents by inverting the funnel several times. If the two layers separate fairly easily, then the contents can be shaken more thoroughly. If the layers do not separate easily, be careful not to shake the funnel too vigorously because intractable emulsions could form.

Remove the stopper from the funnel, clamp it, and then, grasping the valve with two hands, empty the bottom layer into an Erlenmeyer flask or other container. If the top layer is desired, pour it out through the top of the separatory funnel—don't drain it through the valve, which may have a drop of the lower layer remaining in it.

STEP 3. DRYING THE ORGANIC LAYER

Dichloromethane dissolves a very small quantity of water, and microscopic droplets of water are suspended in the organic layer, often making it cloudy. To remove the water, a drying agent, for example, anhydrous calcium chloride pellets, is added to the dichloromethane solution.

FIG. 7.8

An aspirator tube being used to remove solvent vapors.

Record the tare of the final container.

How Much Drying Agent Should Be Used?

When a small quantity of the drying agent is added, the crystals or pellets become sticky with water, clump together, and fall rapidly as a lump to the bottom of the reaction tube. There will come a point when a new small quantity of drying agent no longer clumps together, but the individual particles settle slowly throughout the solution. As they say in Sweden, "Add drying agent until it begins to snow." The drying process takes about 10–15 minutes, during which time the tube contents should be mixed occasionally by flicking the tube. The solution should no longer be cloudy, but clear (although it may be colored).

Once drying is judged complete, the solvent is removed by forcing a Pasteur pipette to the bottom of the reaction tube and pulling the solvent in. Air is expelled from the pipette as it is being pushed through the crystals or pellets so that no drying agent will enter the pipette. It is very important to wash the drying agent left in the reaction tube with several small quantities of pure solvent to transfer all the extract.

STEP 4. REMOVING THE SOLVENT

If the quantity of extract is relatively small, say 3 mL or less, then the easiest way to remove the solvent is to blow a stream of air (or nitrogen) onto the surface of the solution from a Pasteur pipette (Fig. 7.8). Be sure that the stream of air is very gentle before inserting it into the reaction tube. The heat of vaporization of the solvent will cause the tube to become rather cold during the evaporation and, of course, slow down the process. The easiest way to add heat is to hold the tube in your hand.

Another way to remove the solvent is to attach the Pasteur pipette to an aspirator and pull air over the surface of the liquid. This is not quite as fast as blowing air onto the surface of the liquid, and runs the danger of sucking up the liquid into the aspirator.

If the volume of liquid is more than about 3 mL, put it into a 25-mL filter flask, put a plastic Hirsch funnel in place, and attach the flask to the aspirator. By placing your thumb in the Hirsch funnel, the vacuum can be controlled, and heat can be applied by holding the flask in the other hand while swirling the contents (Fig. 7.9).

The reaction tube or filter flask in which the solvent is evaporated should be tared (weighed empty), and this weight recorded in your notebook. In this way, the weight of material extracted can be determined by again weighing the container that contains the extract.

EXPERIMENT

PARTITION COEFFICIENT OF BENZOIC ACID

> IN THIS EXPERIMENT, you will shake a solution of benzoic acid in water with the immiscible solvent dichloromethane. The benzoic acid will distribute (partition) itself between the two layers. By removing the organic layer, drying, and evaporating it, the weight of benzoic acid in the dichloromethane can be determined and thus the ratio in the two layers. This ratio is a constant known as the partition coefficient.

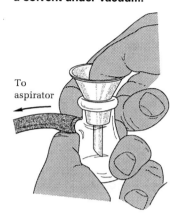

FIG. 7.9
The apparatus for removing a solvent under vacuum.

To aspirator

In a reaction tube, place about 100 mg of benzoic acid (weighed to the nearest milligram) and add exactly equal volumes of water followed by dichloromethane (about 1.6 mL each). While making this addition, note which layer is organic and which is aqueous. Put a septum on the tube and shake the contents vigorously for at least 2 minutes. Allow the tube to stand undisturbed until the layers separate and then carefully draw off, using a Pasteur pipette, *all* of the aqueous layer without removing any of the organic layer. It may be helpful to draw out the tip of the pipette to a finer point in a flame and, using this, to tilt the reaction tube at a 45° angle to make this separation as clean as possible.

Add anhydrous calcium chloride pellets to the dichloromethane in very small quantities until they no longer clump together. Mix the contents of the tube by flicking it, and allow it to stand for about 5 minutes to complete the drying process. Using a dry Pasteur pipette, transfer the dichloromethane to a tared dry reaction tube or a 10-mL Erlenmeyer flask containing a boiling chip. Complete the transfer by washing the drying agent with two more portions of solvent that are added to the original solution, and then evaporate the solvent. This can be done by boiling off the solvent while removing solvent vapors with an aspirator tube, or by blowing a stream of air or nitrogen into the container while warming it in one's hand (*see* Fig. 7.8). This operation should be performed in a hood.

From the weight of the benzoic acid in the dichloromethane layer, the weight in the water layer can be obtained by difference. The ratio of the weight in dichloromethane to the weight in water is the distribution coefficient because the volumes of the two solvents were equal. Report the value of the distribution coefficient in your notebook.

Cleaning Up. The aqueous layer can be flushed down the drain. Dichloromethane goes into the halogenated organic solvents waste container. After allowing the solvent to evaporate from the sodium sulfate in the hood, place the sodium sulfate in the non-hazardous solid waste container. If local regulations do not allow for the evaporation of solvents in a hood, dispose of the wet sodium sulfate in a special waste container.

PART 2: ACID/BASE LIQUID/LIQUID EXTRACTION

Acid/base liquid/liquid extraction involves carrying out simple acid/base reactions to separate strong organic acids, weak organic acids, neutral organic compounds, and basic organic substances. The chemistry involved is given in the following equations, using benzoic acid, phenol, naphthalene, and aniline as examples of the four types of compounds.

Here is the strategy (refer to the flow sheet in Fig. 7.10): The four organic compounds are dissolved in *t*-butyl methyl ether. The ether solution is shaken with a saturated aqueous solution of sodium bicarbonate, a weak base. This will react only with the strong acid, benzoic acid (**1**), to form the ionic salt, sodium benzoate (**5**), which dissolves in the aqueous layer and is removed. The ether solution now contains just phenol (**2**), naphthalene (**4**), and aniline (**3**). A 3 *M* aqueous solution of sodium hydroxide is added, and the mixture is shaken. The hydroxide, a strong base, will react only with the phenol (**2**), a weak acid, to form sodium phenoxide (**6**), an ionic compound that dissolves in the aqueous layer and is removed. The ether now contains only naphthalene (**4**) and aniline (**3**). Shaking it with dilute

142 *Macroscale and Microscale Organic Experiments*

FIG. 7.10
A flow sheet for the separation of a strong acid, a weak acid, a neutral compound, and a base: benzoic acid, phenol, naphthalene, and aniline (this page). Acid/base reactions of the acidic and basic compounds (opposite page).

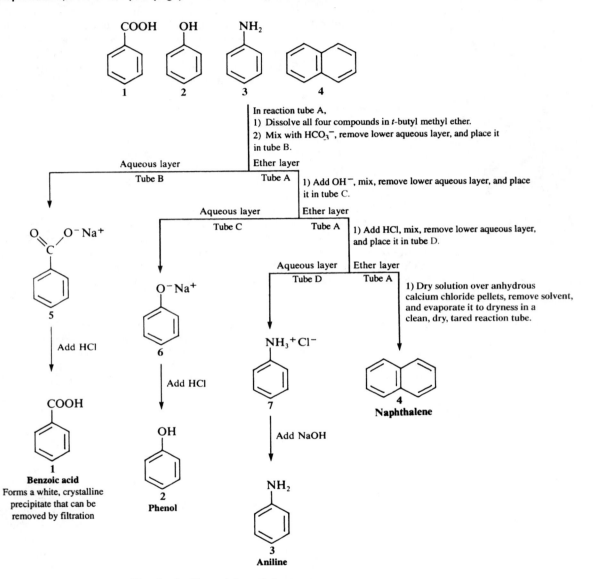

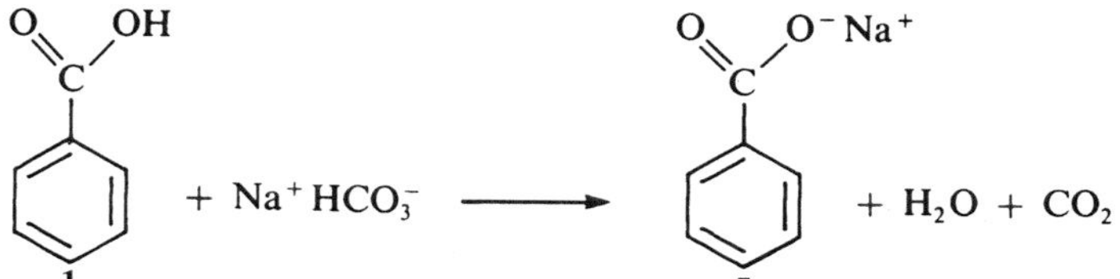

Chapter 7 ■ Extraction

1 Benzoic acid (C₆H₅COOH)
pK_a = 4.17
Covalent, sol. in org. solvents

+ Na⁺HCO₃⁻ ⟶ **5** Sodium benzoate (C₆H₅COO⁻Na⁺) + H₂O + CO₂
Ionic, sol. in water

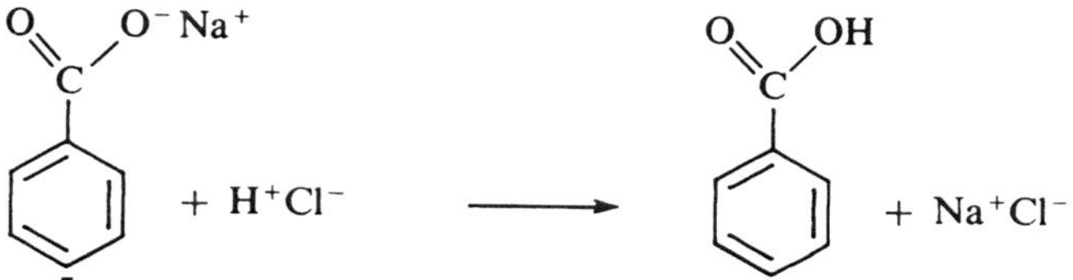

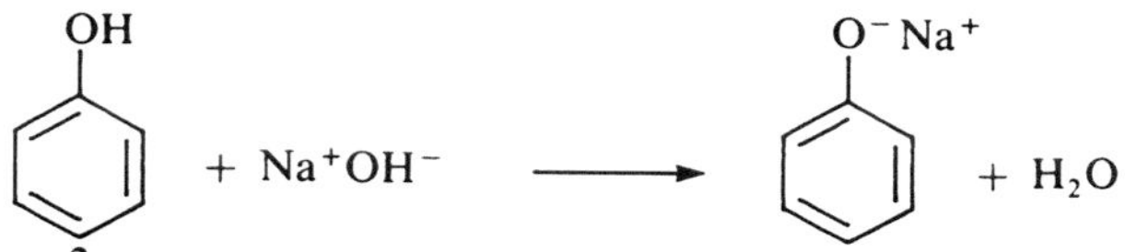

5 Sodium benzoate + H⁺Cl⁻ ⟶ **1** Benzoic acid + Na⁺Cl⁻

2 Phenol (C₆H₅OH)
pK_a = 10
Covalent, sol. in org. solvents

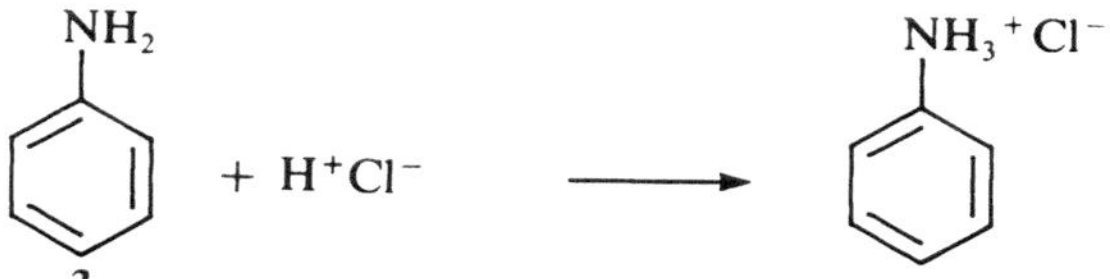

+ Na⁺OH⁻ ⟶ **6** Sodium phenoxide (C₆H₅O⁻Na⁺) + H₂O
Ionic, sol. in water

6 Sodium phenoxide + H⁺Cl⁻ ⟶ **2** Phenol + Na⁺Cl⁻

3 Aniline (C₆H₅NH₂)
pK_b = 9.30
Covalent, sol. in org. solvents

+ H⁺Cl⁻ ⟶ **7** Anilinium chloride (C₆H₅NH₃⁺Cl⁻)
Ionic, sol. in water

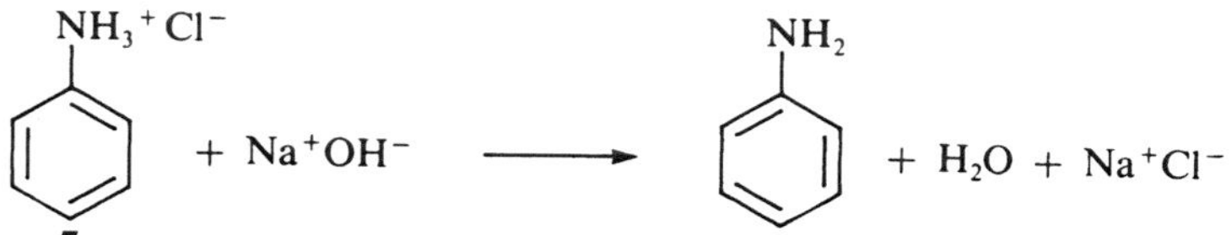

7 Anilinium chloride + Na⁺OH⁻ ⟶ **3** Aniline + H₂O + Na⁺Cl⁻

hydrochloric acid removes the aniline, a base, as the ionic anilinium chloride (**7**). The aqueous layer is removed. Evaporation of the *t*-butyl methyl ether now leaves naphthalene (**4**), the neutral compound. The other three compounds are recovered by adding acid to the sodium benzoate (**5**) and sodium phenoxide (**6**) and base to the anilinium chloride (**7**) to regenerate the covalent compounds benzoic acid (**1**), phenol (**2**), and aniline (**3**).

The ability to separate strong acids from weak acids depends on the acidity constants of the acids and the basicity constants of the bases. In the first equation, consider the ionization of benzoic acid, which has an equilibrium constant (K_a) of 6.8×10^2. The conversion of benzoic acid to the benzoate anion in equation 4 is governed by the equilibrium constant, K (equation 5), obtained by combining equations 3 and 4.

> The pK_a of carbonic acid, H_2CO_3, is 6.35.

$$C_6H_5COOH + H_2O \rightleftharpoons C_6H_5COO^- + H_3O^+ \quad (1)$$

$$K_a = \frac{[C_6H_5COO^-][H_3O^+]}{[C_6H_5COOH]} = 6.8 \times 10^{-5}, \text{p}K_a = 4.17 \quad (2)$$

$$K_w = [H_3O^+][OH^-] = 10^{-14} \quad (3)$$

$$C_6H_5COOH + OH^- \rightleftharpoons C_6H_5COO^- + H_2O \quad (4)$$

$$K = \frac{[C_6H_5COO^-]}{[C_6H_5COOH][OH^-]} = \frac{K_a}{K_w} = \frac{6.8 \times 10^{-5}}{10^{-14}} = 3.2 \times 10^8 \quad (5)$$

If 99% of the benzoic acid is converted to $C_6H_5COO^-$,

$$\frac{[C_6H_5COO^-]}{[C_6H_5COOH]} = \frac{99}{1} \quad (6)$$

then from equation 5 the hydroxide ion concentration would need to be 6.8×10^{-7} M. Because saturated $NaHCO_3$ has $[OH^-] = 3 \times 10^{-4}$ M, the hydroxide ion concentration is high enough to convert benzoic acid completely to sodium benzoate.

For phenol, with a K_a of 10^{-10}, the minimum hydroxide ion concentration that will produce the phenoxide anion in 99% conversion is 10^{-2} M. The concentration of hydroxide in 10% sodium hydroxide solution is 10^{-1} M, and so phenol in a strong base is entirely converted to the water-soluble salt.

GENERAL CONSIDERATIONS

If acetic acid was used as the reaction solvent, it would also be distributed largely into the aqueous phase. If the reaction product is a neutral substance, however, the residual acetic acid in the ether can be removed by one washing with excess 5% sodium bicarbonate solution. If the reaction product is a higher molecular weight acid, for example, benzoic acid (C_6H_5COOH), it will stay in the ether layer, while acetic acid is being removed by repeated washing with water. The benzoic acid can then be separated from neutral byproducts by extraction with sodium bicarbonate or sodium hydroxide solution and acidification of the extract. Acids of high molecular weight are extracted only slowly by sodium bicarbonate, so sodium carbonate is used in its place; however, carbonate is more prone than bicarbonate to produce emulsions. Sometimes an emulsion in the lower layer can be settled by twirling the separatory funnel by its stem. An emulsion in the upper layer can be broken by grasping the funnel at the neck and swirling it. Because the tendency to emulsify increases with the removal of electrolytes and solvents, a little sodium chloride solution is added with each portion of wash water. If the layers are largely clear but an emulsion persists at the interface, the clear part of the water layer can be drawn off, and the emulsion run into a second funnel and shaken with fresh ether.

Liquid/liquid extraction and acid/base extraction are employed in the majority of organic reactions because it is unusual to have the product crystallize from the reaction mixture or to be able to distill the reaction product directly from the reaction mixture. In the research literature, one will often see the statement "the reaction mixture was worked up in the usual way," which implies an extraction process of the type described here. Good laboratory practice dictates, however, that the details of the process be written out.

8 CHAPTER

Thin-Layer Chromatography: Analyzing Analgesics and Isolating Lycopene from Tomato Paste

> **PRELAB EXERCISE:** Based on the number and polarity of the functional groups in aspirin, acetaminophen, ibuprofen, and caffeine, whose structures are shown on page 184, predict which of these four compounds has the highest R_f value and which has the lowest.

Chromatography is the separation of two or more compounds or ions caused by their molecular interactions with two phases—one moving and one stationary. These two phases can be a solid and a liquid, a liquid and a liquid, a gas and a solid, or a gas and a liquid. You very likely have seen chromatography carried out on paper towels or coffee filters, which showed various colors denoting the separation of inks and food dyes. In the laboratory, cellulose paper is the stationary or solid phase, and a propanol-water mixture is the mobile or liquid phase. The samples are spotted near one edge of the paper, and this edge is dipped into the liquid phase. The solvent is drawn through the paper by capillary action, and the molecules are separated based on how they interact with the paper. Although there are several different forms of chromatography, the principles are essentially the same.

Thin-layer chromatography (TLC) is a sensitive, fast, simple, and inexpensive analytical technique that you will use repeatedly while carrying out organic experiments. It is a micro technique; as little as 10^{-9} g of material can be detected, although the usual sample size is from 1×10^{-6} g to 1×10^{-8} g. The stationary phase is normally a polar solid adsorbent, and the mobile phase can be a single solvent or a combination of solvents.

TLC requires micrograms of material.

USES OF THIN-LAYER CHROMATOGRAPHY

1. **To determine the number of components in a mixture.** TLC affords a quick and easy method for analyzing such things as a crude reaction mixture, an extract from a plant substance, or the ingredients in a pill. Knowing the

number and relative amounts of the components aids in planning further analytical and separation steps.

2. **To determine the identity of two substances.** If two substances spotted on the same TLC plate give spots in identical locations, they *may* be identical. If the spot positions are not the same, the substances cannot be the same. It is possible for two or more closely related but not identical compounds to have the same positions on a TLC plate. Changing the stationary or mobile phase will usually effect their separation.

3. **To monitor the progress of a reaction.** By sampling a reaction at regular intervals, it is possible to watch the reactants disappear and the products appear using TLC. Thus, the optimum time to halt the reaction can be determined, and the effect of changing such variables as temperature, concentrations, and solvents can be followed without having to isolate the product.

4. **To determine the effectiveness of a purification.** The effectiveness of distillation, crystallization, extraction, and other separation and purification methods can be monitored using TLC, with the caveat that a single spot does not guarantee a single substance.

5. **To determine the appropriate conditions for a column chromatographic separation.** In general, TLC is not satisfactory for purifying and isolating macroscopic quantities of material; however, the adsorbents most commonly used for TLC—silica gel and alumina—are also used for column chromatography, which is discussed in Chapter 9. Column chromatography is used to separate and purify up to 1 g of a solid mixture. The correct adsorbent and solvent to use for column chromatography can be rapidly determined by TLC.

6. **To monitor column chromatography.** As column chromatography is carried out, the solvent is collected in a number of small flasks. Unless the desired compound is colored, the various fractions must be analyzed in some way to determine which ones have the desired components of the mixture. TLC is a fast and effective method for doing this.

THE PRINCIPLES OF CHROMATOGRAPHY

To thoroughly understand the process of TLC (and other types of chromatography), we must examine the process at the molecular level. All forms of chromatography involve a dynamic and rapid equilibrium of molecules between the liquid and the stationary phases. For the chromatographic separation of molecules A and B shown in Figure 8.1, there are two states:

1. **Free**—dissolved in the liquid or gaseous mobile phase.
2. **Adsorbed**—sticking to the surface of the solid stationary phase.

Molecules A and B are continuously moving back and forth between the dissolved (free) and adsorbed states, with billions of molecules adsorbing and billions of other molecules desorbing from the solid stationary phase each second. The equilibrium between the free and adsorbed states depends on the relative strength of the attraction of A and B to the liquid phase molecules *versus* the

FIG. 8.1

The mixture of molecules A and B is in a dynamic equilibrium between the free and adsorbed states.

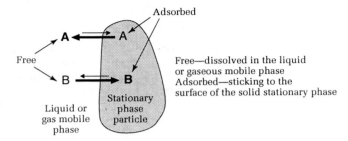

FIG. 8.2

The mixture of molecules A and B is in a dynamic equilibrium between the stationary adsorbent and a *flowing* mobile phase.

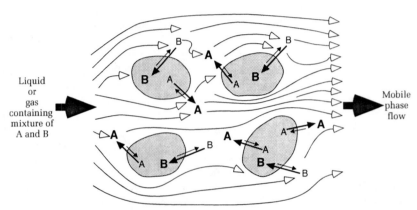

strength of attraction of A and B to the stationary phase structure. As discussed in the introduction to Chapter 3, the strength of these attractive forces depends on the following factors:

- Size, polarity, and hydrogen bonding ability of molecules A and B
- Polarity and hydrogen bonding ability of the stationary phase
- Polarity and hydrogen bonding ability of the mobile phase solvent

Molecules distribute themselves, or *partition*, between the mobile and stationary phases depending on these attractive forces. As implied by the equilibrium arrows in Figure 8.1, the A molecules are less polar and are only weakly attracted to a polar stationary phase, spending most of their time in the mobile phase. In contrast, equilibrium for the more polar B molecules lies in the direction of being adsorbed onto the polar stationary phase. The equilibrium constant k (also called the *partition coefficient*) is a measure of the distribution of molecules between the mobile phase and the stationary phase, and is similar to the distribution coefficient for liquid/liquid extraction. This constant changes with structure.

Simply adding a mixture to a combination of a liquid phase and a stationary phase will not separate it into its pure components. For separation to happen, the liquid phase must be mobile and be flowing past the stationary phase, as depicted in Figure 8.2. Because the A molecules spend more time in the mobile phase, they will be carried through the stationary phase and be eluted faster and move farther in a given amount of time. Because the B molecules are adsorbed

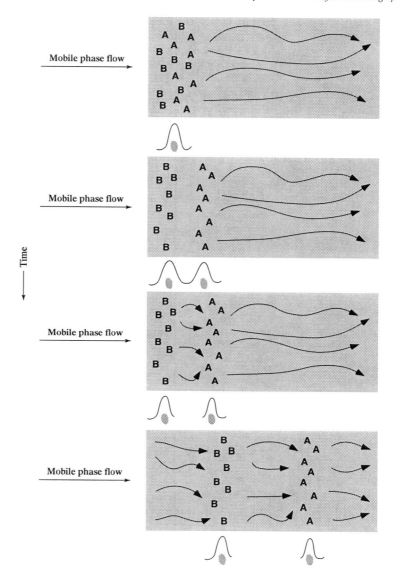

FIG. 8.3
A chromatographic separation. Over time, the mobile phase carries the less weakly adsorbed A molecules ahead of the more strongly adsorbed B molecules.

on the stationary phase more than A molecules, the B molecules spend less time in the mobile phase and therefore migrate through the stationary phase more slowly and are eluted later. The B molecules do not migrate as far in the same amount of time. The consequence of this difference is that A is gradually separated from B by moving ahead in the flowing mobile phase as time passes, as shown in Figure 8.3.

A simple analogy may help to illustrate these concepts. Imagine a group of hungry and not-so-hungry people riding a moving sidewalk (the mobile phase in this analogy) that moves beside a long buffet table covered with all sorts of delicious food (the stationary phase). Hungry people, attracted to the food, will step off and on the moving belt many times in order to fill their plates. The not-so-hungry people will step off and on to get food far less often. Consequently, the more strongly attracted hungry people will lag behind, while the not-so-hungry ones will move ahead on the belt. The two types of people are thus separated based on the strength of their attraction for the food.

Stationary Phase Adsorbents

In TLC, the stationary phase is a polar adsorbent, usually finely ground alumina $[(Al_2O_3)_x]$ or silica $[(SiO_2)_x]$ particles, coated as a thin layer on a glass slide or plastic sheet. Silica, commonly called *silica gel* in the laboratory, is simply very pure white sand. The extended covalent network of these adsorbents creates a very polar surface. Partial structures of silica and alumina are shown below. The silicon or aluminum atoms are the smaller, darker spheres:

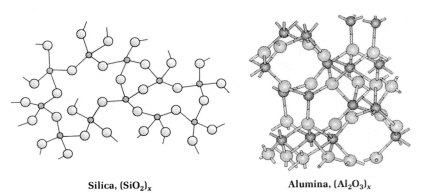

Silica, $(SiO_2)_x$ Alumina, $(Al_2O_3)_x$

FIG. 8.4
A partial silica structure showing polar Si—O bonds.

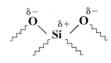

The electropositive character of the aluminum or silicon and the electronegativity of oxygen create a very polar stationary phase (Fig. 8.4). Therefore, the more polar the molecules to be separated, the stronger the attraction to the stationary phase. Nonpolar molecules will tend to stay in the mobile phase. In general, the more polar the functional group, the stronger the adsorption on the stationary phase and the more slowly the molecules will move. In an extreme situation, the molecules will not move at all. This problem can be overcome by increasing the polarity of the mobile phase so that the equilibrium between the free and adsorbed state is shifted toward the free state.

Although silica is the most common stationary phase used for TLC, many other types are used, ranging from paper to charcoal, nonpolar to polar, and reverse phase to normal phase. Several different types of stationary phases are listed according to polarity in Table 8.1.

Silica gel and alumina are commonly used in column chromatography for the purification of macroscopic quantities of material (*see* Chapter 9). Of the two, alumina, when anhydrous, is the more active; that is, it will adsorb substances more strongly. It is thus the adsorbent of choice for the separation of relatively nonpolar substrates, such as hydrocarbons, alkyl halides, ethers, aldehydes, and ketones. To separate more polar substrates, such as alcohols, carboxylic acids, and amines, the less active adsorbent, silica gel, is often used.

TABLE 8.1 *Common Stationary Phases Listed by Increasing Polarity*

Increasing polarity ↓

- Polydimethyl siloxane*
- Methyl- or Phenylsiloxane*
- Cyanopropyl siloxane*
- Carbowax [poly(ethyleneglycol)]*
- Reverse phase (hydrocarbon-coated silica, e.g., C_{18})
- Paper
- Cellulose
- Starch
- Calcium sulfate
- Silica (silica gel)
- Florisil (magnesium silicate)
- Magnesium oxide
- Alumina (aluminum oxide; acidic, basic, or neutral)
- Activated carbon (charcoal or Norit pellets)

*Stationary phase for gas chromatography

Elution sequence is the order in which the components of a mixture move during chromatography.

Ethyl acetoacetate

Ethyl pentanoate

Molecular Polarity and Elution Sequence

Assuming we are using a polar adsorbent, how can we determine how rapidly the compounds in our particular mixture move, that is, their elution sequence? Because the more polar compounds will adsorb more strongly to the polar stationary phase they will move the slowest and the shortest distance on a TLC plate. Non-polar compounds will move rapidly, and will elute first or move the greatest distance on the TLC plate. Table 8.2 lists several common compound classes according to how they move or elute on silica or alumina.

You should be able to look at a molecular structure, identify its functional group(s), and easily determine whether it is more or less polar than another structure with different functional groups. Note that the polarity of a molecule increases as the number of functional groups in that molecule increases. Thus, ethyl acetoacetate, with both ketone and ester groups, is more polar than ethyl pentanoate, which has only an ester group. However, it should be noted that chromatography is not an exact science. The rules discussed here can be used to help predict the order of elution; however, only performing an experiment will give definitive answers.

Mobile Phase Solvent Polarity

The key to a successful chromatographic separation is the mobile phase. You normally use silica gel or alumina as the stationary phase. In extreme situations, very polar substances chromatographed on alumina will not migrate very far from the starting point (i.e., give low R_f values), and nonpolar compounds chromatographed on silica gel will travel with the solvent front (i.e., give high R_f values). These extremes of behavior are markedly affected, however, by the solvents used to carry out the chromatography. A polar solvent will carry along with it polar substrates,

TABLE 8.2 *Elution Order for Some Common Functional Groups with a Silica or Alumina Stationary Phase*

Highest/fastest (elute with nonpolar mobile phase)
Alkane hydrocarbons
Alkyl halides (halocarbons)
Alkenes (olefins)
Dienes
Aromatic hydrocarbons
Ethers
Esters
Ketones
Aldehydes
Amines
Alcohols
Phenols
Carboxylic acids
Sulfonic acids
Lowest/slowest (need polar mobile phase to elute)

↕ Increasing polarity of functional group

Avoid using benzene, carbon tetrachloride, and chloroform. Benzene is known to be a carcinogen when exposure is prolonged; the others are suspected carcinogens.

and nonpolar solvents will do the same with nonpolar compounds—another example of the generalization "like dissolves like." You cannot change the polarities of the compounds in your mixture, but by using different solvents, either alone or as mixtures, you can adjust the polarity of the mobile phase and affect the equilibria between the free and adsorbed states. Changing the polarity of the mobile phase can optimize the chromatographic separation of mixtures of compounds with a wide variety of polarities.

Table 8.3 lists, according to increasing polarity, some solvents that are commonly used for both TLC and column chromatography. Because the polarities of benzene, carbon tetrachloride, or chloroform can be matched by other, less toxic solvents, these three solvents are seldom used. Carbon tetrachloride and chloroform are suspected carcinogens and benzene is known to be a carcinogen when exposure is prolonged. In general, the solvents for TLC and column chromatography are characterized by having low boiling points that allow them to be easily evaporated and low viscosities that allow them to migrate rapidly. A solvent more polar than methanol is seldom needed. Often, two solvents are used in a mixture of varying proportions; the polarity of the mixture is a weighted average of the two. Hexane and ether mixtures are often employed.

Finding a good solvent system is usually the most critical aspect of TLC. If the mobile phase has not been previously determined, start with a nonpolar solvent such as hexanes and observe the separation. If the mixture's components do not move very far, try adding a polar solvent such as ether or ethyl acetate to the hexanes. Compare the separation to the previous plate. In most cases, a combination of two solvents is the best choice. If the spots stay at the bottom of the plate, add more of the polar solvent. If they run with the solvent front (move to the top), increase the proportion of the nonpolar solvent. Unfortunately, some trial and error is

TABLE 8.3 *Common Mobile Phases Listed by Increasing Polarity*

Increasing polarity →

- Helium
- Nitrogen
- Pentanes (petroleum ether)
- Hexanes (ligroin)
- Cyclohexane
- Carbon tetrachloride*
- Toluene
- Chloroform*
- Dichloromethane (methylene chloride)
- *t*-Butyl methyl ether
- Diethyl ether
- Ethyl acetate
- Acetone
- 2-Propanol
- Pyridine
- Ethanol
- Methanol
- Water
- Acetic acid

*Suspected carcinogens

usually involved in determining which solvent system is the best. There is a large amount of literature on the solvents and adsorbents used in the separation of a wide variety of substances.

THE SIX STEPS OF THIN-LAYER CHROMATOGRAPHY

The process of thin-layer chromatography (TLC) can be broken down into six main steps: (1) preparing the sample; (2) spotting the TLC plate; (3) picking a solvent; (4) developing the TLC plate; (5) visualizing the TLC plate; (6) calculating R_f values. A detailed description of each of these steps is given in the following sections.

Step 1. Preparing the Sample

Too much sample is a frequent problem. Use a 1% solution of the mixture. Apply very small spots.

You need to dissolve only a few milligrams of material (a 1% solution) because one can detect a few micrograms of compound on a TLC plate. Choose a volatile solvent such as diethyl ether. Even if the material is only partially soluble, you will normally be able to observe the compound because only low concentrations are needed. If the prepared sample is too concentrated, streaking can occur on the plate. Streaking may lead to the overlap of two or more compounds, thus skewing results. See Figure 8.5 for an example of a streaking spot.

FIG. 8.5
A streaking spot.

FIG. 8.6
A marked and spotted TLC plate.

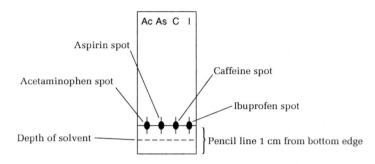

FIG. 8.7
(a) A spotting capillary.
(b) Soften the glass by heating at the base of the flame as shown and then stretch to a fine capillary.

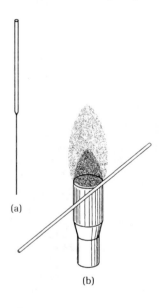

CAUTION: Bunsen burners should be used only in lab areas that are far from flammable organic solvents.

Step 2. Spotting the TLC Plate

The use of commercially available TLC plates, poly(ethylene terephthalate) (Mylar) sheets coated with silica gel using polyacrylic acid as a binder, is highly recommended; these fluoresce under ultraviolet (UV) light.[1] The TLC plates must be handled gently and by the edge, or the 100-mm-thick coating of silica gel can be easily scratched off or contamination of the surface can occur. With a pencil, lightly draw a faint line 1 cm from the end and then three or four short hash marks to guide spotting. Lightly write identifying letters at the top of the plate to keep track of the placement of the compound spots (Fig. 8.6). Note that a pencil is always used to mark TLC plates because the graphite (carbon) is inert. If ink is used to mark the plate, it will chromatograph just as any other organic compound, interfering with the samples and giving flawed results.

Once the sample is prepared, a spotting capillary must be used to add the sample to the plate. Spotting capillaries can be made by drawing out open-end melting point tubes or Pasteur pipette stems in a burner flame (Fig. 8.7).[2] The bore of these capillaries should be so small that once a liquid is drawn into them, it will not flow out to form a drop. Practice spotting just pure solvent onto an unmarked TLC plate. Dip the capillary into the solvent and let a 2–3 cm column of solvent flow into it by capillary action. Do this by holding the capillary vertically over the *coated* side of the plate, and lower the pipette until the tip just touches the adsorbent. Only then will liquid flow onto the plate; quickly withdraw the capillary when the spot is about 1 mm in diameter. The center of the letter *o* on this page is more than 1 mm in diameter. The solvent will evaporate quickly, leaving your mixture behind on the plate. You may have to spot the plate a couple of times in the same place to ensure that sufficient material is present; do not spot too much sample because this will lead to a poor separation. It is extremely important that the spots be as small as possible. If the spot is large, then two or more spots of a sample may overlap on the TLC plate, thus causing erroneous conclusions about the separation and/or the sample's purity or content. Practice placing spots a number of times until you develop good spotting technique. You are now ready to spot the solutions of mixtures as described in Experiments 1 and 2.

---

[1]Whatman flexible plates for TLC, cat. no. 4410 222 (Fisher cat. no. 05-713-162); cut with scissors to 1" × 3". Unlike student-prepared plates, these coated sheets give very consistent results. A supply of these plates makes it a simple matter to examine most of the reactions in this book for completeness of reaction, purity of product, and side reactions.
[2]Three-inch pieces of old and unusable gas chromatography capillary columns are also effective spotting capillaries.

FIG. 8.8
A fast method for determining the correct solvent for TLC. See the text for the procedure.

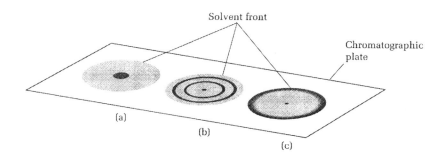

Step 3. Picking a Solvent

One of the most crucial parts of a successful TLC separation is choosing the appropriate mobile phase; see the section "Mobile Phase Polarity" above. Make some spots of the mixture to be separated on a test TLC plate, either a new or a used one. Touch each spot with a different solvent held in a capillary. Referring to Fig. 8.8, note that the mixture did not travel away from the point of origin in Fig. 8.8a, and in Fig. 8.8b, two concentric rings are seen between the origin and the solvent front. This is how a good solvent behaves. In Fig. 8.12c, the mixture traveled with the solvent front.

FIG. 8.9
Using a wide-mouth bottle to develop a TLC plate.

Step 4. Developing the Plate

Once the dilute solution of the mixture has been spotted on the plate, the next step is the actual chromatographic separation, called *plate development*.

Once a mobile phase has been chosen, the marked and spotted TLC plate is inserted into a 4-oz wide-mouth bottle (Fig. 8.9) or beaker (Fig. 8.10) containing 4 mL of the mobile phase, either a solvent or a solvent mixture. The bottle is lined with filter paper that is wet with solvent to saturate the atmosphere within the container. Use tweezers to place the plate in the development chamber; oils from your fingers can sometimes smear or ruin a TLC plate. Also make sure that the origin spots are not below the solvent level in the chamber. If the spots are submerged in the solvent, they are washed off the plate and lost. The top of the container is put in place and the time noted. (If a beaker is used, the beaker is to be covered with aluminum foil.) The solvent travels up the thin layer by capillary action. If the substance is a pure colored compound, one soon sees a spot traveling either along with the solvent front or, more commonly, at some distance behind the solvent front. Once the solvent has run within a centimeter of the top of the plate, remove the plate with tweezers. Immediately, before the solvent evaporates, use a pencil to draw a line across the plate where the solvent front can be seen. The proper location of this solvent front line is needed for R_f calculations.

FIG. 8.10
Using a foil-covered beaker to develop a TLC plate.

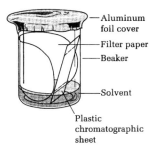

Step 5. Visualizing the Results

If you are fortunate enough to be separating organic molecules that are colored, such as dyes, inks, or indicators, then visualizing the separated spots is easy. However, because most organic compounds are colorless, this is rarely the case.

For most compounds, a UV light works well for observing the separated spots. TLC plates normally contain a fluorescent indicator that makes them glow green under UV light of wavelength 254 nm. Compounds that adsorb UV light at this wavelength will quench the green fluorescence, yielding dark purple or bluish spots on the plate.

FIG. 8.11
A UV lamp used to visualize spots.

Never look into a UV lamp.

R_f is the ratio of the distance the spot travels from the origin to the distance the solvent travels.

Simply hold the plate by its edges under a UV lamp as shown in Figure 8.11, and the compound spots become visible to the naked eye. Lightly circle the spots with a pencil so that you will have a permanent record of their location for later calculations.

Another useful visualizing technique is to use an iodine (I_2) chamber. Certain compounds, such as alkanes, alcohols, and ethers, do not absorb UV light sufficiently to quench the fluorescence of the TLC plate and therefore will not show up under a UV lamp. However, they will adsorb iodine vapors and can be detected (after any residual solvent has evaporated) by placing the plate for a few minutes in a capped 4-oz bottle containing some crystals of iodine. Iodine vapor is adsorbed by the organic compound to form brown spots. These brown spots should be outlined with a pencil immediately after removing the plate from the iodine bottle because they will soon disappear as the iodine sublimes away; a brief return to the iodine chamber will regenerate the spots. Using both the UV lamp and iodine vapor visualization methods will ensure the location of all spots on the TLC plate.

Many specialized spray reagents, also known as TLC stains, have also been developed to give specific colors for certain types of compounds. A detailed listing of different TLC stains that help visualize certain compounds with specific functional groups can be found in *The Chemist's Companion: A Handbook of Practical Data, Techniques, and References*,[3] a valuable classic reference book that contains copious amounts of helpful information.

Step 6. Calculating R_f Values

In addition to qualitative results, TLC can also provide a chromatographic parameter known as an R_f value. The R_f value is the retention factor expressed as a decimal fraction. The R_f value is the ratio of the distance the spot travels from the point of origin to the distance the solvent travels. The R_f value can be calculated as follows:

$$R_f = \frac{\text{distance spot travels}}{\text{distance solvent travels}}$$

This number should be calculated for each spot observed on a TLC plate. Figure 8.12 shows a diagram of a typical TLC plate and how the distances are

FIG. 8.12
A developed TLC plate with spots visualized and R_f values determined.

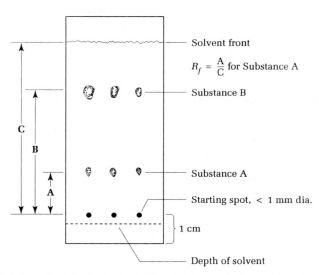

[3] Gordon, A. J., Richard A. Ford, The *Chemist's Companion: A Handbook of Practical Data, Techniques, and References*. John Wiley & Sons, Inc.: New York, 1972.

TABLE 8.4 *Chromatography Terms and Their Definitions with Examples*

Chromatography Term	Definition	Examples
Mixture	A collection of different compounds	Aspirin, ibuprofen, caffeine, and fluorene/fluorenone
Stationary phase	A fixed material that can adsorb compounds	Alumina, silica gel, and silicone gum, et al.
Mobile phase	A moving liquid or gas that dissolves compounds and carries them along	Hexane, CH_2Cl_2, and ethyl acetate (TLC and column chromatography) et al.; helium gas (gas chromatography)
Adsorption	The strength of attraction between the compounds and the stationary phase	London forces, hydrogen bonds, and dipole-dipole attractive forces
Separation	A measure of the elution or migration rate of compounds	R_f (TLC); elution volume (column chromatography); retention time (gas chromatography)

measured to calculate the R_f value. The best separations are usually achieved when the R_f values fall between 0.3 and 0.7.

If two spots travel the same distance or have the same R_f value, then it might be concluded that the two components are the same molecule. Just as many organic molecules have the same melting point and color, many can have the same R_f value, so identical R_f values do not necessarily mean identical compounds. For comparisons of R_f values to be valid, TLC plates must be run under the exact same conditions for the stationary phase, mobile phase, and temperature. Even then, additional information such as a mixed melting point or an IR spectrum should be obtained before concluding that two substances are identical.

COMPARISON OF DIFFERENT TYPES OF CHROMATOGRAPHY

Table 8.4 summarizes the terminology used in chromatography and how these apply to different types of chromatography. All of the chromatographic types involve the same principles but vary in the nature of the stationary phase and the mobile phase, and the measure of separation.

EXPERIMENTS

1. ANALGESICS

Analgesics are substances that relieve pain. The most common of these is aspirin, a component of more than 100 nonprescription drugs. In Chapter 41, the history of this most popular drug is discussed. In this experiment, analgesic tablets will be analyzed by TLC to determine which analgesics they contain and whether they contain caffeine, which is often added to counteract the sedative effects of the analgesic.

In addition to aspirin and caffeine, the most common components of the currently used analgesics are acetaminophen and ibuprofen. In addition to one or more of these substances, each tablet contains a binder—often starch, microcrystalline cellulose, or silica gel. And to counteract the acidic properties of aspirin, an inorganic buffering agent is added to some analgesics. An inspection of analgesic labels will reveal that most cold remedies and decongestants contain both aspirin and caffeine in addition to the primary ingredient.

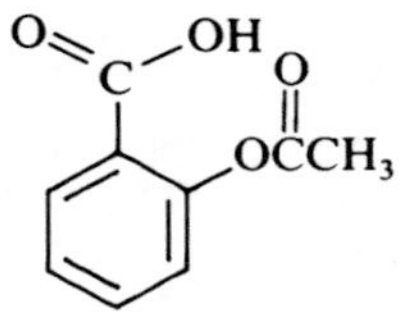

Aspirin
Acetylsalicylic acid

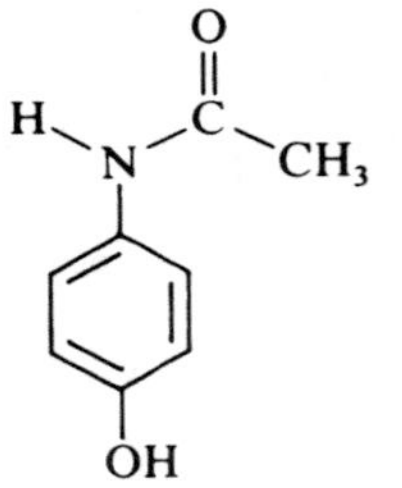

Acetaminophen
4-Acetamidophenol

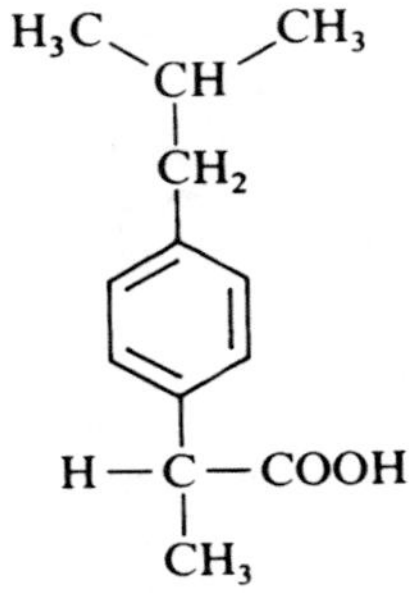

Ibuprofen
2-(4-Isobutylphenyl)propionic acid

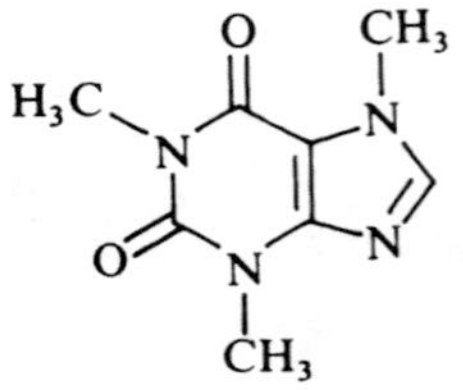

Caffeine

To identify an unknown by TLC, the usual strategy is to run chromatograms of known substances (the standards) and the unknown at the same time. If the unknown has one or more spots that correspond to spots with the same R_f values as the standards, then those substances are probably present.

Proprietary drugs that contain one or more of the common analgesics and sometimes caffeine are sold under some of the following brand names: Bayer Aspirin, Anacin, Datril, Advil, Excedrin, Extra Strength Excedrin, Tylenol, and Vanquish. Note that ibuprofen has a chiral carbon atom. The S-(+)-enantiomer is more effective than the other.

Procedure

Before proceeding, practice the TLC spotting technique described earlier. Following that procedure, draw a light pencil line about 1 cm from the end of a chromatographic plate. On this line, spot aspirin, acetaminophen, ibuprofen, and caffeine, which are available as reference standards. Use a separate capillary for each standard (or rinse the capillary carefully before reusing). Make each spot as small as possible, preferably less than 0.5 mm in diameter. Examine the plate under UV light to see that enough of each compound has been applied; if not, add more. On a separate plate, run the unknown and one or more of the standards.

Because of the insoluble binder, not all of the unknown will dissolve.

The unknown sample is prepared by crushing a part of a tablet, adding this powder to a test tube or small vial along with an appropriate amount of ethanol, and then mixing the suspension. Not all of the crushed tablet will dissolve, but enough will go into solution to spot the plate. The binder—starch or silica—will not dissolve. Weigh out only part of the tablet to try to prepare a 1% solution of the unknown. Typically, ibuprofen tablets contain 200 mg of the active ingredient, aspirin tablets contain 325 mg, and acetaminophen tablets contain 500 mg.

To the developing jar or beaker (*see* Fig. 8.9 or Fig. 8.10 on page 173), add 4 mL of the mobile phase, a mixture of 95% ethyl acetate and 5% acetic acid. Insert the spotted TLC plates with tweezers. After the solvent has risen nearly to the top of the plate, remove the plate from the developing chamber, mark the solvent front with a pencil, and allow the solvent to dry. Examine the plate under UV light to see the components as dark spots against a bright green-blue background. Outline the spots with a pencil. The spots can also be visualized by putting the plate in an iodine chamber made by placing a few crystals of iodine in the bottom of a capped 4-oz jar. Calculate the R_f values for the spots and identify the components in the unknown.

Cleaning Up. Solvents should be placed in the organic solvents waste container; dry, used chromatographic plates can be discarded in the nonhazardous solid waste container.

CHAPTER 9

Column Chromatography: Fluorenone, Cholesteryl Acetate, Acetylferrocene, and Plant Pigments

> **PRELAB EXERCISE:** Compare column chromatography and thin-layer chromatography (TLC) with regard to the (1) quantity of material that can be separated, (2) time needed for the analysis, (3) solvent systems used, and (4) ability to separate compounds.

Column chromatography is one of the most useful methods for the separation and purification of both solids and liquids when carrying out microscale experiments. It becomes expensive and time consuming, however, when more than about 10 g of material must be purified. Column chromatography involves the same chromatographic principles as detailed for TLC in Chapter 8, so be sure that you understand those before doing the experiments in this chapter.

As discussed in Chapter 1, organic chemists obtain new compounds by synthesizing or isolating natural products that have been biosynthesized by microbes, plants, or animals. In most cases, initial reaction products or cell extracts are complex mixtures containing many substances. As you have seen, recrystallization, distillation, liquid/liquid extraction, and sublimation can be used to separate and purify a desired compound from these mixtures. However, these techniques are frequently not adequate for removing impurities that are closely related in structure. In these cases, column chromatography is often used. The broad applicability of this technique becomes obvious if you visit any organic chemistry research lab, where chromatography columns are commonplace.

Three of the five experiments in this chapter involve synthesis and may be your first experience in running an organic reaction. Experiments 1 and 2 involve the synthesis of a ketone. In Experiment 3, an ester of cholesterol is prepared. Experiment 4 demonstrates the separation of colored compounds. Experiment 5 involves the isolation and separation of natural products (plant pigments), which is analogous to Experiment 2 in Chapter 8 but on a larger scale.

The most common adsorbents for column chromatography—silica gel and alumina—are the same stationary phases as used in TLC. The sample is dissolved in a small quantity of solvent (the eluent) and applied to the top of the column. The eluent, instead of rising by capillary action up a thin layer, flows down through the column filled with the adsorbent. Just as in TLC, there is an equilibrium established between the solute adsorbed on the silica gel or alumina and the eluting solvent flowing down through the column, with the less strongly absorbed solutes moving ahead and eluting earlier.

Three mutual interactions must be considered in column chromatography: the activity of the stationary adsorbent phase, the polarity of the eluting mobile solvent phase, and the polarity of the compounds in the mixture being chromatographed.

ADDITIONAL PRINCIPLES OF COLUMN CHROMATOGRAPHY

Adsorbents

A large number of adsorbents have been used for column chromatography, including cellulose, sugar, starch, and inorganic carbonates; but most separations employ alumina [$(Al_2O_3)_x$] or silica gel [$(SiO_2)_x$]. Alumina comes in three forms: acidic, neutral, and basic. The neutral form of Brockmann activity grade II or III, 150 mesh, is most commonly employed. The surface area of this alumina is about 150 m^2/g. Alumina as purchased will usually be activity grade I, meaning that it will strongly adsorb solutes. It must be deactivated by adding water, shaking, and allowing the mixture to reach equilibrium over an hour or so. The amount of water needed to achieve certain activities is given in Table 9.1. The activity of the alumina on TLC plates is usually about III. Silica gel for column chromatography, 70–230 mesh, has a surface area of about 500 m^2/g and comes in only one activity.

TABLE 9.1 *Alumina Activity*

Brockmann activity grade	I	II	III	IV	V
Percent by weight of water	0	3	6	10	15

TABLE 9.2 *Elutropic Series for Solvents*

n-Pentane (least polar)
Petroleum ether
Cyclohexane
Hexanes
Carbon disulfide
t-Butyl methyl ether
Dichloromethane
Tetrahydrofuran
Dioxane
Ethyl acetate
2-Propanol
Ethanol
Methanol
Acetic acid (most polar)

Solvents

Solvent systems for use as mobile phases in column chromatography can be determined from TLC, the scientific literature, or experimentally. Normally, a separation will begin with a nonpolar or low-polarity solvent, allowing the compounds to adsorb to the stationary phase; then the polarity of the solvent is *slowly* increased to desorb the compounds and allow them to move with the mobile phase. The polarity of the solvent should be changed gradually. A sudden change in solvent polarity will cause heat evolution as the alumina or silica gel adsorbs the new solvent. This will vaporize the solvent, causing channels to form in the column that severely reduce its separating power.

Several solvents are listed in Table 9.2, arranged in order of increasing polarity (elutropic series), with n-pentane being the least polar. The order shown in the table reflects the ability of these solvents to dislodge a polar substance adsorbed onto either silica gel or alumina, with n-pentane having the lowest solvent power.

As a practical matter, the following sequence of solvents is recommended in an investigation of unknown mixtures: elute first with petroleum ether (pentanes); then hexanes; followed by hexanes containing 1%, 2%, 5%, 10%, 25%, and 50% ether; pure ether; ether and dichloromethane mixtures; followed by dichloromethane and methanol mixtures. Either diethyl ether or *t*-butyl methyl ether can be used, but *t*-butyl methyl ether is recommended. Solvents such as methanol and water are normally not used because they can destroy the integrity of the stationary phase by dissolving some of the silica gel. Some typical solvent combinations are hexanes-dichloromethane, hexanes-ethyl acetate, and hexanes-toluene. An experimentally determined ratio of these solvents can sufficiently separate most compounds.

Petroleum ether: mostly isometric pentanes.

TABLE 9.3 *Elution Order for Solutes*

| Alkanes (first) |
| Alkenes |
| Dienes |
| Aromatic hydrocarbons |
| Ethers |
| Esters |
| Ketones |
| Aldehydes |
| Amines |
| Alcohols |
| Phenols |
| Acids (last) |

Compound Mobility

The ease with which different classes of compounds elute from a column is indicated in Table 9.3. Molecules with nonpolar functional groups are least adsorbed and elute first, while more polar or hydrogen-bonding molecules are more strongly adsorbed and elute later. The order is similar to that of the eluting solvents—another application of "like dissolves like."

Sample and Column Size

Chromatography columns can be as thin as a pencil for milligram quantities to as big as a barrel for the industrial-scale separation of kilogram quantities. A microscale column for the chromatography of about 50 mg of material is shown in Figure 9.1; columns with larger diameters, as shown in Figures 9.2 and 9.3, are used for macroscale procedures. The amount of alumina or silica gel used should generally weigh at least 30 times as much as the sample, and the column, when packed, should have a height at least 10 times the diameter. The density of silica gel is 0.4 g/mL, and the density of alumina is 0.9 g/mL, so the optimum size for any column can be calculated.

Packing the Column

Microscale Procedure

Before you pack the column, tare several Erlenmeyer flasks, small beakers, or 20-mL vials to use as receivers. Weigh each one carefully and mark it with a number on the etched circle.

Uniform packing of the chromatography column is critical to the success of this technique. Two acceptable methods for packing a column are dry packing and slurry packing, which normally achieve the best results. Assemble the column as depicted in Figure 9.1. To measure the amount of adsorbent, fill the column one-half to two-thirds full; then pour the powder out into a small beaker or flask. Clamp the column in a vertical position and close the valve. Always grasp the valve with one hand while turning it with the other. Fill the column with a non-polar solvent such as hexanes almost to the top.

Photo: Column Chromatography; Video: Column Chromatography

Dry Packing Method. This is the simplest method for preparing a microscale column. Slowly add the powdered alumina or silica gel through the funnel while gently tapping the side of the column with a pencil. The solid should "float" to the bottom of the column. Try to pack the column as evenly as possible; cracks, air bubbles, and channels in the powder will lead to a poor separation.

Slurry Packing Method. To slurry pack a column, add about 8 mL of hexanes to the adsorbent in a flask or beaker, stir the mixture to eliminate air bubbles, and

FIG. 9.1
A microscale chromatographic column.

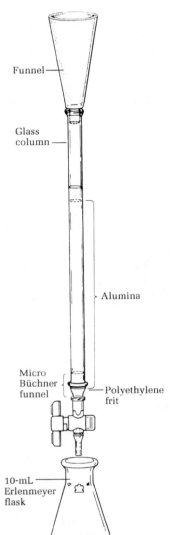

FIG. 9.2
A macroscale chromatographic column.

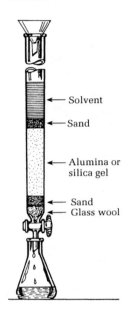

FIG. 9.3
A chromatographic tube on a ring stand.

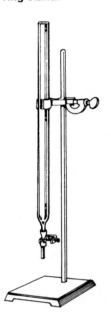

then (this is the hard part) swirl the mixture to get the adsorbent suspended in the solvent and immediately pour the entire slurry into the funnel. Open the valve, drain some solvent into the flask that had the adsorbent in it and finish transferring the slurry to the column. Place an empty flask under the column and allow the solvent to drain to about 5 mm above the top surface of the adsorbent. Tap the column with a pencil until the packing settles to a minimum height. Try to pack the column as evenly as possible; cracks, air bubbles, and channels in the packed column will lead to a poor separation.

The slurry method normally gives the best column packing, but it is also the more difficult technique to master. Whether the dry packing or slurry packing method is chosen, the most important aspect of packing the column is creating an evenly distributed and packed stationary phase. The slurry method is often used for macroscale separations.

Once the column is loaded with solvent and adsorbent, place a flask under it, open the stopcock (use two hands for the microscale column), and allow the solvent level to drop to the *top* of the packing. Avoid allowing the solvent level to go below the stationary phase (known as letting the column "run dry") because this allows air bubbles and channel formation to occur, which leads to a poor separation.

Macroscale Procedure

Before you pack the column, prepare several small Erlenmeyer flasks to use as receivers by taring (weighing) each one carefully, recording the weight in your laboratory notebook, and marking each with a number on the etched circle.

Extinguish all flames; work in the laboratory hood.

Dry Packing Method

The column can be prepared using a 50-mL burette such as the one shown in Figure 9.2 or using the less expensive and equally satisfactory chromatographic tube shown in Figure 9.3, in which the flow of solvent is controlled by a screw pinchclamp. Weigh the required amount of silica gel (12.5 g in the first experiment), close the pinchclamp on the tube, and fill about half full with a 90:10 mixture of hexanes and ether. With a wooden dowel or glass rod, push a small plug of glass wool through the liquid to the bottom of the tube, dust in through a funnel enough sand to form a 1-cm layer over the glass wool, and level the surface by tapping the tube. Unclamp the tube. With your right hand, grasp both the top of the tube and the funnel so that the whole assembly can be shaken to dislodge silica gel that may stick to the walls; with your left hand pour in the silica gel slowly (Fig. 9.4) while tapping the column with a rubber stopper fitted on the end of a pencil. If necessary, use a Pasteur pipette full of a 90:10 mixture of hexanes and ether to wash down any silica gel that adheres to the walls of the column above the liquid. When the silica gel has settled, add a little sand to provide a protective layer at the top. Open the pinchclamp, let the solvent level fall until it is just slightly above the upper layer of sand, and then stop the flow.

FIG. 9.4

A useful technique for filling a chromatographic tube with silica gel.

Slurry Packing Method

Alternatively, the silica gel can be added to the column (half-filled with hexanes) by slurrying the silica gel with a 90:10 mixture of hexanes and ether in a beaker. The powder is stirred to suspend it in the solvent and it is immediately poured through a wide-mouth funnel into the chromatographic tube. Rap the column with a rubber stopper to cause the silica gel to settle and to remove bubbles. Add a protective layer of sand to the top. The column is now ready for use.

Cleaning Up. After use, the tube is conveniently emptied by pointing the open end into a beaker, opening the pinchclamp, and applying gentle air pressure to the tip. If the plug of glass wool remains in the tube after the alumina leaves, wet it with acetone and reapply air pressure. Allow the adsorbent to dry in the hood and then dispose of it in the nonhazardous waste container.

Adding the Sample

Dissolve the sample completely in a very minimum volume of dichloromethane (just a few drops) in a small flask or vial. Add to this solution 300 mg of the adsorbent, stir, and evaporate the solvent completely by heating the slurry *very gently* with *constant* stirring to avoid bumping. Remember that dichloromethane boils at 41°C. Pour this dry powder into the funnel of the chromatography column, wash it down onto the column with a few drops of hexane, and then tap the column to remove air bubbles from the layer of adsorbent-solute mixture just added. Open the valve and carefully add new solvent in such a manner that the top surface of the column is not disturbed. A thin layer of fine sand can be added to the column after the sample to avoid disturbance of the column surface when the solvent is being added. Run the solvent down near to the surface several times to apply the sample as a narrow band at the top of the column.

Eluting the Column

Fill the column with solvent, open the stopcock, and continue to add more solvent while collecting 1–3 mL fractions in small previously tared Erlenmeyer flasks. Collecting small fractions is important to the success of your column separation. Fractions that are too small can always be pooled together; however, if the collected fractions are too large, you may get more than one compound in any particular fraction. If this occurs, the only way to attain separation is to redo the chromatography. Column chromatography is a lengthy process, so collecting large fractions is discouraged.

Isolating the Separated Compounds

If the mixture to be separated contains colored compounds, then monitoring the column is very simple. The colored bands will move down the column along with the solvent, and as they approach the end of the column, you can collect the separated colors in individual containers. However, most organic molecules are colorless. In this case, the separation must be monitored by TLC. Spot each fraction on a TLC plate (Fig. 9.5). Four or five fractions can be spotted on a single plate. Before you develop the plate, do a quick examination under UV light to see if there is any compound where you spotted. If not, you can spot the next fraction in that location. Note which fraction is in which lane. Develop the plate and use the observed spot(s) to determine which compound is in each of the collected fractions. Spotting some of the starting material or the product (if available) on the TLC plate as a standard will help in the identification.

The colors of the fractions or the results from analyzing the fractions by TLC will indicate which fractions contain the compound(s) you are interested in isolating. Combine fractions containing the same compound and evaporate the solvent. Recrystallization may be used to further purify a solid product. However, on a milligram scale, there is usually not enough material to do this.

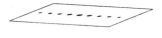

FIG. 9.5
Spot each fraction on a TLC plate. Examine under UV light to see which fractions contain the compound.

OTHER TYPES OF CHROMATOGRAPHY

Flash Chromatography

Relying only on gravity, liquid flow through a column can be quite slow, especially if the column is tightly packed. One method to speed up the process is *flash chromatography*. This method uses a pressure of about 10 psi of air or nitrogen on top of the column to force the mobile phase through the column. Normally, doing this would give a poorer separation. However, it has been found that with a finer mesh of alumina or silica gel, flash chromatography can increase the speed without lowering the quality of the separation. Go to this book's web site for an illustrated set of instructions for packing and using a flash chromatography column.

High-performance liquid chromatography.

High-performance liquid chromatography (HPLC) is a high-tech version of column chromatography, which is capable of separating complex mixtures with dozens of components. A high-pressure pump forces solvent at pressures up to 10,000 psi through a stainless steel tube packed tightly with extremely small adsorbent particles. The eluent flows from the column to a detector, such as a tiny UV absorbance cell or a mass spectrometer that is able to detect extremely small amounts of separated components, as little as a picogram (10^{-12} g).

Reverse-phase chromatography.

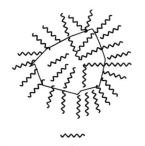

$-\text{O}-\underset{\underset{\text{CH}_3}{|}}{\overset{\overset{\text{CH}_3}{|}}{\text{Si}}}-(\text{CH}_2)_{17}\text{CH}_3$

C$_{18}$ reverse-phase packing

Chiral chromatography

HPLC is used extensively in biochemistry to separate cellular components such as proteins, lipids, and nucleic acids. Mixtures of these types of compounds can be dissolved only in a predominantly aqueous mobile phase such as methanol-water or acetonitrile-water, and normal silica gel or alumina stationary phases do not work well with high concentrations of water. Rather than polar stationary phases, highly nonpolar ones called *reverse-phase packings* are used. These are manufactured by bonding lots of hydrocarbon molecules to the surfaces of silica gel particles, which convert the particles into highly nonpolar grease balls. With this packing, the order of elution is the reverse of that observed for a normal silica gel phase. On a reverse-phase column, the more nonpolar compounds will adhere to the nonpolar stationary phase more strongly, and the polar compounds will elute first.

Is it possible to separate two enantiomers (optical isomers), both of which have the same intermolecular attractive forces? Chiral stationary phases can be used to separate enantiomers. Giving the stationary phase an asymmetry or handedness allows one enantiomer to be specifically retained on the column. Such columns are quite expensive and are limited to a particular type of separation, but they have led to great achievements in separation science. This separation technique is of great importance in the pharmaceutical industry because the U.S. Food and Drug Administration (FDA) specifies the amounts of impurities, including enantiomers, that can be found in drugs. For example, thalidomide, a drug prescribed as a sedative and an antidepressant in the 1960s, was found to be a potent teratogen that caused birth defects when pregnant women took the drug. It was quickly pulled from the market. Thalidomide has two enantiomers, and further research demonstrated that only one of the enantiomers caused the birth defects.

CHAPTER 10

Gas Chromatography: Analyzing Alkene Isomers

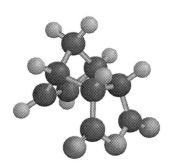

> **PRELAB EXERCISE:** If the dehydration of 2-methyl-2-butanol occurred on a purely statistical basis, what would be the relative proportions of 2-methyl-1-butene and 2-methyl-2-butene?

Gas chromatography (GC) is a rapid and sensitive method of separating and analyzing mixtures of gaseous or liquid compounds. The information it provides can tell you:

1. if you have successfully synthesized your product;
2. whether your product contains unreacted starting material or other impurities;
3. whether your product is a mixture of isomers; and
4. the relative amounts of different materials or isomers in a mixture.

Gas chromatography is one of several *instrumental* analysis methods used by organic chemists. Unlike the apparatus for thin-layer (TLC) or column chromatography, the apparatus used for gas chromatography—a modern gas chromatograph like that shown in Figure 10.1—costs thousands of dollars and is therefore shared among many users. There are two major reasons for spending this much money for this instrument. First, very complex mixtures containing hundreds of components can be separated. Second, separated components can be detected in very small amounts, 10^{-6} to 10^{-15} g. The chromatogram in Figure 10.2 is a good example of the power of gas chromatography; it shows the separation of 1 mg of crude oil into the hundreds of compounds that are present: alkanes, alkenes, and aromatics, including many isomers. The exceptional sensitivity of GC instruments is the major reason they are often used in environmental and forensic chemistry labs, where the detection of trace amounts is necessary.

Gas chromatography involves the same principles that apply to all forms of chromatography, which were covered in Chapter 8. The mobile phase is a gas, usually helium. The stationary phase is often a fairly nonpolar polymer such as polydimethylsiloxane or poly(ethylene glycol) (Carbowax) that is stable at temperatures as high as 350°C. The most commonly used phase is polydimethyl-siloxane in which 5% of the methyl groups have been replaced by phenyl groups.

There are two types of GC columns:

1. *Packed columns* in which the stationary phase consists of solid particles similar to the alumina or silica gel used in column chromatography, but coated with a nonpolar polymer (e.g., polydimethylsiloxane or Carbowax) and packed into 3–6 mm diameter metal or glass tubes that are 1–3 m long and rolled into a compact coil.

FIG. 10.1
A typical modern gas chromatograph with a data acquisition analysis computer.

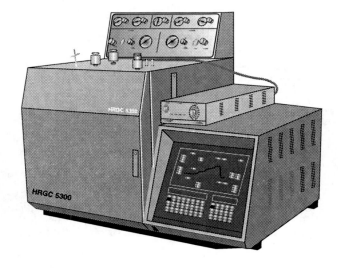

FIG. 10.2
A chromatogram from the GC analysis of 1 mg of crude petroleum. The larger peaks represent the series of *n*-alkanes from C_5 to >C_{30}. The column was temperature programmed from 40°C to 300°C at 20°C/minutes.

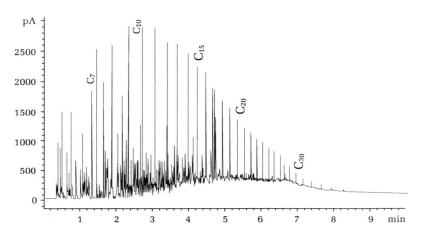

$$\left[\begin{array}{c} CH_3 \\ | \\ Si-O \\ | \\ CH_3 \end{array}\right]_n$$

Polydimethylsiloxane

$$+CH_2-CH_2-O+_n$$

Carbowax

2. *Capillary columns* are long, thin, flexible quartz tubes with a very thin coating of a nonpolar stationary phase polymer on the inside wall. They can be 15–60 m long with internal diameters of 0.1–0.5 mm. Their successful development came about because of the invention of fiber optics and the discovery that coating the outside of hollow quartz fibers with a polyimide polymer prevented them from being easily scratched and broken.

Packed columns can separate milligram quantities of materials, while capillary columns work best with microgram quantities or less. On the other hand, capillary columns have much better separating power than packed columns.

The GC instrument has the following components (Fig. 10.3). Helium at a pressure of 10–30 psi from a compressed gas cylinder flows through the heated injector port, the column (located in a temperature-controlled oven), and the heated detector at a flow rate of 10–60 mL/minutes for a packed column and about 1 mL/minutes for a capillary column. A microliter syringe is used to inject 1–25 μL of sample

FIG. 10.3

A diagram of a gas chromatograph.

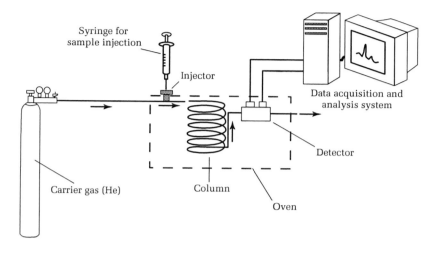

through a rubber septum into the hot (~250°C) injector, where any liquid compounds are instantly vaporized to the gaseous state. As the vaporized molecules are swept through the column, they interact with the stationary phase and are adsorbed and desorbed many times each second in a dynamic equilibrium. As discussed in Chapter 8, the equilibria between adsorbed and free states depends on each molecule's size (London forces), polarity, and ability to hydrogen bond. Molecules spending more time absorbed to the stationary phase will not be carried through the column as rapidly as those that are weakly bound and free. A molecular view of this differential adsorption and flow is depicted in Figure 10.4.

The separated components pass one after the other into the detector. In column chromatography (*see* Chapter 9), human eyes are the detectors, either seeing colored compounds flow out of the column or visualizing the presence of colorless compounds on a fluorescent TLC plate. Gas chromatographs have electronic detectors that produce a signal voltage proportional to the number of molecules passing through them at any instant in time. A record of this voltage versus time is a gas chromatogram, with peak areas representing the amount of each individual component passing through the detector. Figure 10.5 is the chromatogram produced by the injection of a mixture of two compounds that were separated on the column and detected to produce peaks A and B. The amount of each is proportional to the area under its peak, so we can say that there appears to be about twice as much B as A. If an integrating recorder or computer data acquisition system is used, the exact areas under all peaks are automatically determined and can be used to quantify the amount of each component in a mixture.

Most GC instruments use one of the following types of detectors:

- The *thermal conductivity detector* (TCD) consists of an electrically heated wire or thermistor. The temperature of the sensing element depends on the thermal conductivity of the gas flowing around it. Changes in thermal conductivity, such as when organic molecules displace some of the carrier gas, cause a temperature rise in the element that is sensed as a change in resistance. The TCD is the least sensitive (detecting micrograms per second, 10^{-6} g/s, of material) of the four detectors described here, but it is quite rugged. It will detect all types of molecules, not just those containing C—H bonds, and compounds passing through it are unchanged and can be collected.

FIG. 10.4

The differential adsorption of two compounds in a flowing gas leads to the separation of the compounds over time.

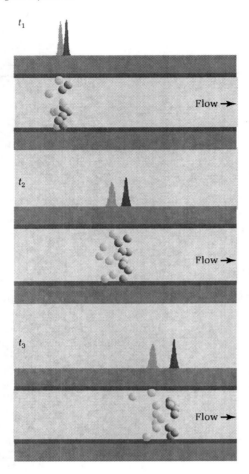

FIG. 10.5

A gas chromatogram of a mixture of two compounds, A and B.

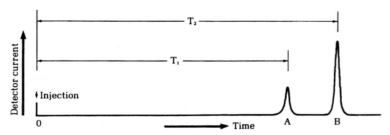

- The *flame ionization detector* (FID) is at least a thousand times more sensitive than the TCD, easily detecting nanograms per second (10^{-9} g/s) of material. The molecules, however, must contain C—H bonds because the sample is actually burned to form ions. These ions carry a tiny current between two electrodes, which is greatly amplified to produce the signal output.

- The *electron capture detector* (ECD) contains a tiny amount of a radioactive substance, such as ^{63}Ni, that emits high-energy electrons (β^- particles).

The sample molecules can capture these electrons, become charged, and carry a current between two electrodes, as in the FID detector. Halocarbons have exceptionally good cross sections for electron capture because of the electronegativity of the halogen atoms, and these types of compounds can be detected at picogram per second (10^{-12} g/s) levels. For this reason, ECD is used to measure levels of chlorocarbon pollutants, such as chloroform, DDT, and dioxin in the environment.

- A *mass spectrometer detector* is the only one of these four detectors that provides information about the structure of the molecules that pass into it. For this reason, combination gas chromatography-mass spectrometry (GC–MS) systems are often used in organic research, forensic science, and environmental analysis in spite of their high cost ($40,000 to $250,000). Sensitivity is at the nanogram to picogram per second level. Chapter 13 discusses mass spectrometry in detail.

The instant you inject your sample into the GC instrument, you normally press a button that starts a clock in the data recorder, either an electronic integrator or a computer data system. Then, as each eluting component produces a peak in the detector, the top of the peak is detected, and the elapsed time since injection, called the *retention time*, is recorded and stored. In Figure 10.5, the retention time of compound A is T_1 and of B is T_2. Which compound is more strongly absorbed by the stationary phase?

Temperature programming

One powerful variable in gas chromatography that is not available in TLC or column chromatography is temperature. Because the GC column is in an oven whose temperature is programmable—that is, the temperature can be raised from near room temperature to as high as 350°C at a constant and reproducible rate—we are able to separate much more complex mixtures than if the oven could be set to only one temperature. At room temperature, a mixture containing lower-boiling, weakly adsorbed components (such as ether and dichloromethane) and higher-boiling, strongly adsorbed components (such as butanol and toluene) might take hours to separate because the latter would move significantly more slowly through the column at low temperature. If the column were kept at a high temperature, the higher-boiling compounds would come out in a shorter time, but the lower-boiling ones would not be retained at all and therefore would not be separated. By programming the oven temperature, we can inject such a mixture at a low initial temperature so that weakly adsorbed components are separated. Then, as the temperature slowly rises, the more strongly retained components will move through the column faster and also have reasonable retention times. Temperature programming allows the separation of very complex mixtures containing both low-boiling small molecules and high-boiling large ones, as demonstrated by the GC analysis of crude petroleum in Figure 10.2 on page 206.

THE GENERAL GC ANALYSIS PROCEDURE

Most GC separations are of gaseous and liquid mixtures. Although some solids will pass through a gas chromatograph at higher temperatures, it is best if all of the samples are distilled or vacuum transferred before injection into the GC instrument to ensure that only volatile compounds are present. If samples contain materials that cannot be vaporized and swept through the GC column, even at high temperatures, these will remain on the column and can ultimately ruin its performance. GC columns are expensive and can cost up to $400 to replace.

The guidelines given here are for GC analysis using a capillary column and an FID detector, which require very dilute solutions to avoid overwhelming the column, even if 90%–95% of the sample is split away so that it does not enter the column, a common practice in capillary chromatography. Samples analyzed with a packed column and a TCD detector can be much more concentrated or even undiluted (neat).

Sample Preparation

There are different ways of preparing a GC sample, depending on whether your product is a solid, a liquid (usually obtained by distillation), or a gaseous mixture; on the type of column (capillary or packed); and on the type of detector (FID, ECD, or TCD) being used.

Solid Samples

Low-melting, sublimable solids can be analyzed at higher column temperatures. Run a GC analysis on solids only if your experimental procedure explicitly says that you can do so. To prepare a solid sample for capillary column analysis, put a *small* crystal of the solid into a small vial and dissolve it in 1 mL of dichloromethane.

Liquid Samples

Dilute solutions of liquid organic mixtures in dichloromethane are the most common form of GC samples. Place 1 or 2 drops of your distilled product into a small vial and add 1 mL of dichloromethane. Because GC columns can be damaged by moisture, it is suggested that enough *anhydrous* sodium sulfate be added to cover the bottom of the vial to a depth of 1–2 mm to ensure that all traces of water are removed. The sodium sulfate can be left in the sample because it usually will not go into the fine needle of the microliter syringe used to inject the sample. Injection of 1 µL of a sample prepared in this manner usually provides strong signals in an FID detector. In special cases, undiluted neat or pure liquids can be injected directly without dilution as noted; for example, the methylbutenes in Experiment 1. Also, if preparative gas chromatography is used to separate and collect the components of a mixture, neat liquids are usually injected.

Gaseous Samples

Gases are normally collected over water, as in Experiment 2. Do not remove the septum-capped collection tube from the beaker of water, but bring it and the beaker of water together to the gas chromatograph. A gas syringe is inserted through the rubber septum and 10–500 µL (depending on the type of the gas chromatograph) of the gas sample is withdrawn and injected into the unit.

Sample Analysis

Make sure that the required gases are flowing and that the instrument has been on at least 1 hour so that all zones, the injector, the oven, and the detector have come to the required temperatures. Enter a temperature program: initial temperature, heating rate (in degrees per minutes), and final temperature, as required by the particular experiment. Fill the microliter syringe with the sample solution, normally 1 µL. A 10-µL capacity syringe with a plunger guide is recommended. Handle this expensive syringe carefully; it has a sharp needle, and the glass barrel is easily dropped and broken. Make sure that there are no bubbles in the syringe barrel when you

Do not touch the metal injector cap. It is very HOT.

draw up the sample. If bubbles are present, fill the syringe about halfway and depress the plunger quickly. Draw up the sample again slowly. Repeat as necessary to remove all bubbles. To obtain a quality chromatogram, the needle should be inserted rapidly, the sample should be injected rapidly and the syringe should be removed as soon as the plunger has been depressed and the sample injected. Then quickly press the start analysis button on the gas chromatograph, computer, or integrator to start data acquisition or mark the paper on the pen recorder.

Once all the compounds have eluted and the oven temperature program has finished, the run should be stopped. Plot your chromatogram using the computer or remove the chromatogram chart from the integrator or flat bed recorder. Record the GC analysis parameters: column diameter and length, column packing type, carrier gas and its flow rate, column temperature (initial, final, and heating rate if programmed), detector and injector temperatures, sample injection amount and solvent (if used), signal attenuation, and chart speed; keep these with your gas chromatogram. Write the structures of all identifiable peaks on the chromatogram. If you used dichloromethane to dilute your sample, the largest peak will be due to this substance and will have a short retention time.

Remember that retention times are not constants. If the programmed oven temperature rate is higher or lower than for a prior analysis, then the retention times for all of the peaks will be consistently shorter or consistently longer. Changes in helium flow and the aging of the column also affect retention times. Chemists primarily look for similarities in the pattern of eluting peaks, not for perfect matches in retention times. In general, for simple mixtures chromatographed on nonpolar stationary phases, the retention time order is the same as the order of increasing boiling points. For example, you would expect cyclohexene (bp 83°C) to have a shorter retention time than toluene (bp 110°C). You can often ascertain the identity of a certain peak by adding a small amount of a known standard to the sample and rerunning the GC analysis. The peak corresponding to the standard will increase significantly and can be identified. Often the chromatogram will show a number of small peaks that cannot be assigned to the product, the starting material, or the solvents; these are likely due to byproducts. As long as these impurities are minor, say, less than 10% of the total area (not including the solvent), they can probably be ignored. The ultimate method of identifying every peak is to analyze the sample on a GC–MS instrument (Fig. 10.6).

Cleaning Up. Rinse out the syringe by filling it with dichloromethane and squirting it onto a tissue or paper towel at least four times. Clean and return any sample collection tubes, if used.

Collecting a Sample for an Infrared Spectrum

The small amount of sample analyzed in gas chromatography is an advantage in many cases, but it precludes isolating the separated components. Some specialized chromatographs can separate samples as large as 0.5 mL per injection and automatically collect each fraction in a separate container. At the other extreme, gas chromatographs equipped with FIDs can detect micrograms of substances, such as traces of pesticides in food, or drugs in blood and urine. Clearly, a gas chromatogram gives little information about the chemical nature of the sample being detected. Gas chromatography cannot positively identify most samples, nor can it always detect all substances in a sample. All that a gas chromatogram can truly tell you is that the detector was sensitive to a compound, and it can give you the relative time that a component eluted from a column. However, certain

FIG. 10.6
A combined gas chromatograph/mass spectrometer, fitted with an automatic sample changer.

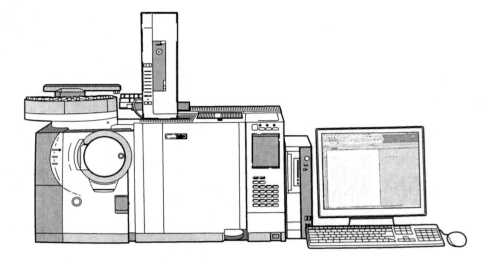

FIG. 10.7
Some gas chromatographs with packed columns allow the collection of small amounts of separated compounds.

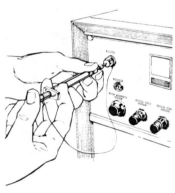

FIG. 10.8
A gas chromatographic collection device. Fill the container with ice or a dry ice–acetone mixture and attach to the outlet port of the gas chromatograph.[1]

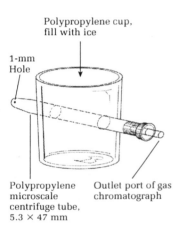

preparative chromatographs, like that shown in Figure 10.7, allow the collection of enough sample at the exit port to obtain an infrared spectrum. About 10–15 μL of a mixture (not diluted in solvent) is injected, and as the peak for the compound of interest appears, a 2-mm-diameter glass tube, 3 in. long and packed with glass wool, is inserted into the rubber septum at the exit port. The sample, if it is not too volatile, will condense in the cold glass tube. Subsequently, the sample is washed out with 1 or 2 drops of solvent, and an infrared spectrum is obtained. This process can be repeated to collect enough sample for obtaining an NMR spectrum (*see* Chapter 12), using a few drops of deuterochloroform ($CDCl_3$) to wash out the tube each time. See Figure 10.8 for another collection device.

---

[1]This apparatus is available from Kimble Kontes (Vineland, NJ).

CHAPTER 11

Infrared Spectroscopy

PRELAB EXERCISE: When an infrared (IR) spectrum is run, it is possible that the chart paper is not properly placed or that the spectrometer is not mechanically adjusted. Describe how you could calibrate an IR spectrum.

The types and molecular environment of functional groups in organic molecules can be identified by infrared (IR) spectroscopy. Like nuclear magnetic resonance (NMR) and ultraviolet (UV) spectroscopy, IR spectroscopy is nondestructive. Moreover, the small quantity of sample needed, the speed with which a spectrum can be obtained, the relatively low cost of the spectrometer, and the wide applicability of the method combine to make IR spectroscopy one of the most common structural elucidation tools used by organic chemists.

IR radiation consists of wavelengths that are longer than those of visible light. It is detected not with the eyes, but by a feeling of warmth on the skin. When absorbed by molecules, radiation of these wavelengths (typically 2.5–5 μm) increases the amplitude of vibrations of the chemical bonds joining atoms.

IR spectra are measured in units of frequency or wavelength. The wavelength is measured in micrometers[1] or microns, μ (1 μm = 1 × 10⁻⁶ m). The positions of absorption bands are measured in frequency units called wavenumbers $\bar{\nu}$, which are expressed in reciprocal centimeters, cm⁻¹, corresponding to the number of cycles of the wave in each centimeter.

> 2.5–25 μm equals 4000–400 cm⁻¹. Wavenumber(cm⁻¹) is proportional to frequency (c is the speed of light).
>
> $$\bar{\nu}\,(\text{cm}^{-1}) = \frac{\nu}{c}$$

$$\text{cm}^{-1} = \frac{10{,}000}{\mu\text{m}}$$

> Examine the scale carefully.

Unlike UV and NMR spectra, IR spectra are inverted, with the strongest absorptions at the bottom (called "peaks" although they look like valleys), and are not always presented on the same scale. Some spectrometers record the spectra on an ordinate linear in microns, but this compresses the low-wavelength region. Other spectrometers present the spectra on a scale linear in reciprocal centimeters, but linear on two different scales: one between 4000 and 2000 cm⁻¹, which spreads out the low-wavelength region; and the other a smaller one between 2000 and 667 cm⁻¹. Consequently, spectra of the same compound run on two different spectrometers will not always look the same.

---

[1] Although micrometers are known as microns, the micron is not the official SI (International System of Units) unit.

IR spectroscopy easily detects:

Hydroxyl groups	—OH
Amines	—NH$_2$
Nitriles	—C≡N
Nitro groups	—NO$_2$

IR spectroscopy is especially useful for detecting and distinguishing among all carbonyl-containing compounds:

$$\text{Acids} \quad -\overset{\overset{O}{\|}}{C}-OH$$

$$\text{Amides} \quad -\overset{\overset{O}{\|}}{C}-NH_2$$

$$\text{Anhydrides} \quad -\overset{\overset{O}{\|}}{C}-O-\overset{\overset{O}{\|}}{C}-$$

$$\text{Aldehydes} \quad -\overset{\overset{O}{\|}}{C}-H$$

$$\text{Ketones} \quad -\overset{\overset{O}{\|}}{C}-$$

$$\text{Esters} \quad -\overset{\overset{O}{\|}}{C}-O-$$

$$\text{Lactones} \quad \boxed{-\overset{\overset{O}{\|}}{C}-O-}$$

To picture the molecular vibrations that interact with IR light, imagine a molecule as being made up of balls (atoms) connected by springs (bonds). The vibration can be described by Hooke's law from classical mechanics, which says that the frequency of a stretching vibration is directly proportional to the strength of the spring (bond) and inversely proportional to the masses connected by the spring. Thus we find that C—H, N—H, and O—H bond-stretching vibrations are high frequency (short wavelength) compared to those of C—C and C—O because of the low mass of hydrogen compared to that of carbon or oxygen. The bonds connecting carbon to bromine and iodine, atoms of large mass, vibrate so slowly that they are beyond the range of most common IR spectrometers. A double bond can be regarded as a stiffer, stronger spring, so we find C=C and C=O vibrations at higher frequencies than C—C and C—O stretching vibrations. And C≡C and C≡N stretch at even higher frequencies than C=C and C=O (but at lower frequencies than C—H, N—H, and O—H). These frequencies are in keeping with the bond strengths of single (~100 kcal/mol), double (~160 kcal/mol), and triple bonds (~220 kcal/mol).

These stretching vibrations are intense and particularly easy to analyze. A nonlinear molecule of n atoms can undergo $3n - 6$ possible modes of vibration, which means cyclohexane with 18 atoms can undergo 48 possible modes of vibration. Each vibrational mode produces a peak in the spectrum because it corresponds to the absorption of energy at a discrete frequency. These many modes of vibration create a complex spectrum that defies simple analysis, but even in very complex molecules, certain functional groups have characteristic frequencies that can easily be recognized. Within these functional groups are the above-mentioned atoms and bonds, C—H, N—H, O—H, C=C, C=O, C≡C, and C≡N. Their absorption frequencies are given in Table 11.1.

When the frequency of IR light is the same as the natural vibrational frequency of an interatomic bond, light will be absorbed by the molecule, and the amplitude of the bond vibration will increase. The intensity of IR absorption bands is proportional to the change in dipole moment that a bond undergoes when it stretches. Thus, the most intense bands (peaks) in an IR spectrum are often from C=O and C—O stretching vibrations, whereas the C≡C stretching band for a symmetrical acetylene is almost nonexistent because the molecule undergoes no net change of dipole moment when it stretches:

TABLE 11.1 *Characteristic IR Absorption Wavenumbers*

Functional Group	Wavenumber (cm^{-1})
O—H	3600–3400
N—H	3400–3200
C—H	3080–2760
C≡N	2260–2215
C≡C	2150–2100
C=O	1815–1650
C=C	1660–1600
C—O	1200–1050

$$\overset{+\longrightarrow}{\underset{/}{\overset{\backslash}{C}}=O} \longleftrightarrow \overset{+\longrightarrow}{\underset{/}{\overset{\backslash}{C}}=O} \qquad \overset{\longleftarrow\quad\longrightarrow}{H_3C-C\equiv C-CH_3} \longleftrightarrow \overset{\longrightarrow\quad\longleftarrow}{H_3C-C\equiv C-CH_3}$$

Change in dipole moment   No change in dipole moment

Unlike proton NMR spectroscopy, where the area of the peaks is strictly proportional to the number of hydrogen atoms causing the peaks, the intensities of IR peaks are not proportional to the numbers of atoms causing them. Given the chemical shifts and coupling constants, it is not too difficult to calculate a theoretical NMR spectrum that is an exact match to the experimental one. For larger molecules, the calculation of all possible stretching and bending frequencies (the IR spectrum) requires large amounts of time on a fast computer. Every peak or group of peaks in an NMR spectrum can be assigned to specific hydrogens in a molecule, but the assignment of the majority of peaks in an IR spectrum with absolute certainty is usually not possible. Peaks to the right (longer wavelength) of 1250 cm^{-1} are the result of combinations of vibrations that are characteristic not of individual functional groups but of the molecule as a whole. This part of the spectrum is often referred to as the *fingerprint region* because it is uniquely characteristic of each molecule. Although two organic compounds can have the same melting points or boiling points and can have identical UV and NMR spectra, they cannot have identical IR spectra (except, as usual, for enantiomers). IR spectroscopy is thus the final arbiter in deciding whether two compounds are identical.

The intensity of absorption is proportional to the change in dipole moment.

ANALYSIS OF IR SPECTRA

Only a few simple rules or equations govern IR spectroscopy. Because it is not practical to calculate theoretical spectra, the analysis is done almost entirely by *correlation* with other spectra. In printed form, these comparisons take the form of lengthy discussions, so detailed analysis of a spectrum is best done with a good reference book at hand.

In a modern analytical or research laboratory, a collection of many thousands of spectra is maintained on a computer. When the spectrum of an unknown compound is run, the analyst picks out five or six of the strongest peaks and asks the computer to list all the known compounds that have peaks within a few reciprocal centimeters of the experimental peaks. From the printout of a dozen or so compounds, it is often possible to pinpoint all the functional groups in the molecule being analyzed. There may be a perfect match of all peaks, in which case the unknown will have been identified.

For relatively simple molecules, a computer search is hardly necessary. Much information can be gained about the functional groups in a molecule from relatively few correlations.

To carry out an analysis, (1) pay most attention to the strongest absorptions; (2) pay more attention to peaks to the left (shorter wavelength) of 1250 cm^{-1}; and (3) pay as much attention to the absence of certain peaks as to the presence of others. The absence of characteristic peaks will definitely exclude certain functional groups. Be wary of weak O—H peaks because water is a common contaminant of many samples. Because potassium bromide is hygroscopic, water is often found in the spectra of samples prepared as KBr pellets.

The Step-by-Step Analysis of IR Spectra

IR spectra are analyzed as follows:

1. Is there a peak between 1820 cm^{-1} and 1625 cm^{-1}? If not, go to Step 2.
   (a) Is there a strong, wide O—H peak between 3200 cm^{-1} and 2500 cm^{-1}? If so, the compound is a carboxylic acid (Fig. 11.1, oleic acid). If not . . .
   (b) Is there a medium-to-weak N—H band between 3520 cm^{-1} and 3070 cm^{-1}? If there are two peaks in this region, the compound is a primary amide; if not, it is a secondary amide. If there is no peak in this region . . .
   (c) Are there two strong peaks, one in the region 1870 cm^{-1} to 1800 cm^{-1} and the other in the region 1800 cm^{-1} to 1740 cm^{-1}? If so, an acid anhydride is present. If not . . .
   (d) Is there a peak in the region of 2720 cm^{-1}? If so, is the carbonyl peak in the region 1715 cm^{-1} to 1680 cm^{-1}? If so, the compound is a conjugated aldehyde; if not, it is an isolated aldehyde (Fig. 11.2, benzaldehyde). However, if there is no peak near 2720 cm^{-1} . . .
   (e) Does the strong carbonyl peak fall in the region 1815 cm^{-1} to 1770 cm^{-1} and the compound give a positive Beilstein test? If so, it is an acid halide. If not . . .
   (f) Does the strong carbonyl peak fall in the region 1690 cm^{-1} to 1675 cm^{-1}? If so, the compound is a conjugated ketone. If not . . .

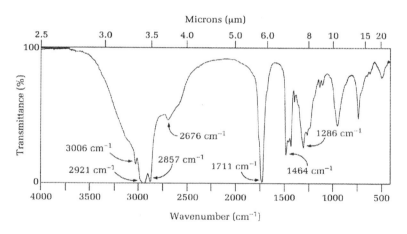

FIG. 11.1
The IR spectrum of oleic acid (thin film).

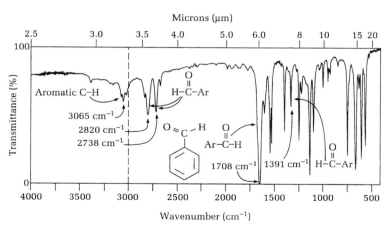

FIG. 11.2
The IR spectrum of benzaldehyde (thin film).

FIG. 11.3
The IR spectrum of *n*-butyl acetate (thin film).

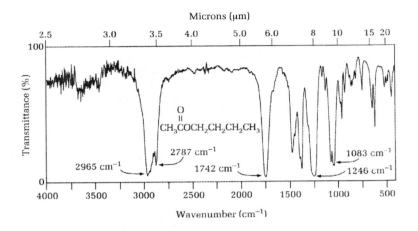

FIG. 11.4
The IR spectrum of cyclohexanol (thin film).

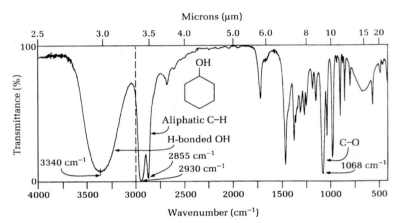

(g) Does the strong carbonyl peak fall in the region 1670 cm^{-1} to 1630 cm^{-1}? If so, the compound is a tertiary amide. If not . . .

(h) Does the spectrum have a strong, wide peak in the region 1310 cm^{-1} to 1100 cm^{-1}? If so, does the carbonyl peak fall in the region 1730 cm^{-1} to 1715 cm^{-1}? If so, the compound is a conjugated ester; if not, the ester is not conjugated (Fig. 11.3, *n*-butyl acetate). If there is no strong, wide peak in the region 1310 to 1100 cm^{-1}, then . . .

(i) The compound is an ordinary nonconjugated ketone (see for example Fig. 22.8, the IR spectrum of cyclohexanone).

2. If the spectrum lacks a carbonyl peak in the region 1820 cm^{-1} to 1625 cm^{-1}, does it have a broad band in the region 3650 cm^{-1} to 3200 cm^{-1}? If so, does it also have a peak at about 1200 cm^{-1}, a C—H stretching peak to the left of 3000 cm^{-1}, and a peak in the region 1600 cm^{-1} to 1470 cm^{-1}? If so, the compound is a phenol. If the spectrum does not meet these latter three criteria, the compound is an alcohol (Fig. 11.4, cyclohexanol). However, if there is no broad band in the region 3650 cm^{-1} to 3200 cm^{-1}, then . . .

(a) Is there a broad band in the region 3500 cm^{-1} to 3300 cm^{-1}, and does the compound smell like an amine, or does it contain nitrogen? If so, are there

FIG. 11.5
The IR spectrum of benzonitrile (thin film).

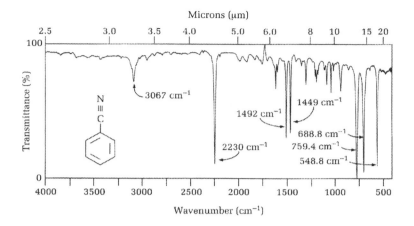

FIG. 11.6
The IR spectrum of 3-methylpentane (thin film).

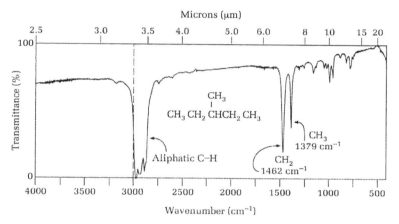

two peaks in this region? If so, the compound is a primary amine; if not, it is a secondary amine. However, if there is no broad band in the region 3500 cm^{-1} to 3300 cm^{-1}, then . . .

(b) Is there a sharp peak of medium-to-weak intensity at 2260 cm^{-1} to 2100 cm^{-1}? If so, is there also a peak at 3320 cm^{-1} to 3310 cm^{-1}? If so, then the compound is a terminal acetylene. If not, the compound is most likely a nitrile (Fig. 11.5, benzonitrile), although it might be an asymmetrically substituted acetylene. If there is no sharp peak of medium-to-weak intensity at 2260 cm^{-1} to 2100 cm^{-1}, then . . .

(c) Are there strong peaks in the region 1600 cm^{-1} to 1540 cm^{-1} and 1380 cm^{-1} to 1300 cm^{-1}? If so, the molecule contains a nitro group. If not . . .

(d) Is there a strong peak in the region 1270 cm^{-1} to 1060 cm^{-1}? If so, the compound is an ether. If not . . .

(e) The compound is either a tertiary amine (odor?), a halogenated hydrocarbon (Beilstein test?), or just an ordinary hydrocarbon (Fig. 11.6, 3-methylpentane, and Fig. 11.7, *t*-butylbenzene).

Many comments can be added to this bare outline. For example, dilute solutions of alcohols will show a sharp peak at about 3600 cm^{-1} for a nonhydrogen-bonded O—H in addition to the usual broad hydrogen-bonded O—H peak.

FIG. 11.7
The IR spectrum of *t*-butylbenzene (thin film).

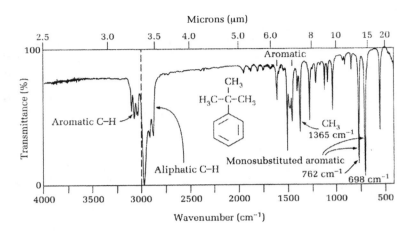

The effect of ring size on the carbonyl frequencies of lactones and esters:

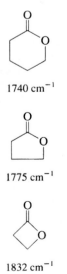

Aromatic hydrogens give peaks just to the left of 3000 cm^{-1}, whereas aliphatic hydrogens appear just to the right of 3000 cm^{-1}. However, NMR spectroscopy is the best method for identifying aromatic hydrogens.

The carbonyl frequencies listed earlier refer to an open chain or an unstrained functional group in a nonconjugated system. If the carbonyl group is conjugated with a double bond or an aromatic ring, the peak will be displaced to the right by 30 cm^{-1}. When the carbonyl group is in a ring smaller than six members or if there is oxygen substitution on the carbon adjacent to an aldehyde or ketone carbonyl, the peak will be moved to the left (refer to the margin notes about the effect of ring size and Table 11.2).

TABLE 11.2 *Characteristic IR Carbonyl Stretching Peaks (Chloroform Solutions)*

Carbonyl-Containing Compounds		Wavenumber (cm^{-1})
RCR (O)	Aliphatic ketones	1725–1705
RCCl (O)	Acid chlorides	1815–1785
R–C=C–C–R (O)	α,β-Unsaturated ketones	1685–1666
ArCR (O)	Aryl ketones	1700–1680
cyclohexanone	Cyclohexanones	1725–1705
–C–CH$_2$–C– (O, O)	β-Diketones	1640–1540

The effect of ring size on carbonyl frequency of ketones:

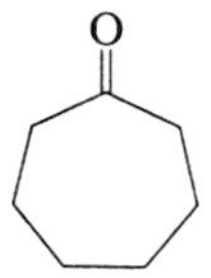

cycloheptanone: 1705 cm^{-1}

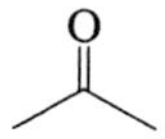

acetone: 1715 cm^{-1}

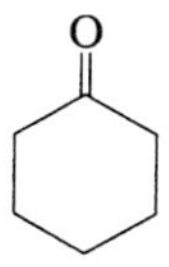

cyclohexanone: 1715 cm^{-1}

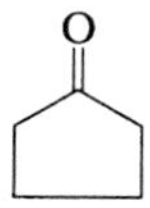

cyclopentanone: 1745 cm^{-1}

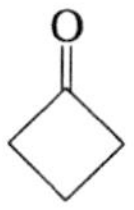

cyclobutanone: 1780 cm^{-1}

cyclopropanone: 1815 cm^{-1}

TABLE 11.2 (*continued*)

Carbonyl-Containing Compounds		Wavenumber (cm^{-1})
RCHO	Aliphatic aldehydes	1740–1720
R—C=C—CHO	α,β-Unsaturated aldehydes	1705–1685
ArCHO	Aryl aldehydes	1715–1695
RCOOH	Aliphatic acids	1725–1700
R—C=C—COOH	α,β-Unsaturated acids	1700–1680
ArCOOH	Aryl acids	1700–1680
RCOOR′	Aliphatic esters	1740
R—C=C—COOR′	α,β-Unsaturated esters	1730–1715
ArCOOR	Aryl esters	1730–1715
HCOOR	Formate esters	1730–1715
CH$_2$=CHOCOCH$_3$ C$_6$H$_5$OCOCH$_3$	Vinyl and phenyl acetate	1776
R—CO—O—CO—R	Acyclic anhydrides (two peaks)	1840–1800 1780–1740
RCONH$_2$	Primary amides	1694–1650
RCONHR′	Secondary amides	1700–1670
RCONR′$_2$	Tertiary amides	1670–1630

FIG. 11.8
The band patterns of toluene and *o-*, *m-*, and *p-*xylene. These peaks are *very weak*. They are characteristic of aromatic substitution patterns in general, not just for these four molecules. Try to find these patterns in other spectra throughout this text.

Methyl groups often give a peak near 1375 cm^{-1}, but NMR is a better method for detecting this group.

The pattern of substitution on an aromatic ring (mono-, *ortho-*, *meta-*, and *para*, di-, tri-, tetra-, and penta-) can be determined from C—H out-of-plane bending vibrations in the region 670–900 cm^{-1}. Much weaker peaks between 1650 cm^{-1} and 2000 cm^{-1} are illustrated in Figure 11.8.

Extensive correlation tables and discussions of characteristic group frequencies can be found in the specialized references listed at the end of this chapter.

CHAPTER 12

Nuclear Magnetic Resonance Spectroscopy

¹H NMR: Determination of the number, kind, and relative locations of hydrogen atoms (protons) in a molecule.

Chemical shift, δ (ppm)

> **PRELAB EXERCISE:** Outline the preliminary solubility experiments you would carry out using inexpensive solvents before preparing a solution of an unknown compound for nuclear magnetic resonance (NMR) spectroscopy using expensive deuterated solvents.

Most organic chemists would agree that the most powerful instrumental method for revealing the structure of organic molecules is nuclear magnetic resonance (NMR) spectroscopy. The 2002 Nobel Prize in Chemistry was awarded, in part, to Kurt Wüthrich for advances in NMR spectroscopy that allowed the determination of the three-dimensional structure of biological macromolecules in solution. Because of such capabilities, introductory organic chemistry courses devote considerable time to the study of NMR. This concise chapter assumes that you have had some prior exposure to the concepts discussed; it focuses on the practical aspects of sample preparation, data acquisition, and interpretation of NMR spectra to elucidate and confirm organic structures.

Let us briefly review the theory of NMR. Certain nuclei such as ^{1}H, ^{13}C, ^{15}N, ^{18}O, ^{19}F, and ^{31}P are said to have a spin, S, of ½ and behave like tiny magnets that can assume two energy states when placed in a magnetic field: aligned (lower energy) and opposed (higher energy). Like the energy states in electronic spectra, the difference in energy between these states is quantized, and energy is absorbed and emitted only at certain radio frequencies. These frequencies depend on the type of nucleus and the magnetic field strength, which is constant for a given instrument. More importantly, however, is the fact that slight differences in the electronegativity and bonding state (sp, sp^2, and sp^3) of surrounding atoms cause small—parts per million (ppm)— variations in the magnetic field felt by each nucleus, causing them to have absorption signals at slightly different frequencies. These small variations, called *chemical shifts*, are plotted versus signal intensity to produce the NMR spectrum. The interpretation of these signals and other spectral features such as splitting patterns and peak areas, as described in the following sections, facilitates organic structure elucidation.

INTERPRETATION OF ¹H NMR SPECTRA

There are two approaches to interpreting proton NMR spectra:

1. The *structure from the spectrum* approach is the strategy of using the information in the NMR spectrum to draw the structure of the molecule based on reference tables and rules. This approach is used if the compound's structure is unknown. In this case, the NMR spectrum alone is often insufficient to "solve" the complete structure, and must be combined with knowledge about the compound's source (synthetic reaction or natural product) and complementary spectral data (infrared, ultraviolet, and/or mass spectrometric).

2. The *spectrum from the structure* approach might be thought of as the reverse of the first approach and is commonly used when verifying products of known reactions using known starting materials. A hypothetical NMR spectrum is created, based on the known compound's structure (or related structural possibilities), using the same reference tables and rules as in the first approach. Computer programs are available that allow the calculation of a hypothetical NMR spectrum for any molecule simply by entering its structure. (Refer to the supplemental information for this chapter on this book's web site.) Even better, there may be a published spectrum of the compound available in a collection of reference spectra. Either way, the sample spectrum should be compared to the hypothetical or reference spectrum, and if the two are closely matched, one can be confident that the correct compound has been obtained.

Ethyl iodide

NMR spectra contain considerably more structural information than do infrared spectra, and this additional information should be used in interpreting them. The following four most informative features are the principal ones to look for when interpreting spectra. We will use the example of ethyl iodide as a specific illustration of each feature.

1. The Number of Signals Due to Equivalent Hydrogen Nuclei or Protons

The three hydrogen atoms of the methyl group are said to be chemically and magnetically equivalent. Replacing any of these hydrogens with another atom of hydrogen will give the same compound. Similarly, the two methylene protons are magnetically and chemically equivalent. This is because there is very fast rotation about the carbon-carbon bond. Individual hydrogens or each set of *equivalent* hydrogens will experience slightly different magnetic fields depending on their chemical environment and, therefore, will produce a signal at different places in the spectrum. Thus, we would predict that ethyl iodide would produce two signals, one for the three equivalent CH_3 protons, and the other from the equivalent protons on the CH_2 bonded to the electronegative iodine. The NMR spectra of ethyl iodide, regardless of whether it is obtained at low field strength (Fig. 12.1) or high field strength (Fig. 12.2), shows three signals, or peaks, at 0.00 ppm, 1.83 ppm, and 3.20 ppm, with the latter two split into patterns of lines. The peak at 0.00 ppm is due to the addition of a reference compound, tetramethylsilane (TMS), used as the zero reference point to assure that the instrument frequency assignments are accurate. Thus, there are only two sets of signal peaks from ethyl iodide as predicted.

Tetramethylsilane

Chapter 12 ■ Nuclear Magnetic Resonance Spectroscopy   241

FIG. 12.1
The ¹H NMR spectrum of ethyl iodide (60 MHz). The stair-step-like line is the integral. In the integral mode of operation, the recorder pen moves from left to right, and moves vertically a distance proportional to the areas of the peaks over which it passes. Hence, the relative areas of the quartet of peaks at 3.20 ppm and the triplet of peaks at 1.83 ppm are given by the relative heights of the integrals (4 cm is to 6 cm as 2 is to 3). The relative numbers of hydrogen atoms are proportional to the peak areas (2 H and 3 H).

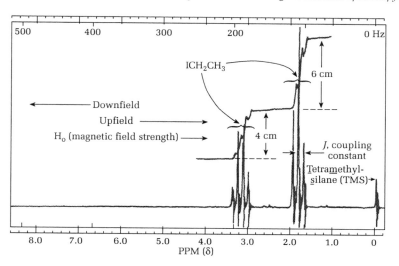

FIG. 12.2
The ¹H NMR spectrum of ethyl iodide (250 MHz). Compare this spectrum to Figure 12.1, run at 60 MHz. The peaks at 1.83 ppm and 3.20 ppm have been expanded and plotted on the left side of the spectrum.

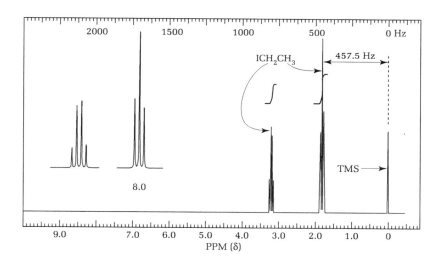

2. The Position of Each Signal on the Horizontal Axis

Each signal's position on the horizontal axis is called the chemical shift, and it indicates the effective magnetic field around a single proton or group of equivalent protons as affected by nearby bonds and atoms. This *chemical environment* depends on structure, that is, the bond types (sp^2, sp^3, sp, or aromatic) and the electronegativities of the atoms one, two, and three bonds away from the particular protons. Chemical shifts, symbolized by δ, are measured in dimensionless parts per million (ppm) to the left of, or *downfield* from, the reference TMS peak, according to the equation:

$$\delta \text{ (ppm)} = \frac{\text{shift of peak downfield TMS (in Hz)}}{\text{spectrometer frequence (in MHz)}}$$

© 2011 Cengage Learning. All Rights Reserved. May not be scanned, copied or duplicated, or posted to a publicly accessible website, in whole or in part.

CH₃—
Methyl

—CH₂—
Methylene

—CH—
  |
Methine

Pascal's triangle

			1			singlet, s
		1		1		doublet, d
	1		2		1	triplet, t
1		3		3		quartet, q
1	4		6		4	1 pentet
1	5	10	10	5	1	sextet

Table 12.1 gives the chemical shifts for protons in different chemical environments, spanning the region from 0 ppm to 12.5 ppm. Each signal in the spectrum of ethyl iodide can be assigned to a particular structural feature using this and similar tables. According to Table 12.1, the hydrogens on CH_2 groups attached to both an alkyl group and iodine as in RCH_2I (where R can be methyl, methylene, or methine) should yield a signal between +2.3 ppm and 3.2 ppm. We see a signal at 3.20 ppm in the spectrum of ethyl iodide, which is in this range. The table shows a range of 0.7 ppm to 1.1 ppm for a methyl group attached to a methylene or methine R group (RCH_3); R is the -CH_2I group for ethyl iodide. In actuality, the other signal in the spectrum of methyl iodide is not in this range; it is at 1.83 ppm, which is further downfield than predicted. This type of inconsistency occurs occasionally and demonstrates that reference tables are only generally representative of the majority of molecules that have been examined. It appears that in this particular case, the electronegative iodine, even though it is three bonds away, still strongly affects the magnetic field and shifts the peak for the methyl protons downfield. An examination of the other features of the spectrum will show that this peak assignment is correct.

3. Splitting Patterns and Coupling Constants

Patterns and coupling constants in the spectrum help define which groups are next to each other in a molecule's carbon skeleton. In an open-chain molecule with free rotation about the bonds and no chiral centers, protons couple with each other over three chemical bonds to give characteristic patterns of lines. If one or more equivalent protons couple to *one* adjacent proton, then the coupling hydrogen(s) appears as a *doublet* of equal intensity lines separated by the coupling *constant* (J) measured in Hertz (Hz). If the coupling proton or protons couple equally to two protons three chemical bonds away, they appear as a *triplet* of lines with relative intensities of 1:2:1, again separated by J. Finally, a *quartet* of lines in the ratio of 1:3:3:1 arises when one or more protons couple to the *three* protons on a methyl group.

In general, chemically equivalent protons give a pattern of lines containing one more line than the number of protons being coupled to, and the intensities of the peaks follow the binomial expansion, conveniently represented by Pascal's triangle. Thus the methylene group of ethyl iodide appears as a quartet of lines at 3.20 ppm with relative intensities of 1:3:3:1 because the two methylene protons are coupled to the three equivalent methyl protons (see Fig. 12.1 and Fig. 12.2). The methyl peak is split into a 1:2:1 triplet centered at 1.83 ppm because the methyl protons are coupled to the two adjacent methylene protons. This quartet-triplet combination is a strong indication of an ethyl group in any molecule. The J indicated in Figure 12.1 is the same 7.6 Hz between all lines in the quartet and triplet because the two groups are adjacent; therefore, the protons are coupled. In the 250 MHz spectrum, Figure 12.2, the splitting patterns are compressed and harder to distinguish because there are more Hz per ppm, but the expansion of these peaks on the left of the spectrum clearly shows a quartet and triplet, and the separation between each peak is still 7.6 Hz.

Other characteristic proton coupling constants are given in Table 12.2. In alkenes, *trans* coupling is larger than cis coupling, and both are much larger than geminal coupling (coupling that takes place between two groups on the same carbon). *Ortho*, *meta*, and *para* couplings in aromatic rings range from 0 Hz to 9 Hz. The couplings in a rigid system of saturated bonds are strongly dependent on the dihedral angle between the coupling protons, as seen in cyclohexane.

TABLE 12.1 *Proton Chemical Shifts*

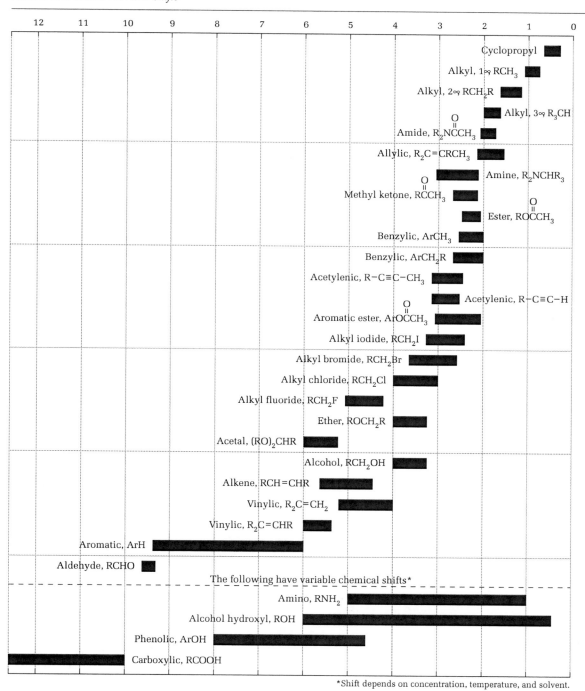

TABLE 12.2 *Spin-Spin Coupling Constants for Various Geometries*

Fragment	J(Hz)	Fragment	J(Hz)
H₂C=CH₂ (cis)	7–12	H-C-H (geminal sp³)	12–15
H₂C=CH₂ (trans)	13–18	H-C-C-H	0–10
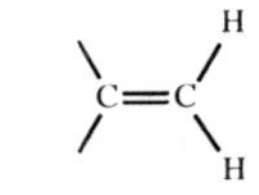 C=C (geminal)	0.5–3	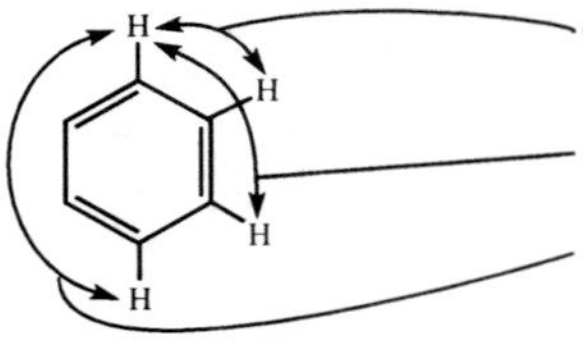 benzene ortho	6–9
		benzene meta	1–3
		benzene para	0–1
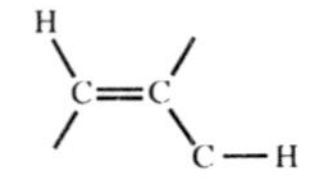 C=C-C-H (allylic)	0.5–2.5	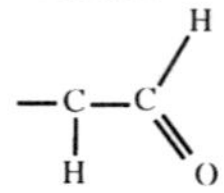 H-C-C=O	1–3
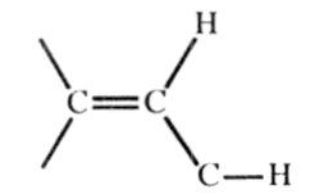 C=C-C-H	4–10	CH₃—CH₂—	6.5–7.5
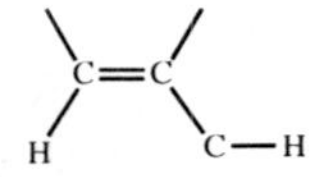 C=C-C-H	0–3	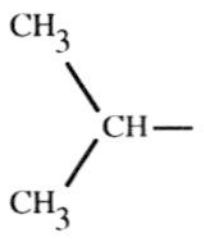 (CH₃)₂CH—	5.5–7
C=C-C=C	9–13	cyclohexane ax-ax	5–9
		cyclohexane ax-eq / eq-eq	2–4

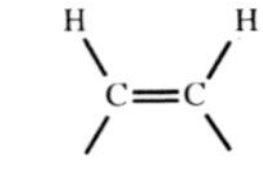

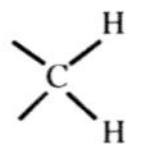

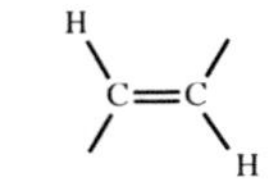

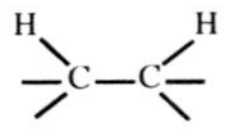

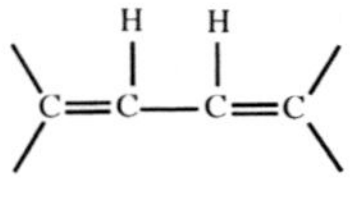

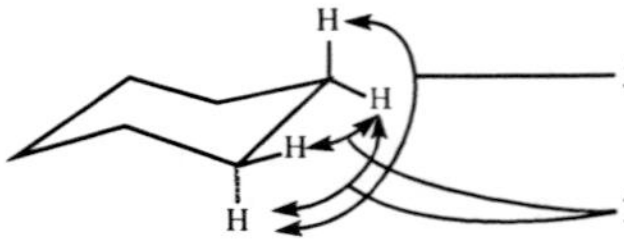

4. The Integral

The relative numbers of distinctive hydrogen atoms (protons) in the molecule of ethyl iodide are determined from the *integral*, the stair-step line over the peaks shown in Figures 12.1 and 12.2. The height of the step is proportional to the area under the NMR signal for each group of equivalent protons. In NMR spectroscopy (contrasted with infrared spectroscopy, for instance), the area for each signal, including all splitting peaks, is directly proportional to the number of hydrogen atoms causing that signal. The methyl protons give a triplet of peaks, and the methylene protons give a quartet of peaks. Both ethyl iodide spectra in Figures 12.1 and 12.2 have a ratio of 3 to 2 for the area of the methyl triplet relative to the area of the methylene quartet, which is further evidence for the presence of an ethyl group. Integrators are part of all NMR spectrometers, and running the integral takes little more time than running the spectrum. Most spectrometers print out a numerical value for the integral (with many more significant figures than are justified!).

FIG. 12.3
The ¹H NMR spectrum of 4-iodotoluene (90 MHz in CDCl₃).

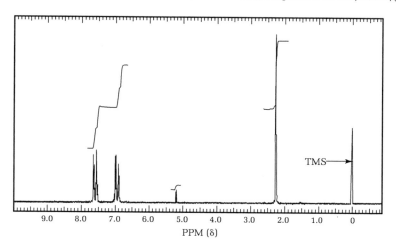

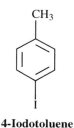

4-Iodotoluene

In actuality, when chemists interpret NMR spectra, they use a combination of the *structure from the spectrum* and the *spectrum from the structure* approaches. A typical mental analysis of the NMR spectrum of 4-iodotoluene (Fig. 12.3) might go like this:

> This molecule should produce three signals: a singlet for the methyl protons, a doublet for the two equivalent aromatic protons ortho to the methyl group, and a doublet for the other two equivalent aromatic protons ortho to the iodo group. The methyl group is attached to an aromatic ring, and according to the table of proton chemical shifts, this should be detected as a singlet peak between 1.8 ppm and 2.8 ppm. The NMR spectrum of my product shows a singlet at 2.28 ppm, which is consistent with this. The aromatic protons show up as doublets in the right range, 6 ppm to 9 ppm according to the table. The spectrum has doublets at 6.95 ppm and 7.60 ppm. It also shows a singlet at 5.2 ppm and, according to the table, this implies that my product contains hydrogen on an alkene or in an acetal. Bad news! From other information I know, my product has neither of these functional groups. But wait! If I look at spectra of common sample impurities, I see that dichloromethane, the solvent I extracted the product with, appears at 5.25 ppm as a singlet. So this peak is not from my product and can be ignored . . .

This back-and-forth analysis correlating structure and NMR data, including coupling constants and integrals, is continued until all NMR peaks, splitting patterns, and integrals can be accounted for, and the identity of the product is verified.

When NMR data are reported in the literature, it is usually in a concise numerical form. For example, the NMR data for ethyl iodide derived from its spectrum (Fig. 12.2) would be reported as ¹H NMR (CDCl₃): 1.83 (3H, t, J = 7.6 Hz), 3.20 (2H, q, J = 7.6 Hz), where CDCl₃ (deuterochloroform) is the solvent, 1.83 and 3.20 are the chemical shifts in ppm, 3H and 2H are integrals, and t = triplet, q = quartet, and J is the coupling constant in Hz.

Some NMR spectra are not as easily analyzed as the spectrum for ethyl iodide. Consider the spectra shown in Figure 12.4. The proton spectrum of this unsaturated chloroester has been run at 500 MHz. Each chemically and magnetically nonequivalent

FIG. 12.4

The 500-MHz ¹H and 75-MHz ¹³C NMR spectra of ethyl (3-chloromethyl)-4-pentenoate.[1] *I. ¹³C DEPT[2] spectrum.* The CH₃ and CH peaks are upright, and the CH₂ peaks are inverted. The quaternary carbon (the carbonyl carbon) does not appear. *II. The normal 75 MHz noise-decoupled ¹³C spectrum.* Note the small size of the carbonyl peak. *III. Expansions of each group of proton NMR peaks.* Protons E and F as well as protons H and I are not equivalent to each other. These pairs of protons are diastereotopic because they are on a carbon adjacent to a chiral carbon atom. The frequencies of all peaks are found on this book's web site. *IV. The integral.* The height of the integral is proportional to the number of protons under it. *V. The 500-MHz ¹H spectrum.*

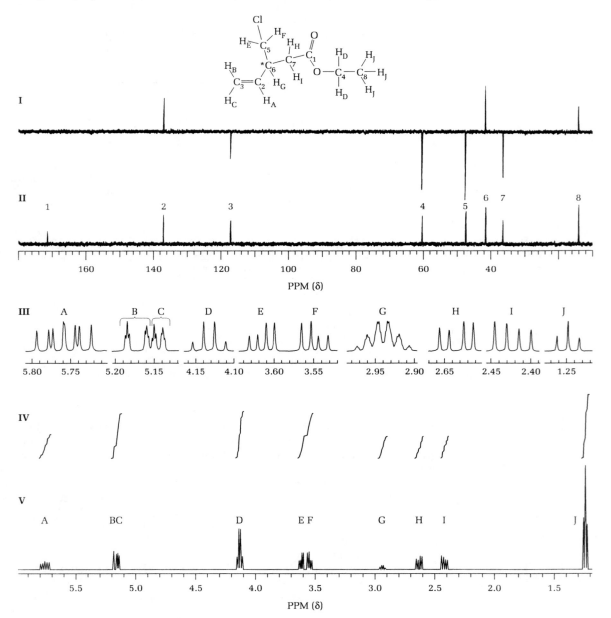

[1]Spectra courtesy of Professor Scott Virgil.
[2]DEPT: distortionless enhancement by polarization transfer.

proton is well resolved so that all of the couplings can be seen. Only the quartet of peaks at 4.1 ppm (relative area 2) and the triplet at 1.2 ppm (relative area 3) follow the simple first-order coupling rules outlined earlier. This quartet/triplet pattern is very characteristic of the commonly encountered ethyl group.

Because of the chiral carbon (marked with an asterisk), the protons on carbons 5 and 7 are diastereotopic, have different chemical shifts, and couple with each other and with adjacent protons to give the patterns seen in the spectrum. Many of the peaks can be assigned to specific hydrogens based simply on their chemical shifts. The coupling patterns then confirm these assignments.

A more complex spectrum.

CARBON-13 SPECTROSCOPY

The element carbon consists of 98.9% carbon atoms with mass 12 (^{12}C) and spin 0 (NMR inactive) and only 1.1% carbon atoms with mass 13 (^{13}C) and spin 1/2 (NMR active). Carbon, with such a low concentration of spin 1/2 nuclei, gives a very small signal when run under the same conditions as used for a proton spectrum. Carbon resonates at 75 MHz in a spectrometer where protons resonate at 300 MHz. Because only 1 in 100 carbon atoms has mass 13, the chances of a molecule having two ^{13}C atoms adjacent to one another are small. Consequently, coupling of one ^{13}C with another is not observed.

Each ^{13}C atom couples to hydrogen atoms over one, two, and three bonds. Because the coupling constants are large, there is a high probability of peak overlap. To simplify the spectra as well as to increase the signal-to-noise ratio, a special technique is routinely used in obtaining ^{13}C spectra: *broadband noise decoupling*. Decoupling has the effect of collapsing all multiplets (quartets, triplets, etc.) into a single peak. Further, the energy put into decoupling the protons will appear in the carbon spectrum in the form of an enhanced peak. This Nuclear Overhauser Enhancement (NOE) effect makes the peak appear three times larger than it would otherwise be. The result of decoupling is that every chemically and magnetically distinct carbon atom will appear as a single sharp line in the spectrum. Because the NOE effect is somewhat variable and does not affect carbons bearing no protons, one cannot do carbon counting from peak integrals in the same way that hydrogen counting is done from proton NMR spectra.

^{13}C spectra: Broadband noise decoupling gives a single line for each carbon.

Carbon Chemical Shifts

The range of carbon chemical shifts is 200 ppm compared to the 10-ppm range for protons. It is not common to have an accidental overlap of carbon peaks. In the spectrum for the unsaturated chloroester in Figure 12.4 (spectrum II), there are eight sharp peaks corresponding to the eight carbon atoms in the molecule.

The generalities governing carbon chemical shifts are very similar to those governing proton shifts, as seen by comparing Table 12.1 to Table 12.3. Most of the downfield peaks are due to those carbon atoms near electron-withdrawing groups. In Fig. 12.4 (spectrum II), the furthest downfield peak is that from the carbonyl carbon of the ester. The attached electronegative oxygens make the peak appear at about 172 ppm. The peak is smaller than any other in the spectrum because it does not have an attached proton, and thus does not benefit from the NOE effect.

The ^{13}C spectrum of sucrose in Figure 12.5 displays a single line for each carbon atom; in Figure 20.3 (on page 345), a single line is seen for each of the 27 carbon atoms in cholesterol.

TABLE 12.3 *Carbon Chemical Shifts*

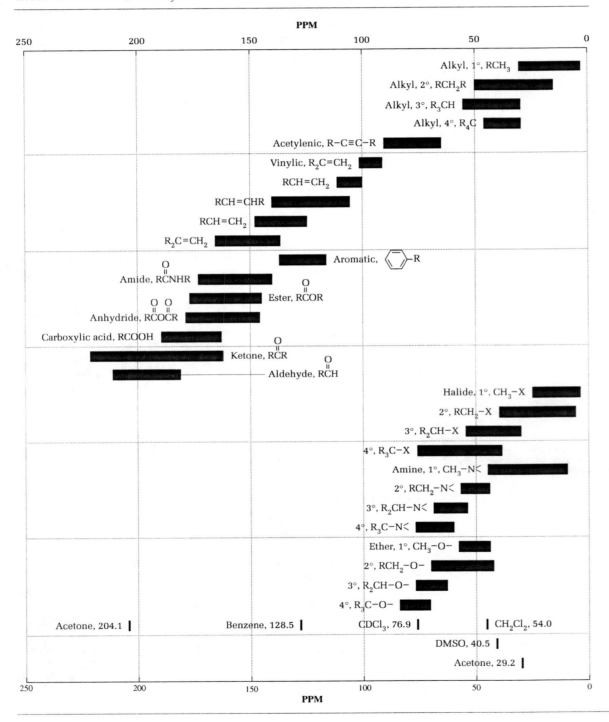

FIG. 12.5
The ¹³C NMR spectrum of sucrose (22.6 MHz). Not all lines have been assigned to individual carbon atoms.

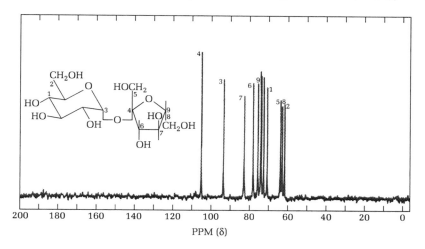

CHAPTER 14

Ultraviolet Spectroscopy, Refractive Indices, and Qualitative Instrumental Organic Analysis

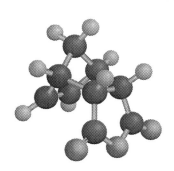

> **PRELAB EXERCISE:** In the identification of an unknown organic compound, certain procedures are more valuable than others. For example, far more information is obtained from an IR or NMR spectrum than from a UV spectrum or a refractive index measurement. Outline, in the order of priority, the steps you will employ in identifying an organic unknown.

ULTRAVIOLET SPECTROSCOPY

UV: Electronic transitions within molecules

Ultraviolet (UV) spectroscopy produces data about electronic transitions within molecules. Whereas absorption of low-energy infrared (IR) radiation causes bonds in a molecule to stretch and bend, the absorption of short-wavelength, high-energy ultraviolet (UV) radiation causes electrons to move from one energy level to another with energies that are capable of breaking chemical bonds. Many molecules produce only a major absorption signal (*see* Figs. 14.2 to 14.7 on pages 280–284) due to a single, extended structural feature involving π bonded systems: carbon-carbon single bonds, carbon-oxygen or carbon-nitrogen double bonds, triple bonds, and aromatic rings. Because many different combinations of these π systems can absorb at nearly the same wavelength, the structure is usually elucidated by other techniques, and its UV spectrum is interpreted based on this structure.

We are most concerned with transitions of π electrons in conjugated and aromatic ring systems. These transitions occur in the wavelength region of 200–800 nm (nanometers, 10^{-9} meters). Most common UV spectrometers cover the region of 200–400 nm, as well as the visible spectral region of 400–800 nm. Below 200 nm, air (oxygen) absorbs UV radiation; spectra in that region must therefore be obtained in a vacuum or in an atmosphere of pure nitrogen.

Consider ethylene, even though it absorbs UV radiation in the normally inaccessible region at 163 nm. The double bond in ethylene has two *s* electrons in a π molecular orbital and two, less tightly held, *p* electrons in a π molecular orbital. Two unoccupied, high-energy-level, antibonding orbitals are associated with these orbitals. When ethylene absorbs UV radiation, one electron moves up from the

Band spectra

Beer-Lambert law

λ_{max} = wavelength of maximum absorption

ε, extinction coefficient

The path length (*l*) is the distance (in centimeters) that light travels through a sample.

bonding π molecular orbital to the antibonding π^* molecular orbital (Fig. 14.1). As Figure 14.1 indicates, this change requires less energy than the excitation of an electron from the σ to the σ^* molecular orbital.

By comparison with IR spectra and nuclear magnetic resonance (NMR) spectra, UV spectra are mostly featureless (Fig. 14.2). This condition results as molecules in a number of different vibrational states undergo the same electronic transition to produce a band spectrum instead of a line spectrum.

Unlike IR spectroscopy, UV spectroscopy lends itself to precise quantitative analysis of substances. The intensity of an absorption band is usually given by the molar extinction coefficient, ε, which, according to the Beer-Lambert law, is equal to the absorbance (A) divided by the product of the molar concentration (c) and the path length (l) in centimeters.

$$\varepsilon = \frac{A}{cl}$$

The wavelength of maximum absorption (the tip of the peak) is given by λ_{max}. Because UV spectra are so featureless, it is common practice to describe a spectrum like that of cholesta-3,5-diene (Fig. 14.2) as λ_{max} = 234 nm (ε = 20,000) and not bother to reproduce the actual spectrum.

The extinction coefficients of conjugated dienes and enones are in the range of 10,000–20,000, so only very dilute solutions are needed for spectra. In the example in Figure 14.2, the absorbance at the tip of the peak is 1.2, and the path length is the usual 1 cm; so the molar concentration, c, needed for this spectrum is 6×10^{-5} mol/L, which is 0.221 mg/10 mL of solvent. The usual laboratory balance cannot accurately weigh such small quantities; therefore sample preparation usually requires the quantitative serial dilution of more concentrated solutions.

FIG. 14.1

The electronic energy levels of ethylene.

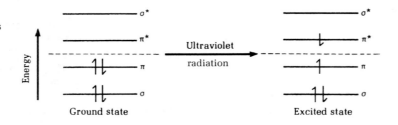

FIG. 14.2

The UV spectrum of cholesta-3,5-diene in ethanol.

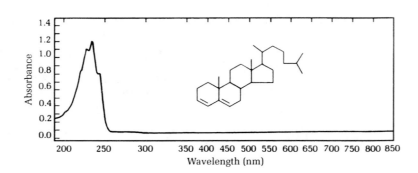

Spectro **grade solvents**

UV quartz cells are expensive; handle with care.

Woodward and Fieser rules for dienes and dienones.

The usual solvents for UV spectroscopy are 95% ethanol, methanol, water, and also saturated hydrocarbons such as hexane, trimethylpentane, and isooctane. The three hydrocarbons are often especially purified to remove impurities that absorb in the UV region. Any transparent solvent can be used for spectra in the visible region.

Sample cells for spectra in the visible region are made of glass or clear plastic, but UV cells must be composed of the more expensive fused quartz because glass absorbs UV radiation. The cells and solvents must be clean and pure because very little of a substance produces a UV spectrum. A single fingerprint will give a spectrum!

Ethylene has $\lambda_{max} = 163$ nm ($\varepsilon = 15{,}000$), and butadiene has $\lambda_{max} = 217$ nm ($\varepsilon = 20{,}900$). As the conjugated system is extended, the wavelength of maximum absorption moves to longer wavelengths (toward the visible region). For example, lycopene, with 11 conjugated double bonds, has $\lambda_{max} = 470$ nm ($\varepsilon = 185{,}000$; Fig. 14.3). Because lycopene absorbs blue visible light at 470 nm, the substance appears bright red. It is responsible for the color of tomatoes; its isolation and analysis are described in Chapters 8 and 9.

The wavelengths of maximum absorption of conjugated dienes and polyenes and conjugated enones and dienones are given by the Woodward and Fieser rules (Tables 14.1 and 14.2). The application of the rules is demonstrated by the spectra of pulegone (1) and carvone (2) in Figure 14.4. The solvent correction is given in Table 14.3. The calculations are given in Tables 14.4 and 14.5.

FIG. 14.3

The UV/Vis spectrum of lycopene in isooctane.

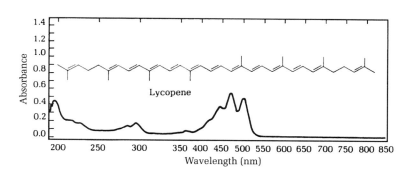

TABLE 14.1 *Rules for Predicting the λ_{max} for Conjugated Dienes and Polyenes*

	Increment (nm)
Parent acyclic diene (butadiene)	217
Parent heteroannular diene	214
Double bond extending the conjugation	30
Alkyl substituent or ring residue	5
Exocyclic location of double bond to any ring	5
Groups: OAc, OR	0
Solvent correction (see Table 14.3)	( )
$\lambda_{max}^{EtOH} =$	Total

TABLE 14.2 Rules for Predicting λ_{max} for Conjugated Enones and Dienones:

$$\overset{\beta}{\beta}-\overset{\alpha}{C}=\overset{R}{C}-\overset{}{C}=O \quad \text{and} \quad \overset{\delta}{\delta}-\overset{\gamma}{C}=\overset{\beta}{\gamma}-\overset{\alpha}{C}=\overset{R}{C}-\overset{}{C}=O$$

	Increment (nm)
Parent α,β-unsaturated system	215
Double bond extending the conjugation	30
R (alkyl or ring residue), OR, OCOCH$_3$	
α	10
β	12
γ, δ and higher	18
α-Hydroxyl, enolic	35
α-Cl	15
α-Br	23
Exocyclic location of double bond to any ring	5
Homoannular diene component	39
Solvent correction (see Table 14.3)	( )
$\lambda_{max}^{EtOH} =$	Total

FIG. 14.4
The UV spectra of (1) pulegone and (2) carvone in hexane.

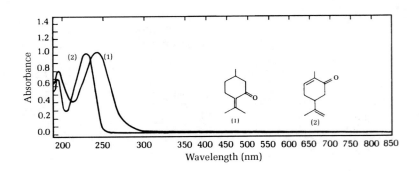

TABLE 14.3 Solvent Correction

Solvent	Factor for Correction to Ethanol
Hexane	+11
Ether	+7
Dioxane	+5
Chloroform	+1
Methanol	0
Ethanol	0
Water	−8

TABLE 14.4 Calculation of λ_{max} for Pulegone (See Fig. 14.4)

Parent α,β-unsaturated system	215 nm
α-Ring residue, R	10
β-Alkyl group (two methyls)	24
Exocyclic double bond	5
Solvent correction (hexane)	−11
Calculated $\lambda_{max} =$	243 nm; found = 244 nm

TABLE 14.5 Calculation of λ_{max} for Carvone (See Fig. 14.4)

Parent α,β-unsaturated system	215 nm
α-Alkyl group (two methyls)	10
β-Ring residue	12
Solvent correction (hexane)	−11
Calculated $\lambda_{max} =$	226 nm; found = 229 nm

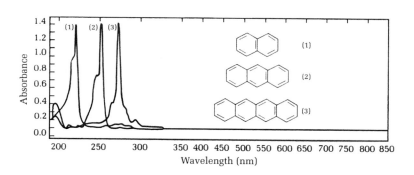

FIG. 14.5
The UV spectra of (1) naphthalene, (2) anthracene, and (3) tetracene.

Effect of acid and base on λ_{max}.

No simple rules exist for the calculation of aromatic ring spectra, but several generalizations can be made. From Figure 14.5, it is obvious that as polynuclear aromatic rings are extended linearly, λ_{max} shifts to longer wavelengths.

As alkyl groups are added to benzene, λ_{max} shifts from 255 nm for benzene to 261 nm for toluene to 272 nm for hexamethylbenzene. Substituents bearing nonbonding electrons also cause shifts of λ_{max} to longer wavelengths—for example, from 255 nm for benzene to 257 nm for chlorobenzene, 270 nm for phenol, and 280 nm for aniline (ε = 6200–8600). That these effects are the result of the interaction of the π-electron system with nonbonded electrons is seen dramatically in the spectra of vanillin and the anion derived by deprotonation of its phenolic OH (Fig. 14.6). The two additional nonbonding electrons in the anion cause λ_{max} to shift from 279 nm to 351 nm and ε to increase. Protonation of the non-bonding electrons on the nitrogen of aniline to give the anilinium cation causes λ_{max} to decrease from 280 nm to 254 nm (Fig. 14.7). These changes of λ_{max} as a function of pH have obvious analytical applications.

FIG. 14.6
The UV spectra of (1) neutral vanillin and (2) the anion of vanillin.

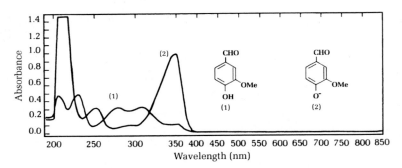

FIG. 14.7
The UV spectra of (1) aniline and (2) aniline hydrochloride.

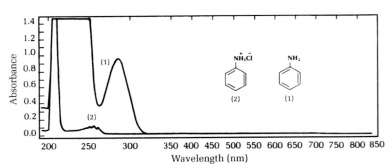

Intense bands result from π-π conjugation of double bonds and carbonyl groups with the aromatic ring. Styrene, for example, has λ_{max} = 244 nm (ε = 12,000), and benzaldehyde has λ_{max} = 244 nm (ε = 15,000).

EXPERIMENT

UV SPECTRUM OF AN UNKNOWN ACID, BASE, OR NEUTRAL COMPOUND

Determine whether an unknown compound obtained from your instructor is acidic, basic, or neutral from UV spectra in a neutral solvent such as pure ethanol or methanol, as well as under acidic conditions (add 1 drop of 5% HCl to the solution in the cuvette and mix) and basic conditions (add 1 drop of 1 M NaOH to the acidic solution in the cuvette, mix, and check that the pH is basic).

Cleaning Up. Because UV samples are extremely dilute solutions in ethanol or methanol, they can normally be flushed down the drain.

REFRACTIVE INDICES

The *refractive index*, symbolized by n, is a physical constant that, like the boiling point, can be used to characterize liquids. It is the ratio of the velocity of light traveling in air to the velocity of light moving in the liquid (Fig. 14.8). It is also

FIG. 14.8
The refraction of light.

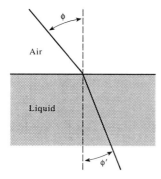

equal to the ratio of the sine of the angle of incidence ϕ to the sine of the angle of refraction ϕ':

$$n = \frac{\text{velocity of light in air}}{\text{velocity of light in liquid}} = \frac{\sin \phi}{\sin \phi'}$$

The angle of refraction is also a function of temperature and the wavelength of light (consider the dispersion of white light by a prism). Because the velocity of light in air (strictly speaking, a vacuum) is always greater than that through a liquid, the refractive index is a number greater than 1, for example, hexane, $n_D^{20} = 1.3751$; diiodobenzene, $n_D^{20} = 1.7179$. The superscript 20 indicates that the refractive index was measured at 20°C, and the subscript D refers to the yellow D-line from a sodium vapor lamp, which produces light with a wavelength of 589 nm.

The measurement is made on a refractometer using a few drops of liquid. Compensation is made within the instrument for the fact that white light, not sodium vapor light, is used, and a temperature correction must be applied to the observed reading by using the following equation which automatically compensates for temperatures higher or lower than 20°C:

$$n_D^{20} = n_D^t + 0.00045(t - 20°C)$$

The refractive index can be determined to 1 part in 10,000, but because the value is quite sensitive to impurities, full agreement in the literature regarding the last figure does not always exist. For this reason, the refractive indices in this book have been rounded to the nearest 1 part in 1000, as have the refractive indices reported in the Aldrich catalog of chemicals. To master the technique of using the refractometer, measure the refractive indices of several known, pure liquids before measuring an unknown.

Specialized hand-held refractometers that read over a narrow range are used to determine the concentration of sugar, salt, or alcohol in water.

USING A REFRACTOMETER

Refractometers come in many designs. In the most common, the Abbé design (Fig. 14.9), two or three drops of the sample are placed on the measuring prism using a polyethylene Beral pipette (to avoid scratching the prism face). The illuminating prism is closed, and the lamp is turned on and positioned for maximum brightness as seen through the eyepiece. If the refractometer is set to a nearly correct value, then a partially gray image will be seen, as shown in Figure 14.10a. Turn the index knob so that the line separating the dark and light areas is at the crosshairs, as shown in Figure 14.10b. Sometimes the line separating the dark and light areas is fuzzy and colored (Fig. 14.10c). Turn the chromatic adjustment until the demarcation line is sharp and colorless. Then read the refractive index. On a newer instrument, press a button or hold down the on/off switch to light up the scale in the field of vision or activate the digital readout. On older models, read the refractive index through a separate eyepiece. Read the temperature on the thermometer attached to the refractometer, and make the appropriate temperature correction to the observed index of refraction. When the measurement is completed, open the prism and wipe off the sample with lens paper, using ethanol, acetone, or hexane only as necessary.

For most organic liquids, the index of refraction decreases approximately 0.00095 ± 0.0001 for every °C increase in temperature.

FIG. 14.9
An Abbé refractometer. The sample block can be thermostatted.

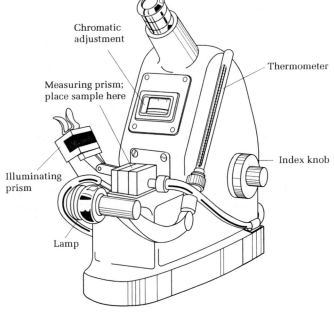

FIG. 14.10
(a) The view into a refractometer when the index knob is out of adjustment. (b) The view into a refractometer when properly adjusted. (c) The view when the chromatic adjustment is incorrect.

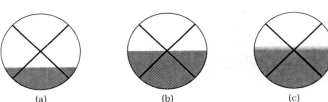

QUALITATIVE INSTRUMENTAL ORGANIC ANALYSIS

Qualitative Organic Analysis

As indicated in many of the previous chapters, physical characterization and structural elucidation are major activities in organic chemistry. In many areas, such as drug metabolism studies and forensic or environmental chemistry, only milligram or microgram quantities of organic compounds are available. Fortunately, today's instrumental methods have the requisite sensitivity to meet this challenge. It may be possible to establish the structure of a compound on the basis of spectra alone (IR, NMR, MS [mass spectrometry], and/or UV), but often these spectra must be supplemented with other information about the unknown: physical state, solubility, and confirmatory tests for functional groups. Before spectra are run, other information about the unknown must be obtained. Is it pure or a mixture (test by thin-layer [TLC], gas, or liquid chromatography)? Once a substance is known to be homogeneous, it can often be identified by spectroscopy alone. What are its physical properties: melting point, boiling point, and color and solubility in various solvents, such as those commonly used in NMR? A mass spectrum can determine a compound's molecular weight and, if measured with sufficient accuracy, identify the elements present and the molecular formula.

For the millions of organic compounds that have been synthesized or isolated from nature, spectral data are included when they are reported in the chemical literature. Thousands of the more common chemicals that can be purchased or easily synthesized have had their IR, NMR, MS, and UV spectra printed in multivolume collections that can be searched by compound class, formula, or spectral features. Today, these collections have been converted into digital form that can be searched by computer and compared with the spectra obtained for an unknown, yielding a list of all closely matching compounds. If two substances have identical IR spectra, they can be regarded as identical. Such is not always true of other spectra. Many substances can have identical or nearly identical UV spectra, and it is possible for the MS or NMR spectra for two different substances to be almost identical. When new substances are encountered in research laboratories, their spectra can be compared with those in commercial databases. Even though the particular new compound is not represented in those databases, a list of very similar substances can be generated, which will guide the determination of the structure of the new substance.

EXPERIMENT

IDENTIFYING AN UNKNOWN COMPOUND

Most students consider the identification of an unknown organic compound to be one of the most enjoyable and challenging organic lab activities. In this laboratory course you may receive an "unknown" substance, which, of necessity, is usually a commercially available compound. At least initially, you may not be given access to a commercial spectral database for searching and comparison, but you will use the skills you have learned to interpret the spectral data you acquire. After you have done the best you can with your IR, NMR, and UV spectra and have arrived at a short list of possible compounds, you will be provided with the mass spectrum of your unknown. Interpretation of this spectrum usually eliminates most of the candidate structures and should allow you to complete the identification of the unknown.

Physical State

Check for Sample Purity

Distill or recrystallize as necessary. A constant boiling point and sharp melting point are indicators of purity, but beware of azeotropes and eutectic substances. Check homogeneity by TLC, GC, or HPLC.

Note the Color

Common colored compounds include nitro and nitroso compounds (yellow), diketones (yellow), quinones (yellow to red), azo compounds (yellow to red), and polyconjugated olefins and ketones (yellow to red). Phenols and amines are often brown to dark purple because of traces of air oxidation products.

Note the Odor

Some liquid and solid amines are recognizable by their fishy odors; esters are often pleasantly fragrant. Alcohols, ketones, aromatic hydrocarbons, and aliphatic olefins have characteristic odors. On the unpleasant side in terms of odor are thiols, isonitriles, and low-molecular-weight carboxylic acids.

Ignition Test

Heat a small sample on a spatula; first hold the sample near the side of a microburner to see if it melts normally and then burns. Then heat directly in the flame. If a large ashy residue is left after ignition, the unknown is probably a metal salt. Aromatic compounds often burn with a smoky flame.

Beilstein Test for Halogens

Although the presence of a halogen can usually be determined by mass spectrometry, this test is so simple that it can easily be run to confirm the MS data. Heat the tip of a copper wire in a burner flame until no further coloration of the flame is noticed. Allow the wire to cool slightly, then dip it into the unknown (solid or liquid), and again heat it in the flame. A green flash is indicative of chlorine, blue-green of bromine, and blue of iodine; fluorine is not detected because copper fluoride is not volatile. The Beilstein test is very sensitive; halogen-containing impurities may give misleading results. Run the test on a compound known to contain a halogen for comparison to your unknown.

Spectra

Obtain IR and NMR spectra following the procedures in Chapters 11 and 12. If these spectra indicate the possible presence of conjugated double bonds, aromatic rings, or conjugated carbonyl compounds, obtain the UV spectrum following the procedures in this chapter. Interpret the spectra as fully as possible by investigating the reference sources cited at the end of the spectroscopy chapters. Once you have interpreted these spectra to the best of your ability, you may obtain or be provided with a mass spectrum (Chapter 13), which should make a secure identification of the unknown possible.

Solubility Tests

There is a logical sequence for determining the solubility of an organic compound in order, for example, to dissolve it for spectroscopic analysis. In the process of determining the solubility of the unknown, much information about the nature of the compound can be obtained.

Like dissolves like.

Like dissolves like; a substance is most soluble in that solvent to which it is most closely related in structure. This statement serves as a useful classification scheme for all organic molecules. The solubility measurements are done at room temperature with 1 drop of a liquid or 5 mg of a solid (finely crushed) and 0.2 mL of solvent. The mixture should be rubbed with a rounded stirring rod and agitated vigorously. Lower members of a homologous series are easily classified; higher members become more like the hydrocarbons from which they are derived. If a very small amount of the sample fails to dissolve when added to some of the solvent, it can be considered insoluble. Conversely, if several portions dissolve readily in a small amount of the solvent, the substance is obviously soluble. If an unknown seems to be more soluble in dilute acid or base than in water, the observation can be

Chapter 14 ■ *Ultraviolet Spectroscopy* 289

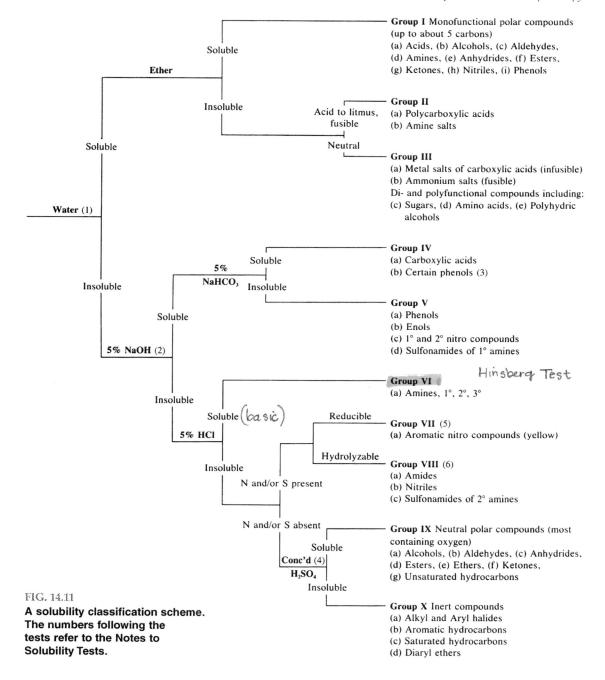

FIG. 14.11
**A solubility classification scheme.
The numbers following the
tests refer to the Notes to
Solubility Tests.**

confirmed by neutralizing the solution; the original material will precipitate if it is less soluble in a neutral medium. If both acidic and basic groups are present, the substance may be amphoteric and therefore soluble in both acid and base. Aromatic aminocarboxylic acids are amphoteric, like aliphatic ones, but they do not exist as zwitterions. They are soluble in both dilute hydrochloric acid and sodium hydroxide, but not in bicarbonate solution. Aminosulfonic acids exist as zwitterions; they are soluble in alkali but not in acid. The solubility tests are not infallible, and many borderline cases are known. Carry out the tests according to the scheme in Figure 14.11 and the following Notes to Solubility Tests, and tentatively assign the unknown to one of the groups I–X.

Notes to Solubility Tests

1. Groups I, II, III (soluble in water). Test the solution with pH paper. If the compound is not easily soluble in cold water, treat it as water insoluble but test with indicator paper.
2. If the substance is insoluble in water but dissolves partially in 5% sodium hydroxide, add more water; the sodium salts of some phenols are less soluble in alkali than in water. If the unknown is colored, be careful to distinguish between the *dissolving and the reacting* of the sample. Some quinones (colored) *react* with alkali and give highly colored solutions. Some phenols (colorless) *dissolve and then* become oxidized to give colored solutions. Some compounds (e.g., benzamide) are hydrolyzed with such ease that careful observation is required to distinguish them from acidic substances.
3. Nitrophenols (yellow), aldehydophenols, and polyhalophenols are sufficiently strongly acidic to react with sodium bicarbonate.
4. Oxygen- and nitrogen-containing compounds form oxonium and ammonium ions in concentrated sulfuric acid and dissolve.
5. On reduction in the presence of hydrochloric acid, nitro compounds form water-soluble amine hydrochlorides. Dissolve 250 mg of tin(II) chloride in 0.5 mL of concentrated hydrochloric acid, add 50 mg of the unknown, and warm. The material should dissolve with the disappearance of the color and give a clear solution when diluted with water.
6. Most amides can be hydrolyzed by short boiling with a 10% sodium hydroxide solution; the acid dissolves with evolution of ammonia. Reflux 100 mg of the sample and a 10% sodium hydroxide solution for 15–20 minutes. Test for the evolution of ammonia, which confirms the elementary analysis for nitrogen and establishes the presence of a nitrile or amide.

Cleaning Up. Because the quantities of material used in these tests are extremely small, and because no hazardous substances are handed out as unknowns, it is possible to dilute the material with a large quantity of water and flush it down the drain, unless it is very smelly or is dissolved in water-insoluble nonhalogenated or halogenated solvents, in which case it should be disposed of in the appropriate containers.

QUESTIONS

1. Calculate the UV absorption maximum for 2-cyclohexene-1-one.
2. Calculate the UV absorption maximum for 3,4,4-trimethyl-2-cyclohexene-1-one.

3. Calculate the UV absorption maximum for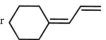

4. What concentration, in g/mL, of a substance with MW 200 should be prepared to give an absorbance value equal to 0.8 if the substance has ε = 16,000 and a cell with a path length of 1 cm is employed?

5. When borosilicate glass (Kimax, Pyrex), n_D^{20} 1.474, is immersed in a solution having the same refractive index, it is almost invisible. Soft glass with n^{20} 1.52 is quite visible. This is an easy way to distinguish between the two types of glass. Calculate the mole percents of toluene and heptane that will have a refractive index of 1.474, assuming a linear relationship between the refractive indices of the two.

REFERENCES

See Web Links

Gillam, Albert. E., Edward S. Stern, and Christopher J. Timmons, *Gillam and Stern's Introduction to Electronic Absorption Spectroscopy in Organic Chemistry*, 3rd. ed. London: Edward Arnold, 1970.

Jaffe, Hans H., and Milton Orchin, *Theory and Applications of Ultraviolet Spectroscopy*. New York: John Wiley, 1962.

Lambert, Joseph B., Herbert F. Shurvell, David A. Lightner, and Robert G. Cooks, *Organic Structural Spectroscopy*. Upper Saddle River, NJ: Prentice-Hall, 1998.

Rao, Chintamani N. R., *Ultraviolet and Visible Spectroscopy: Chemical Applications*, 3rd ed. London: Butterworths, 1975.

Silverstein, Robert M., Francis X. Webster, and David Kiemle, *Spectrometric Identification of Organic Compounds*, 7th ed.[1] New York: Wiley, 2005.

Williams, Dudley H., and Ian Fleming, *Spectroscopic Methods in Organic Chemistry*, 5th ed. New York: McGraw-Hill, 1995.

---

[1]This book includes IR, UV, and NMR spectra.

19 CHAPTER

Alkenes from Alcohols: Cyclohexene from Cyclohexanol

PRELAB EXERCISE: Prepare a detailed flow sheet for the preparation of cyclohexene, indicating at each step which layer contains the desired product.

$$\underset{\substack{\text{Cyclohexanol} \\ \text{mp 25°C, bp 161°C} \\ \text{den. 0.96, MW 100.16}}}{\text{C}_6\text{H}_{11}\text{OH}} \xrightarrow{\text{H}_3\text{PO}_4} \underset{\substack{\text{Cyclohexene} \\ \text{bp 83°C} \\ \text{den. 0.81, MW 82.14}}}{\text{C}_6\text{H}_{10}} + \text{H}_2\text{O}$$

The dehydration of cyclohexanol to cyclohexene can be accomplished by pyrolysis of the cyclic secondary alcohol with an acid catalyst at a moderate temperature, or by distillation over alumina or silica gel. The procedure selected for these experiments involves catalysis by phosphoric acid; sulfuric acid is not more efficient, causes charring, and gives rise to sulfur dioxide. When a mixture of cyclohexanol and phosphoric acid is heated in a flask equipped with a fractionating column, the formation of water soon becomes evident. On further heating, the water and the cyclohexene that form distill together by the principle of steam distillation, and any higher-boiling cyclohexanol that may volatilize is returned to the flask. However, after dehydration is complete and the bulk of the product has distilled, the fractionating column remains saturated with a water-cyclohexene mixture that merely refluxes and does not distill. Hence, for the recovery of otherwise lost reaction product, a chaser solvent is added, and distillation is continued. A suitable chaser solvent is the water-immiscible, aromatic solvent toluene (bp 110°C); as it steam distills, it carries over the more volatile cyclohexene. When the total water-insoluble layer is separated, dried, and redistilled through the dried column, the chaser again drives the cyclohexene from the column; the difference in boiling points is such that a sharp separation is possible. The holdup in a metal sponge-packed column is so great (about 1.0 mL) that if a chaser solvent is not used in the procedure to free it from the column, the yield will be much lower.

The mechanism of this reaction involves initial rapid protonation of the hydroxyl group by the phosphoric acid:

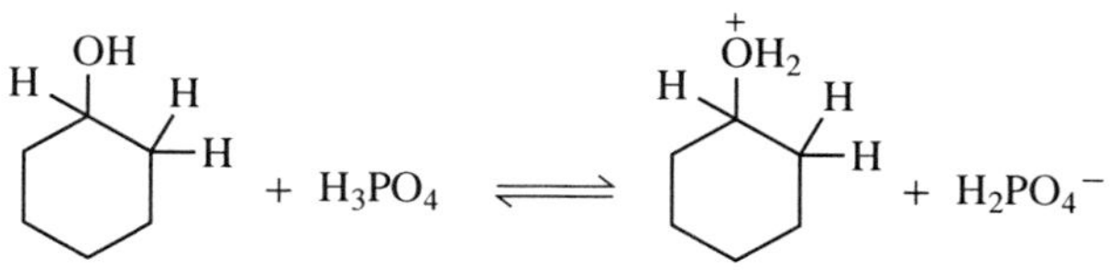

This is followed by loss of water to give the unstable secondary carbocation, which quickly loses a proton to water to give the alkene:

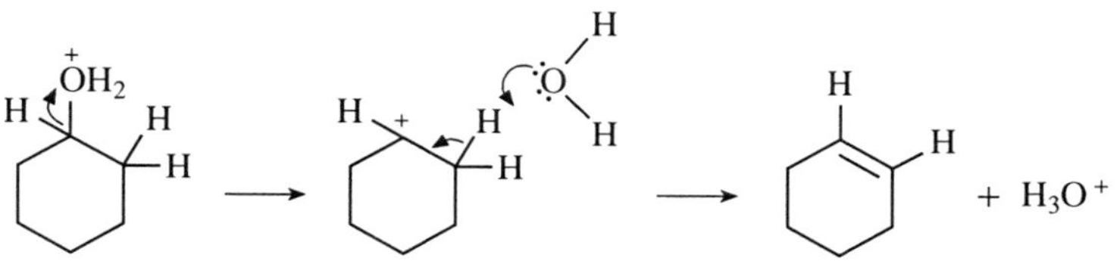

FIG. 19.1
The apparatus for synthesizing cyclohexene from cyclohexanol. The column is packed with a copper sponge.

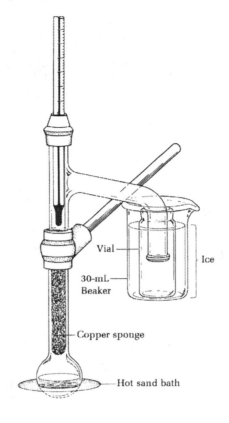

The Preparation of Cyclohexene

Adapted from Kenneth Williamson, Macroscale and Microscale Organic Experiments, 4th ed., Houghton Mifflin, Boston, MA, 2003, Ch 19, pp. 287-293, by M. J. Lusch and B. L. Groh

Overall Reaction Equation:

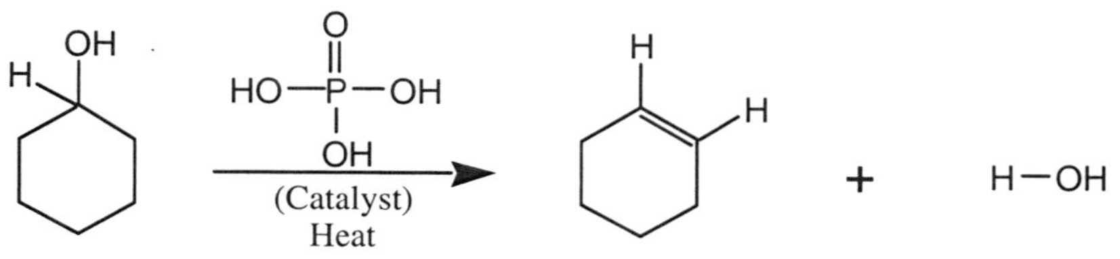

Detailed Mechanism:

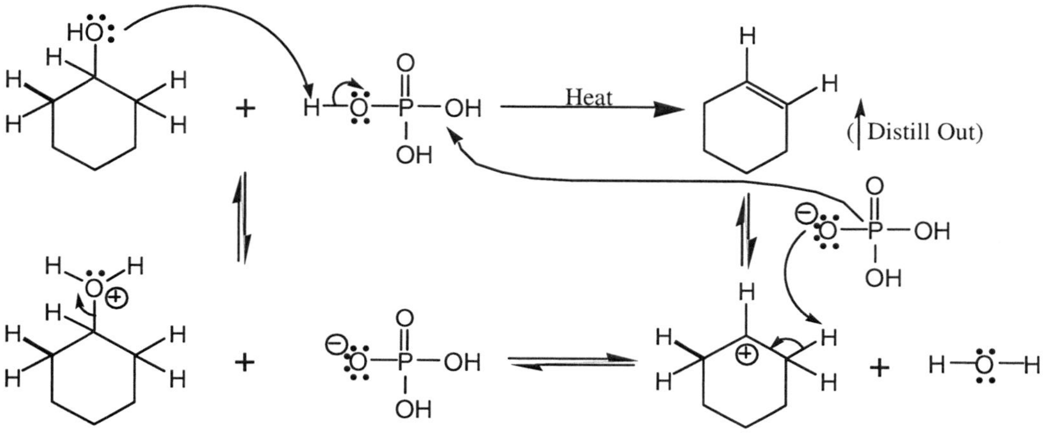

Possible Side Reactions:

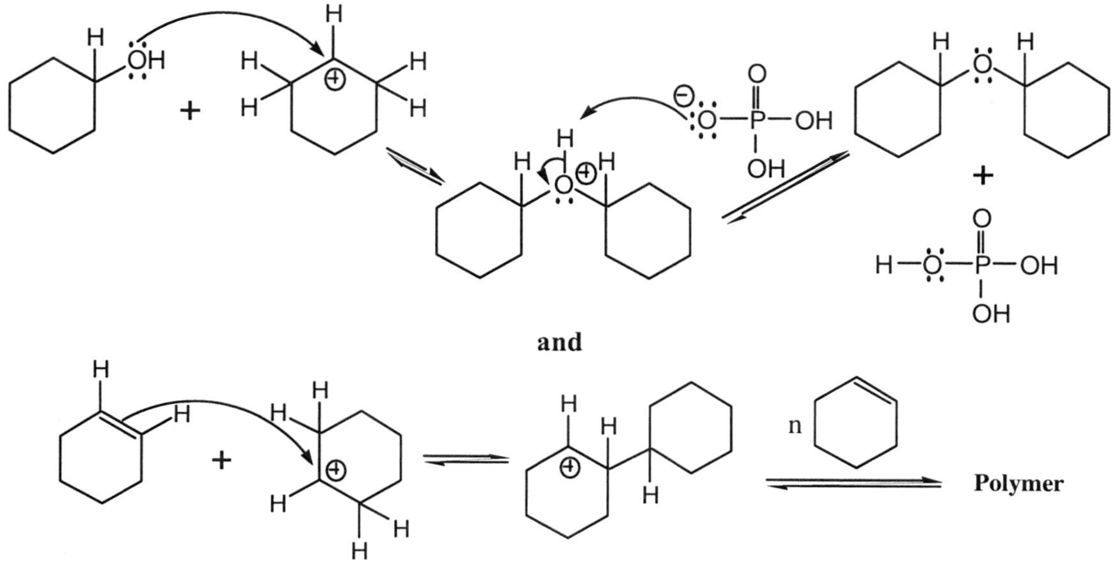

Procedure:

1. Place **0.5 mL of 85% phosphoric acid** (*caution!*) into a <u>long-necked</u> **5 mL round bottomed flask** [note the volume this occupies for deciding when the reaction has distilled to 0.5 mL of residue]. Add **2 g of cyclohexanol** and a boiling chip. Place Copper turnings (used) in the neck of the flask for fractional

distillation. Mix well. Set up for distillation as in Fig. 19.1. Use the **black** (Viton rubber) connector to join the flask to the distillation head. Wrap the distillation column with a paper towel held in place with a test tube holder. Heat *gently* to boiling for 10 minutes (without distilling anything) and then distill until ca. 0.5 mL remains in the distillation flask (the volume of the nonvolatile phosphoric acid catalyst). **Cool** and then add 2 mL of xylenes through the top of the distilling head by first removing the thermometer and thermometer adapter [mix well and note volume], and finally distill about 1 mL of xylenes out of the pot.

2. Transfer the **distillate** (two phases) to a centrifuge tube and extract (thoroughly mix) the organic layer with an approx. equal volume of saturated (sat'd) aqueous NaCl solution. Separate the organic layer by removing the aqueous layer from the bottom of the centrifuge tube with a glass pipette, and then transfer it by pipette to a dry small test tube and dry it with anhydrous $CaCl_2$. Transfer the dry organic layer to a clean and **dry** *fractional* distillation apparatus (Fig. 19.1). Distill and collect the product in a pre-weighed conical vial until the temperature begins to rise above 83°C. **Record the actual temperature *range* over which your product distilled.** *Typical* (NOT theoretical) yield is about 1g.

3. Submit a sample of 0.5–1.0 mL of your distilled cyclohexene product for gas chromatographic analysis (gives a quantitative measure of the amount of cyclohexene, xylenes, and cyclohexanol in the product).

Technical Notes: (These notes will help you be more efficient and safe in the lab. Please read them carefully.)
1. Be sure to use the *long*-necked 5 mL round-bottomed flask.
2. Measure the phosphoric acid using the 1 mL / 0.25 mL graduation pipettes in the hood.
3. Use **recycled** (used) copper turnings for the first distillation. Place the used turnings in the beaker in the hood.
4. Be sure to add a boiling chip or two.
5. Use the **black** (Viton rubber) connector to join the flask to the distillation head (to avoid swelling caused by contact with xylenes (an *aromatic* solvent).
6. Wrap the distillation apparatus with a piece of folded paper towel and hold it in place with your test tube holder (not rack). This will insulate the column and distillation head.
7. The cyclohexene and water distill together. (This first distillation is actually a steam distillation.) Collect the distillate in a 10 mL Erlenmeyer immersed in an ice-water filled 50mL beaker.
8. Addition of **xylenes** serves to "chase" the remaining product from the distillation apparatus. This entire mixture will be extracted, dried, and redistilled.
9. Transfer the mixture to a centrifuge tube and extract, rinsing the 10 mL Erlenmeyer with a small amount of additional **xylenes**.
10. After separating the product layer (top or bottom?), dry it using a minimal amount of calcium chloride. You will have to determine the proper amount. It will likely take less than you could pile up on a nickel. Using too much will considerably lower your yield.
11. Clean and **dry** your distillation apparatus before the second distillation (see text p. 279; rinse out the remaining phosphoric acid catalyst with water into a 100 mL beaker (for later neutralization), and then rinse with acetone and blow the flask dry with air). Pack the distillation column with *new*, **dry** copper turnings.
12. There is solid sodium carbonate (Na_2CO_3) in the hood for neutralizing the acidic waste.

Waste Disposal:
1. **First distillation pot residue** (mainly phosphoric acid and xylenes): Dilute with water and neutralize the acid with **solid sodium carbonate**. Separate the aqueous layer from the xylenes layer. Rinse the neutral aqueous layer down the drain with water. Dispose of the xylenes layer in the **Non-Halogenated Organics** waste bottle.
2. **Aqueous sodium chloride wash solution**, and any ethanol used for washing and drying the distillation apparatus prior to the final distillation: can be flushed down the drain.
3. **Acetone washings and all xylenes containing solutions** and pot residues should be placed in the **Non-Halogenated Organics** waste bottle.
4. **Used Calcium Chloride**: Once free of volatile liquids, can be washed down the drain or placed in the nonhazardous solid waste container in the hood.
5. **Used Boiling Chips** – White: Dry and save in container for used boiling chips. Black: Trash.

22 CHAPTER

Oxidation: Cyclohexanol to Cyclohexanone; Cyclohexanone to Adipic Acid

When you see this icon, sign in at this book's premium web site at www.cengage.com/login to access videos, Pre-Lab Exercises, and other online resources.

PRELAB EXERCISE: Write balanced equations for the dichromate and hypochlorite oxidations of cyclohexanol to cyclohexanone, and for the permanganate oxidation of cyclohexanone to adipic acid.

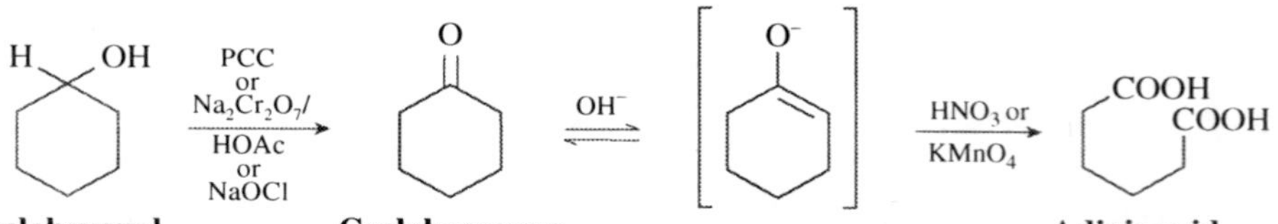

Cyclohexanol
bp 161.5°C, den. 0.96
MW 100.16

Cyclohexanone
bp 157°C, den. 0.95
MW 98.14
solubility 1.5g/100mL $H_2O^{10°}$

Adipic acid
mp 153°C, MW 146.14
solubility 1.4g/100g $H_2O^{15°}$

The oxidation of a secondary alcohol to a ketone is accomplished by many different oxidizing agents, including sodium dichromate, pyridinium chlorochromate, and sodium hypochlorite (household bleach). The ketone can be oxidized further to the dicarboxylic acid, producing adipic acid. Both of these oxidations can be carried out by the permanganate ion to give the diacid. Nitric acid is a powerful oxidizing agent that can oxidize cyclohexane, cyclohexene, cyclohexanol, or cyclohexanone to adipic acid.

THE OXIDATION OF AN ALCOHOL TO A KETONE

Dichromate oxidation

The dichromate mechanism of oxidizing an alcohol to a ketone appears to be the following:

Chapter 22 ■ *Oxidation: Cyclohexanol to Cyclohexanone; Cyclohexanone to Adipic Acid* 357

$$H_2O + Cr_2O_7^{2-} \rightleftharpoons 2\ HCrO_4^-$$

A number of intermediate valence states of chromium are involved in this reaction—the orange Cr^{6+} ion is ultimately reduced to the green Cr^{3+} ion. The course of the oxidation can be followed by these color changes.

In microscale Experiment 1, cyclohexanol is oxidized to cyclohexanone using pyridinium chlorochromate in dichloromethane. The progress of the reaction can be followed by thin-layer chromatography (TLC). In Experiment 2, the macroscale version of this reaction is carried out using sodium dichromate in acetic acid because the reagents are less expensive, the reaction is faster, and much less solvent is required.

Chromium(VI) is probably the most widely used and versatile laboratory oxidizing agent; it is used in a number of different forms to carry out selective oxidations in this text. From an environmental standpoint, however, it is far from ideal. The inhalation of the dust from insoluble Cr(VI) compounds may cause cancer of the respiratory system. The product of the reaction [Cr(III)] should not be flushed down the drain because it is toxic to aquatic life at extremely low concentrations. Therefore, as stated in the "Cleaning Up" section of Experiments 1 and 2, the Cr(III) must be precipitated as insoluble $Cr(OH)_3$, and this material is considered a hazardous waste.

Experiments 3 and 4 use an alternative oxidant for secondary alcohols that is just as efficient and much safer from an environmental standpoint: 5.25% (0.75 *M*) sodium hypochlorite solution, which is available in the grocery store as household bleach.[1] The mechanism of the reaction is not clear. It is not a free radical reaction, the reaction is much faster in acid than in base, elemental chlorine is presumably the oxidant, and hypochlorous acid must be present. It may form an intermediate alkyl hypochlorite ester, which by an E_2 elimination gives the ketone and chloride ion.

Sodium hypochlorite oxidation

[1] Mohrig JR, Nienhuis DM, Linck CF, Van Zoeren C, Fox BG. *J Chem Educ.* **1985**;62:519.

Excess hypochlorite is easily destroyed with bisulfite; the final product is the chloride ion, which is far less toxic to the environment than Cr(III).

THE OXIDATION OF A KETONE TO A CARBOXYLIC ACID

Nitric acid oxidation

In microscale Experiment 5, nitric acid is the oxidant. The balanced equation for the oxidation of cyclohexanone to adipic acid is as follows:

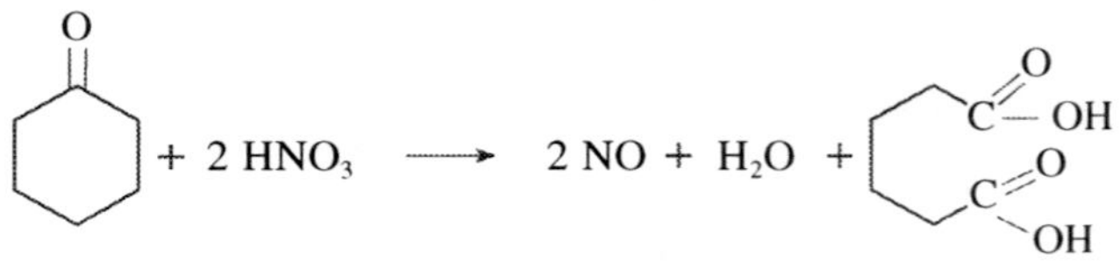

$$\text{cyclohexanone} + 2\,HNO_3 \longrightarrow 2\,NO + H_2O + \text{adipic acid}$$

In this reaction, nitric acid is reduced to nitric oxide.

Permanganate oxidation

Experiment 6 involves the permanganate oxidation of a ketone to a dicarboxylic acid. The reaction can be followed as the bright purple permanganate solution reacts to give a brown precipitate of manganese dioxide. A possible mechanism for this oxidation starts by the reaction of MnO_4^- with the enol form of the ketone and continues as shown here:

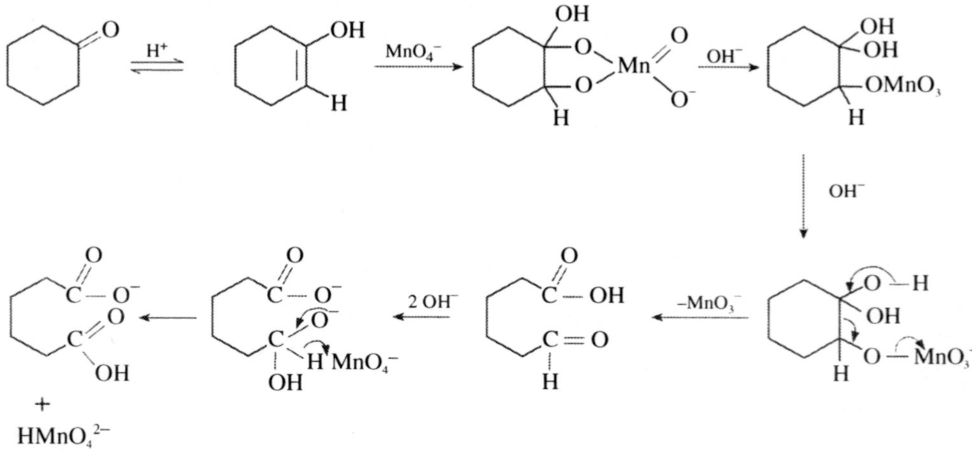

$$3\,HMnO_4^{2-} + H_2O \longrightarrow 2\,MnO_2 + MnO_4^- + 5\,OH^-$$

Oxidation of Cyclohexanol to Cyclohexanone using Household Bleach (Sodium Hypochlorite)

From Joy of Organic Chemistry by Berton C. Weberg and John E. McCarty
Mankato State University, Mankato
Adapted by B. L. Groh and M. J. Lusch

$$\text{Cyclohexanol} + \text{NaOCl} \xrightarrow{\text{CH}_3\text{CO}_2\text{H (Acetic Acid)}} \text{Cyclohexanone} + \text{NaCl} + \text{H}_2\text{O}$$

Procedure:
Place **2.00 g (2.08 mL) of cyclohexanol, 1.00 mL of acetic acid**, and a 1 inch magnetic stirring bar in a 50-mL Erlenmeyer flask. While stirring the reaction mixture rapidly, add **32 mL of 6.00% (0.875 M) sodium hypochlorite** solution (<u>Ultra</u> bleach), in portions (2-3 mL at a time) over a period of 5–10 minutes. DO NOT allow the reaction mixture to become warm; cool in ice/water if necessary. Continue stirring the reaction mixture for a further **20 minutes** after the addition of NaOCl is complete.

Work-Up: At the end of the reaction period, test the reaction mixture for excess hypochlorite by placing a drop of the reaction solution on a piece of *wet* starch-iodide indicator paper. The appearance of a blue-black color from the formation of the triiodide-starch complex on the indicator paper signifies the presence of excess hypochlorite. Add 0.5 mL of saturated sodium bisulfite solution, swirl the flask, and again test a drop of the reaction mixture with starch-iodide paper. If necessary, continue adding bisulfite solution in 0.5 mL increments and testing with starch-iodide paper until the pale yellow color of the solution disappears and the test for excess oxidant is finally negative (excess oxidant has been destroyed).

Add **2.0 mL of 6 M sodium hydroxide** solution to the reaction mixture (What is the purpose of this step – what acid is being neutralized?). Check the pH by placing a drop of the reaction solution on a piece of pH paper. Continue adding 6 M NaOH dropwise until the reaction solution shows a **pH of 6-8**, or *slightly* higher.

Extraction: Pour the reaction mixture into a **60 mL separatory funnel** (obtain from instructor) and extract the aqueous layer with **5 mL of dichloromethane** (agitate vigorously) and then transfer the organic (lower) layer to a 16 x 100 mm (small) test tube. Extract the aqueous layer again with another **5 mL of dichloromethane** (agitate vigorously) and again transfer the lower organic layer to the test tube containing the first dichloromethane extract. Add **anhydrous potassium carbonate** drying agent, cap the tube with a *cork*, and swirl the contents briefly. Dry the product solution for at least 10 min.

Evaporation: Using a Pasteur pipet, transfer the dried solution containing your product away from the drying agent into a *tared* 50-mL Erlenmeyer flask. <u>In a hood</u>, evaporate all of the dichloromethane with a stream of air. [Warming the flask in warm tap water, 40-45°C will greatly speed up the process.] When all dichloromethane has been evaporated and only cyclohexanone product remains, weigh the flask and calculate your percent yield.

Calculations: Show sample calculations in your notebook and set up a formula so that you can rapidly enter the numbers and calculate the yield of your product.

Cleanup: Wash the aqueous solution remaining in the separatory funnel from the extraction down the sink. Place the drying agent in the container for nonhazardous solid waste or the inorganic waste container.

"Green" Catalytic Oxidation of Cyclohexanone to Adipic Acid using Hydrogen Peroxide[1]

Brian Groh, Jason Pendleton and Duane Anderson
Department of Chemistry and Geology
Minnesota State University, Mankato, MN 56001

Introduction:

This experiment was developed by undergraduate students at MSU,M as a research project with the intent of creating a new experiment that would afford adipic acid using greener chemistry (see http://www.epa.gov/greenchemistry/ or search "green chemistry" on the web). Typical typical oxidations of cyclohexanone to adipic acid use hazardous metals (chromium IV) or create large amounts of waste as with permanganate oxidation. Of interest in this reaction is a tungsten catalyzed oxidation using hydrogen peroxide. Any excess hydrogen peroxide breaks down into water and oxygen. The reaction is quite clean and proceeds in very good yield making it easy to isolate the product.

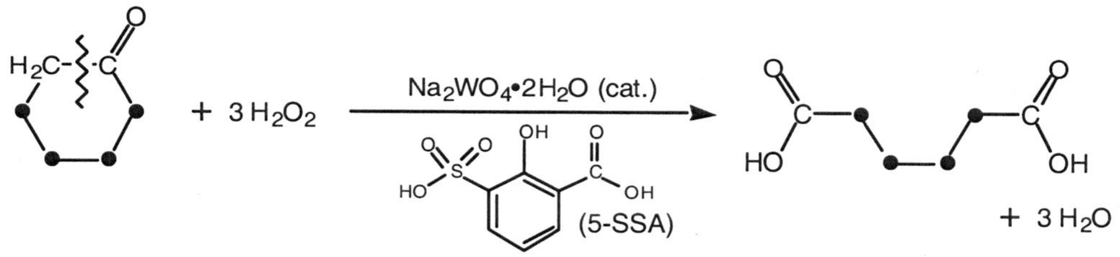

Mechanism: (Partial)

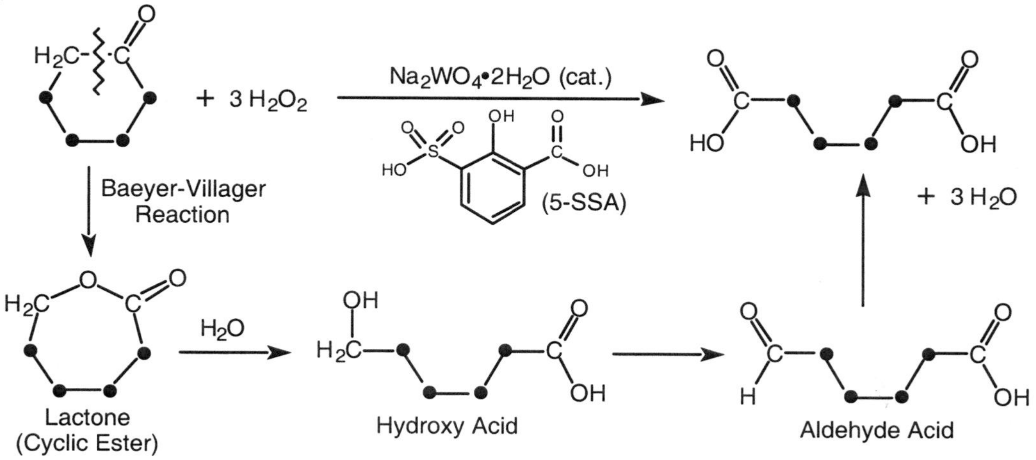

Procedure (Overnight Heating):

Weigh out **83 mg (0.083 g) of sodium tungstate ($Na_2WO_4 \cdot 2H_2O$)** into an 18 x 150 mm (large) test tube that has been labeled with your name. Next, weigh out **27 mg of 5-sulfo-salicylic acid [5-SSA, $HO_3S-C_6H_3(OH)CO_2H \cdot 2H_2O$]** into a 16 x 100 mm (small) test tube. (Use the small funnel from the microscale kit to facilitate these transfers/weighings.) Add the 5-sulfo-salicylic acid *to* the sodium tungstate in the large test tube. Add a small (4 x 12 mm) magnetic stir bar to the test tube. *In the Hood, carefully* add **5.0 mL of 30% (w/w) hydrogen peroxide (H_2O_2)** to the catalyst mixture and loosely cork the test tube. **Caution!** *30% hydrogen peroxide is extremely corrosive to the skin! Should any hydrogen peroxide come into contact with your skin, you should <u>immediately</u> wash it off with lots of water!*

At your bench, stir this mixture with the magnetic stirrer until the solution is homogeneous (no solids left). Then, add **1.0 mL of the cyclohexanone *you*** prepared in the first part of this experiment. Stir the mixture with the magnetic stirrer, and then cover the test tube with a piece of aluminum foil and make a small pin hole in the center of the foil. Using a small three-pronged clamp, clamp the test tube to a ring stand and place it into a 250 mL beaker containing about 125 mL of sand and heated to 97°C on the stirring hot plate (control the temperature with the thermocouple wire inserted into the sand next to the test tube. Allow the test tube to heat in the sand bath overnight until the next day. The next day's lab class should remove these test tubes at the beginning of their lab and place them in the designated test tube rack to cool for the previous day's class.

After cooling and crystallizing until the next lab period (or at least overnight), complete the crystallization by cooling in an ice/water bath for an additional 10 minutes or more. Filter the resulting solid crystalline adipic acid using the Hirsch funnel and suction (a vacuum trap, clamped to a ring stand, should be used between the filter flask and the vacuum source). When most of the mother liquor has been filtered off, wash the product with about **1 mL of *ice-cold* distilled water**. (**Note:** Adipic acid is quite soluble in water, even when cold; using an excess of wash water will substantially reduce your yield.)

Transfer the filtered solid to a 10 mL Erlenmeyer flask and recrystallize the product from **1–2 mL of boiling water**. (It will be necessary to heat the water using the Thermowell heating mantle or the stirring hot plate.) Allow the hot solution to cool *slowly* to room temperature and then cool in an ice/water bath for another 10 minutes. Filter the recrystallized adipic acid again using the Hirsch funnel and suction. Rinse the collected solid with about 1 mL of *ice-cold* water. When washing the product, be sure to use a minimum of *ice-cold* water. The product is ***very soluble,*** <u>even</u> in cold water. After drying the washed solid on the Hirsch funnel with suction for 10–15 minutes, place the semi-dry product in a paper envelope and allow to dry until the next lab period. Record the yield and mp *range* and turn in the product to your instructor. Yields as high as 82% have been obtained for this reaction.

Clean Up:

1. While you are waiting for your adipic acid to recrystallize, test the filtrate from the original isolation of the adipic acid for excess hydrogen peroxide or other peroxides using starch-iodide test paper. Wet the test paper and place a drop of the filtrate on the paper strip. A bluish to brownish color indicates the presence of peroxides. If peroxides are present, add some saturated aqueous sodium bisulfite ($NaHSO_3$) to reduce the peroxides and re-test. Continue adding bisulfite until the test for peroxides

is negative. Then place this aqueous solution in the waste bottle for **acidic aqueous waste**.
2. The aqueous filtrate with from the recrystallization of adipic acid (containing traces of adipic acid) can be flushed down the drain.

---

1. This experiment has been adapted from an experiment described in the literature: *Green Catalytic Oxidation of Cyclohexanone to Adipic Acid.* Shi-gang Zhang, Heng Jiang, Hong Gong, and Zhao-lin Sun, **Petroleum Science and Technology, 2003,** *21*(1,2), 275-282.

CHAPTER 24

Oxidative Coupling of Alkynes: 2,7-Dimethyl-3,5-octadiyn-2,7-diol

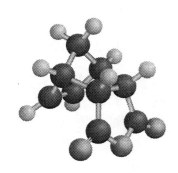

PRELAB EXERCISE: Show the reactions for a two-step method that might be used to convert 2-methyl-3-butyn-2-ol to isoprene.

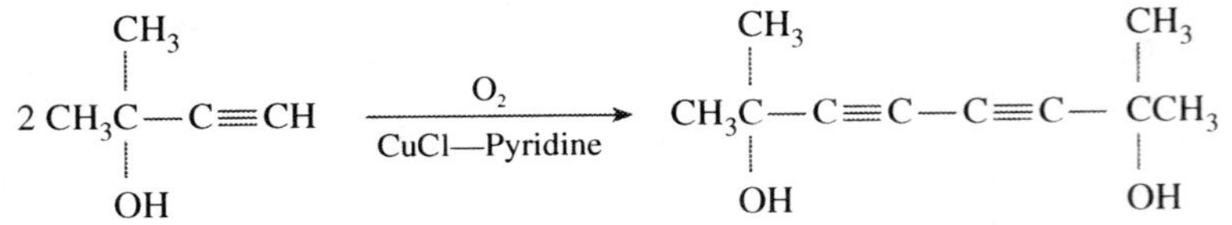

$$2\ CH_3\underset{OH}{\overset{CH_3}{\underset{|}{\overset{|}{C}}}}-C\equiv CH \xrightarrow[CuCl-Pyridine]{O_2} CH_3\underset{OH}{\overset{CH_3}{\underset{|}{\overset{|}{C}}}}-C\equiv C-C\equiv C-\underset{OH}{\overset{CH_3}{\underset{|}{\overset{|}{C}}}}CH_3$$

2-Methyl-3-butyn-2-ol
MW 84.11, den. 0.868, bp 103°C

2,7-Dimethyl-3,5-octadiyn-2,7-diol
MW 166.21, mp 130°C

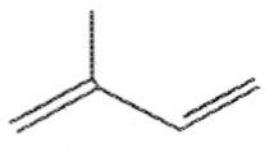

Isoprene

The starting material, 2-methyl-3-butyn-2-ol, is made commercially from acetone and acetylene and is convertible into isoprene. This experiment illustrates the oxidative coupling of a terminal acetylene to produce a diacetylene, commonly known as the Glaser reaction.

Glaser reported the coupling of acetylenes using the cuprous ion in 1869. Using density functional theory, a mechanism was reported only in 2002 by Fomina, et al.[1] The complex mechanism involves Cu^+, Cu^{2+} and Cu^{3+} ions. As noted by Glaser, the cuprous acetylide is oxidized by oxygen. The new work shows that a dicopper-dioxo complex is formed in this oxidation, but Cu^{2+} is the actual oxidizing agent. The ammonium hydroxide seems to be needed to keep the acetylide in solution.

$$C_6H_5C\equiv CH \xrightarrow[NH_4OH]{CuCl} C_6H_5C\equiv CCu \xrightarrow[O_2]{air} C_6H_5C\equiv C-C\equiv CC_6H_5$$

The reaction is very useful in the synthesis of polyenes, vitamins, fatty acids, and the annulenes. Johann Baeyer used the reaction in his historic synthesis of

---

[1] Fomina L, Vazquez B, Tkatchouk E, Fomine S. *Tetrahedron.* 2002; 58:6641–6647.

indigo back in 1882. The reaction allowed the unequivocal establishment of the carbon skeleton in this dye:

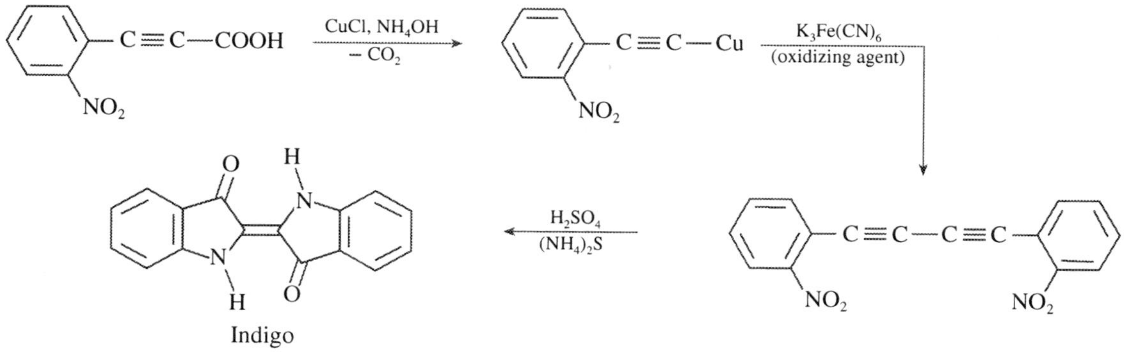

Oxidative Coupling of Alkynes: Synthesis of 2,7-Dimethyl-3,5-octadiyn-2,7-diol

Adapted from *Macroscale and Microscale Organic Experiments,* 4th Ed., K. Williamson, Houghton-Mifflin, 2003, Ch 24; pp. 335-39, by B. L. Groh and M. J. Lusch

Procedure:

1. Equip a **25 mL Filtering Flask** with a **small magnetic stirring bar** and a **small balloon** attached to the side arm and secured with a **small rubber band**. Practice quickly attaching a **large septum stopper** to the top opening of the flask.
2. Add, in the following order, **0.100 g (100 mg) of solid CuCl, 1.0 mL of 95% Ethanol, 1.0 mL of 2-Methyl-3-Butyn-2-ol**, and *finally*, **0.3 mL of N,N,N',N'-Tetramethylethylenediamine (TMEDA)** (In the Hood: Smell!).
3. With the instructor's assistance, degas the solution and flush the flask with Oxygen gas by briskly bubbling oxygen through a needle into the solution for **1 minute**.
4. Quickly cap the flask with a **large white septum stopper**, and secure it with a **large rubber band**.
5. Insert the needle through the septum and allow the balloon to inflate to the size of a medium sized pear or apple. Withdraw the needle, leaving the flask under about 10 psi of oxygen pressure initially.
6. Magnetically stir the reaction mixture briskly but without too much splashing for a full **60 minutes** (solution becomes greenish blue within 15 min or less, balloon deflates, and reaction mixture eventually turns deep midnight blue).
7. If the balloon deflates appreciably during the first 30 min of stirring, reinflate the balloon to the same size by inserting the needle a second time, and then continue the stirring.
8. Open the reaction flask, cool if warm, and add **0.5 mL of concentrated hydrochloric acid** to neutralize the TMEDA and keep the copper compounds in solution. The color will change from blue to yellow-green. If the mixture is not a yellowish color after mixing thoroughly, add a few more drops of acid.
9. Add 2.5 mL of saturated aqueous NaCl solution to precipitate the diol and stir the paste that results.
10. Add **MTBE (methyl t-butyl ether)** in 2 mL increments, mixing well using a Pasteur pipette until all organic solids dissolve (add up to 6 mL MTBE). If a white precipitate of NaCl separates on the bottom of the flask or if the lower aqueous layer looks deep yellow-brown, add a small amount of water to dissolve the NaCl and/or turn the aqueous layer a pale blue color.
11. Allow the two layers to separate, then tip the filter flask slightly and remove the aqueous layer (bottom?) using a Pasteur pipette. Use the same pipette to then transfer the organic layer to a centrifuge tube.
12. Add **3 mL of saturated aqueous NaCl** and mix well using the pipette. Allow the layers to separate and again remove the water layer. The organic layer should be clear and colorless or *light* yellow (IF NOT, wash again with distilled water).
13. Transfer the organic layer to a dry **16 x 100 mm test tube** and dry with **anhydrous sodium sulfate** (use approximately the amount you could pile onto a dime) for at least 10 min.
14. Transfer the organic layer with a pipette to a **10mL Erlenmeyer flask,** leaving the drying agent behind, and then gently boil off all the solvent (use a boiling stick!) on a hot plate **in the hood**. This will leave a residue of the **solid** diol when all of the solvent is gone.
15. Return to your bench and recrystallize the product diol from a minimum amount of **boiling hot toluene (1-3 mL)**. You will need to use a heating mantle or the hot plate to generate sufficient heat to boil the toluene.
16. Cool slowly to crystallize the diol and isolate this product using the Hirsch funnel and suction, rinsing the recrystallization flask and the collected product with a small amount of **ice-cold** toluene.
17. Allow the product to air dry in a sample envelop in your drawer until next week, and then determine yield and the mp range. Submit your product to your instructor next week (label with your name, the sample name, and the mp range).

18. While waiting for the recrystallization solution to cool and the product to crystallize, treat your waste:
    (a) The acidic aqueous layer from the initial extraction of the acidified reaction mixture with MTBE should be placed in the *Acidic* Aqueous Waste bottle (in the hood).
    (b) The filtrate from recrystallization of the final product contains toluene and should be discarded in the non-halogenated organic waste.

Waste Disposal:

1. (a) The acidic aqueous layer from the initial extraction of the acidified reaction mixture with MTBE should be placed in the *Acidic* Aqueous Waste bottle (in the hood).
   (b) The saturated NaCl solution that is used to wash the MTBE solution after removal of the acidic aqueous layer can be flushed down the drain with copious quantities of water.
   (c) The filtrate from recrystallization of the final product contains toluene and should be discarded in the non-halogenated organic waste.

2. Used Sodium Sulfate: Once free of volatile liquids, this neutral salt can be placed in the nonhazardous solid waste container in the hood or dissolved in water and washed down the drain.

3. Used Teflon Boiling Chips – Dry and save in container for used boiling chips.

CHAPTER 25

Catalytic Hydrogenation

> **PRELAB EXERCISE:** Calculate the volume of hydrogen gas generated when 3 mL of 1 M sodium borohydride reacts with concentrated hydrochloric acid. Write a balanced equation for the reaction of sodium borohydride with platinum chloride. Calculate the volume of hydrogen that can be liberated by reacting 1 g of zinc with acid.

Catalytic reduction is a very important and widely used industrial process; usually no harmful wastes are produced in the process. Catalytic hydrogenation and dehydrogenation are carried out on an enormous scale, for example, in the catalytic cracking and reforming of crude oil to make gasoline.

Nitrobenzene can be reduced catalytically to aniline with water as the only byproduct:

$$3 H_2 + \text{Nitrobenzene (PhNO}_2\text{)} \xrightarrow{\text{catalyst}} \text{Aniline (PhNH}_2\text{)} + 2 H_2O$$

Styrene is made by the catalytic dehydrogenation of ethylbenzene at very high temperatures, but styrene can also be very easily hydrogenated back to ethylbenzene. Palladium, as a catalyst, lowers the energy barrier for the reaction in both directions.

$$\text{Ethylbenzene} \underset{\text{Pd}}{\overset{400-500°C}{\rightleftharpoons}} \text{Styrene} + H_2$$

The addition of hydrogen to alkenes is one of the most common reactions. The alkene is more reactive toward this process than is the aromatic ring or functional groups such as esters or ketones.

377

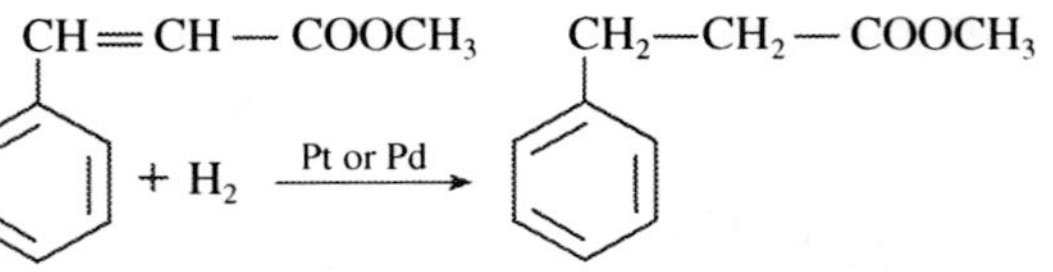

$$\text{Ph-CH=CH-COOCH}_3 + H_2 \xrightarrow{\text{Pt or Pd}} \text{Ph-CH}_2\text{-CH}_2\text{-COOCH}_3$$

Hydrogenation is stereospecific, so alkynes are reduced to *cis*-alkenes. The metal is usually supported on a high-surface-area material such as charcoal. The alkene and the hydrogen probably are both adsorbed onto the surface of the catalyst before the transfer occurs. This heterogeneous reaction, a reaction that involves reactants in the liquid or gas phase and a catalyst in the solid phase, is difficult to study.

In the experiments in this chapter, catalytic hydrogenation is carried out in several different ways. Experiment 1 is a puzzle for you to solve. In Experiment 2, hydrogen gas from an external supply is used to hydrogenate a long-chain unsaturated alcohol to the corresponding saturated alcohol. In Experiment 3, the hydrogen is generated in situ using the Brown hydrogenation technique. Experiment 4 utilizes a process called *transfer hydrogenation* to produce a saturated fat from an unsaturated one (olive oil). In Experiment 5, olive oil is catalytically reduced by using hydrogen gas.

4. TRANSFER HYDROGENATION OF OLIVE OIL[4,5]

> IN THIS EXPERIMENT, olive oil is treated with a palladium-on-carbon catalyst and the hydrogen donor cyclohexene to produce a solid fat by transfer hydrogenation. In Experiment 5, olive oil is catalytically reduced using hydrogen gas. If both experiments are performed, the results can be compared. Analyze the two products by titration with bromine in carbon tetrachloride (or dichloromethane) and also by NMR spectroscopy. Alternatively, you can carry out just one reduction and compare your results with a classmate who used the other procedure.

The metabolism of olive oil, like other vegetable oils containing unsaturated fatty acids, results in an increased production of high-density lipoproteins (HDL) that do

---

[4] Discussions with Gottfried Brieger regarding transfer hydrogenation are gratefully acknowledged.
[5] Barry B. Snider of Brandeis University points out some interesting problems with this experiment. *See* the text website.

not deposit as much cholesterol in the arteries as the low-density lipoproteins (LDL); LDLs contribute to the disease arteriosclerosis (hardening of the arteries).

Olive oil is a triester consisting of a trihydric alcohol, glycerol, and three long-chain fatty (carboxylic) acids. The fatty acid in olive oil is primarily oleic acid, an 18-carbon monounsaturated compound. So olive oil can be regarded as primarily glycerol trioleate, although it contains about 15% saturated fat and an equal quantity of polyunsaturated fat. The double bond in oleic acid has the *cis* configuration; the molecule is bent in the center and does not pack well into a crystal lattice; thus, olive oil is a liquid at room temperature. If the double bonds are saturated, then the triester, glycerol tristearate, is a solid, and melts at almost 70°C. Similarly, a molecule having *trans* double bonds is a solid.

$$\begin{array}{c}
\text{CH}_2\text{OC(CH}_2\text{)}_7\text{CH}=\text{CH(CH}_2\text{)}_7\text{CH}_3 \\
\text{\textit{cis}} \\
| \\
\text{CHOC(CH}_2\text{)}_7\text{CH}=\text{CH(CH}_2\text{)}_7\text{CH}_3 \\
\text{\textit{cis}} \\
| \\
\text{CH}_2\text{OC(CH}_2\text{)}_7\text{CH}=\text{CH(CH}_2\text{)}_7\text{CH}_3 \\
\text{\textit{cis}}
\end{array} \xrightarrow[]{\text{10\% Pd/C, cyclohexene}} \begin{array}{c}
\text{CH}_2\text{OC(CH}_2\text{)}_{16}\text{CH}_3 \\
| \\
\text{CHOC(CH}_2\text{)}_{16}\text{CH}_3 \\
| \\
\text{CH}_2\text{OC(CH}_2\text{)}_{16}\text{CH}_3
\end{array}$$

Glycerol trioleate
mp −5.5°C
MW 885.47

Glycerol tristearate
mp 69.9°C
MW 891.52

To manufacture margarine, a vegetable oil such as olive oil or, more commonly, corn or soybean oil is partially hydrogenated. This gives a mixture of liquid and solid fats that possesses the consistency of butter. In the process, some of the *cis* double bonds are also isomerized to the more stable *trans* isomers. Like saturated fatty acids, the *trans* isomers also contribute to the formation of LDL. When oils such as these are completely hydrogenated, the resulting fat is a solid. Many saturated fats are found in animals. The hard fat (tallow) on a raw steak is made up of saturated fats.

In this procedure, we use 100% extra-virgin olive oil from Italy. The label says that of each 14 g of oil, 10 are monounsaturated (mostly from oleic acid), 2 are polyunsaturated (mostly linoleic acid, an 18-carbon acid with two double bonds), and 2 are saturated (this is a mixture of predominantly 18-carbon saturated stearic acid and a smaller amount of 16-carbon palmitic acid).

From this rough analysis, it can be concluded that complete hydrogenation of all the double bonds in olive oil will give a product containing about 90%–95% stearic acid (C-18) and 5%–10% palmitic acid (C-18) esterified to glycerol. Although the product will not be pure glycerol tristearate, it should be a solid with a melting point above 50°C.

The technique of transfer hydrogenation is employed. In the presence of a catalyst, hydrogen is lost from the cyclohexene donor molecules and transferred to the double bonds in the olive oil. Although it cannot be used in all cases, this technique of hydrogenation is quite convenient because it consists of simply refluxing for a few minutes the substance to be hydrogenated with the hydrogen donor and the catalyst.

A variety of hydrogen donor molecules can be used, among which are cyclohexene, hydrazine, formic acid, cyclohexadiene, and ammonium formate. When ammonium formate decomposes in the presence of a catalyst, it produces hydrogen, ammonia, and carbon dioxide. Not only does reduction of double bonds take place, but the ammonia can react with esters to form amides. Cyclohexene and cyclohexadiene both lose hydrogen to form the same stable end product. In this experiment, cyclohexene is used as the hydrogen donor.

Microscale Procedure

In a reaction tube, place 400 mg of olive oil, 1 mL of cyclohexene, 50 mg of 10% palladium-on-carbon catalyst, and a boiling chip. Reflux the mixture for at least 30 minutes. Be careful not to boil the alkene out of the reaction tube. If necessary, place a coil of damp pipe cleaner around the top of the tube. This will keep the top of the tube cool so the cyclohexene will not escape. The empty distilling column could also be mounted on top of the reaction tube to provide even more condensing area (*see* Fig. 25.1 on page 379); however, with careful adjustment of the heat input (the depth of the tube in the sand), this should not be necessary. The isolation procedure is given in Experiment 5.

Macroscale Procedure

In a 14/20 standard taper flask equipped with a water-cooled condenser, place 0.80 g of olive oil, 2 mL of cyclohexene, 100 mg of 10% palladium-on-carbon catalyst, and a boiling chip (*see* Fig. 25.3 on page 380). Reflux the mixture for at least 15 minutes. The isolation procedure is given in Experiment 5.

5. CATALYTIC HYDROGENATION OF OLIVE OIL

Microscale Procedure

Follow the procedure of Experiment 2, using the apparatus illustrated in Figure 25.5 on page 381. If 9 mL of hydrogen is generated, this will be sufficient to completely hydrogenate 119 mg of olive oil (assuming that the olive oil is 100 percent glycerol trioleate). Demonstrate in your laboratory notebook that this calculation is correct.

Dissolve the olive oil in 0.6 mL of methanol and inject it into a flask containing 1.5 mL of methanol, 20 mg of 10 percent palladium-on-carbon catalyst, and a small magnetic stirring bar. Stir the reaction mixture until the uptake of hydrogen ceases, and then isolate the product.

Macroscale Procedure

Follow the procedure of Experiment 2. Using an apparatus similar to that in Figure 25.5 on page 381 but with a larger beaker and graduated cylinder, generate 45 mL of hydrogen and hydrogenate 590 mg of olive oil. Dissolve the olive oil in 3 mL of methanol and inject it into a 25-mL flask containing 7.5 mL of methanol, 100 mg of 10 percent palladium-on-carbon catalyst, and a small magnetic stirring bar. Stir the reaction mixture until the uptake of hydrogen ceases and then isolate the product.

Isolation of Products

Prepare a Pasteur pipette as a micro filter by forcing a small piece of cotton firmly down to the constriction (*see* Fig. 25.2 on page 379). Add the reaction mixture to the filter pipette and then force the solution through the cotton into a tared, 25-mL

Vacuum evaporation while heating is necessary to ensure that the fat will solidify when cooled on ice.

FIG. 25.9

An apparatus for removing a solvent under vacuum.

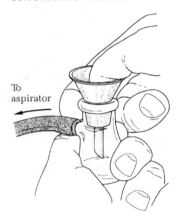

filter flask. Rinse the reaction tube, filter with a few drops of hexane, and then evaporate the solution to dryness (Fig. 25.9). To remove the last traces of volatile liquid from the product, heat the filter flask in a hot sand bath under vacuum. After cooling on ice, the product should solidify to a hard, white fat. Analyze a portion of the product. The bulk of the product can be used for the synthesis of soap (*see* Experiment 6 in Chapter 40).

Analysis of Products

If you performed both Experiments 4 and 5, determine and compare the weights of the products and their physical properties. Count the number of drops of 3% bromine in carbon tetrachloride (or dichloromethane) that can be decolorized by equivalent quantities of each product and of olive oil.

Analyze the NMR spectra of olive oil and of the two products. In olive oil, a complex set of peaks centered at 5.38 ppm are produced by protons on a *cis* double bond. If the catalyst has isomerized the *cis* to the *trans*, double-bond peaks for the *trans* isomer will be found centered at 5.35 ppm. Completely hydrogenated olive oil will, of course, have no peaks in this region of the spectrum. A set of 5 peaks centered at 5.26 ppm will always be present and can be used to calibrate the integration. These peaks arise from the hydrogen on the central carbon of the glycerol part of the triester.

The IR spectra of olive oil (Fig. 25.10) and of the two products show very small differences. A weak peak in the 3013–3011 cm^{-1} region is characteristic of the *cis* double bond, and a weak peak in the 970–967 cm^{-1} region is characteristic of the *trans* double bond.

Computational Chemistry

Calculate, using a semiempirical molecular orbital program such as AM1, the heats of formation of cyclohexene, cyclohexane, and benzene. Use this information to

FIG. 25.10

The IR spectrum (thin film) of olive oil (Bertolli brand, extra virgin), which consists primarily of glycerol trioleate.

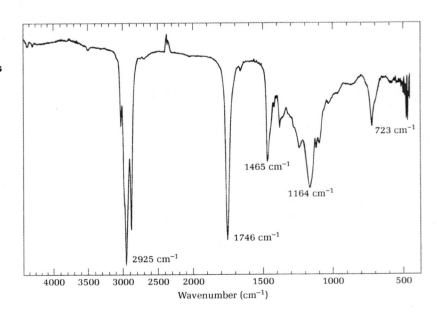

Microscale Catalytic Transfer Hydrogenation of Olive Oil

Adapted from Kenneth Williamson, Macroscale and Microscale Organic Experiments, 4th ed., Houghton Mifflin, Boston, MA, 2003, Ch 25, pp. 340-53.
by M. J. Lusch and B. L. Groh

Reaction Equation:

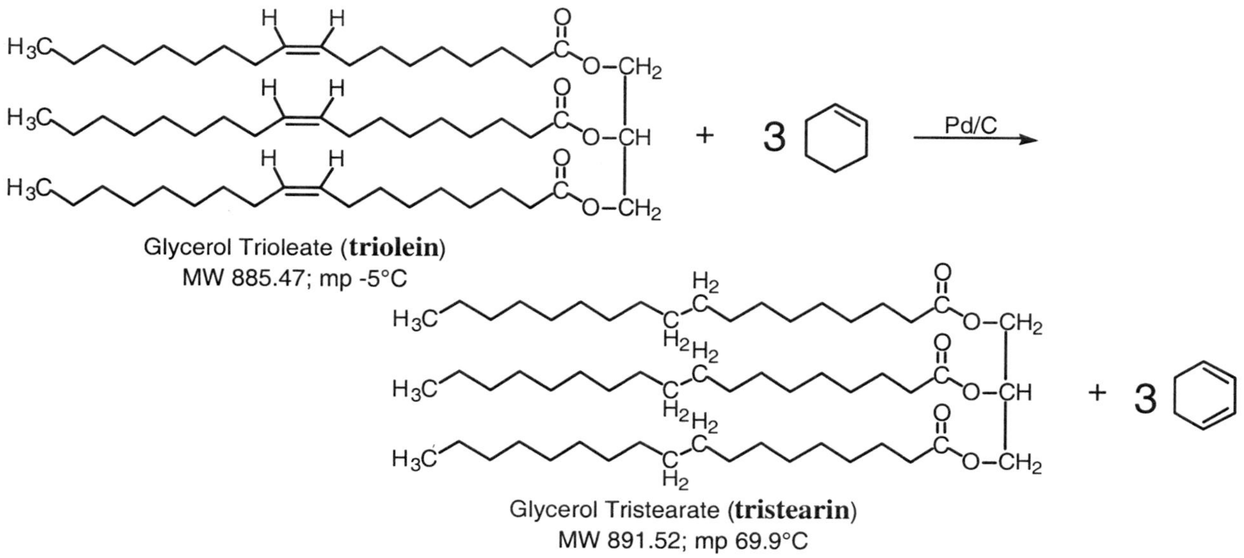

Glycerol Trioleate (**triolein**)
MW 885.47; mp -5°C

Glycerol Tristearate (**tristearin**)
MW 891.52; mp 69.9°C

Possible Side Reactions: Incomplete hydrogenation can result in some of the <u>remaining</u> double bonds being isomerized to their trans isomer (**trans fats**).

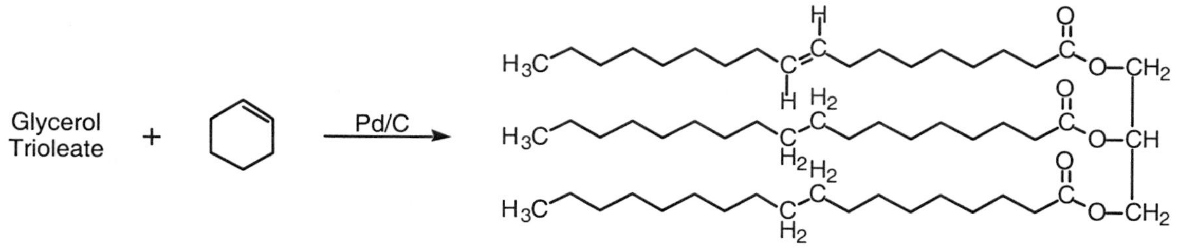

Procedure:

1. Put the **catalyst in first (50 mg; the catalyst has been pre-weighed for you)**, *olive oil* (**400 mg; 450 uL**) *in second*, and *then rinse them in with the cyclohexene* (**1 mL**) that is being added.
2. The catalyst has been pre-weighed for you. You only need to transfer this to your reaction flask.
3. The olive oil will be measured out using an **auto pipettor**. It has been calibrated to deliver the correct volume (and therefore also weight) of oil to your flask. You may pre-weigh your flask to determine you have correctly added the proper amount.
   Use of Autopipettors to measure the olive oil:
   - Pipettor will be preset to **450μL.**
   - Be careful not to get air in the tip, but only put the <u>tip</u> into the olive oil; depress and release the plunger slowly.

- Do not push too hard when the plunger is fully depressed, otherwise the tip will be ejected from the plunger body.
4. Measurement of the Cyclohexene: use the **3 mL disposable plastic pipets**.
5. **Reflux Procedure: Use the water-cooled condenser** instead of just a straight tube.
   - Water should be connected so it goes *in at the bottom* (lower) inlet and *out at the top* (upper) outlet.
   - Thin-wall (black) rubber tubing can be more easily connected if the glass inlet and outlet nipples and the inside of the rubber tubing are **lubricated with water**, and the tubing is fitted on the glass connections with a **twisting, screw-like motion.**
   - Turn on the flow of water to give only a steady but mild flow of water – you DON'T want it to be a high-pressure fire hose flow of water.
   - Heat the solution sufficiently to *boil* the liquid smoothly (**boiling chips!**), but only so the condensing vapors go no higher than the **middle** of jacketed condenser. However, if cyclohexene can be smelled, the condenser is not cooled well enough.
6. Pack a glass pipet with cotton as shown on the board by the reagent shelves.
   - Use an applicator stick and a *small* piece of cotton.
   - *Pack the cotton firmly but not too tightly or you will not be able to filter your sample; too loosely and it will not remove all the catalyst!*
   - Add a small amount of *Celite (about 5mm)* to the pipette, then carefully filter your sample.
   - Pressurize the pipet filter by connecting the top to the air line on the bench and push all of the liquid through the filter. Disconnect the air hose, rinse the pipet and filter bed with a few drops of **hexanes**, and push the hexane rinse through the filter with air pressure also.
7. *Isolate the product in a tared 25 mL filtering flask to obtain the weight of the product and its yield.*
8. **For Solvent removal:** use the filter adapter and a centrifuge tube to seal the top of the filter flask (don't use the Hirsch funnel filter).
   - Apply house vacuum (use the filter trap apparatus to prevent liquid back-up into the vacuum line), and evaporate volatile materials (excess cyclohexene, hexanes, and 1,3-butadiene by-product) until no further evaporation takes place.
   - The flask can be warmed in the heating mantle sand bath, which will melt the waxy fat to a liquid so that it can be swirled in the flask to encourage evaporation.
   - When the evaporation is finished, disconnect the suction and cool the product in an ice bath to cause it to solidify.
9. **Show** the final waxy product to your instructor, then continue with the experiment.
10. Try the test for unsaturation on your product using a **5% solution of Bromine in CH_2Cl_2** Do this by dissolving one drop of your oil in 0.5 mL of CH_2Cl_2. Add the 5% bromine solution dropwise with stirring. Compare this result with the result using a sample of the starting olive oil (will react) and a sample of just bromine in CH_2Cl_2 (no reaction - called a "blank" or "control"). Addition of more than 2 drops of bromine solution indicates unsaturation.

Waste Disposal:

1. The **crude hydrogenated fat product** is used in the next part of the experiment.

2. The Celite/cotton/glass pipet filter can be disposed of in the **waste glass** container.

3. **Used Boiling Chips** – Dry and save in the ceramic dish in the hood for used boiling chips.

4. **Halogenated solvents:** Deposit in the **Halogenated Hazardous Organic Waste** bottle (Hood).

29 CHAPTER

Friedel–Crafts Alkylation of Benzene and Dimethoxybenzene; Host-Guest Chemistry

PRELAB EXERCISE: Prepare a flow sheet for the alkylation of benzene and the alkylation of dimethoxybenzene, indicating how the catalysts and unreacted starting materials are removed from the reaction mixture.

The Friedel–Crafts[1] alkylation of aromatic rings most often employs an alkyl halide and a strong Lewis acid catalyst. Some of the catalysts that can be used, in order of decreasing activity, are the halides of aluminum, antimony, iron, titanium, tin, bismuth, and zinc. Although useful, the reaction has several limitations. The aromatic ring must be unsubstituted or bear activating groups; because the product—an alkylated aromatic molecule—is more reactive than the starting material, multiple substitution usually occurs. Furthermore, primary halides will rearrange under the reaction conditions:

Reaction Temperature	−6°C: 60%	40%
	+35°C: 40%	60%

In this reaction, a tertiary halide and the most powerful Friedel–Crafts catalyst, $AlCl_3$, are allowed to react with benzene. (If you prefer not to work with benzene, you can carry out alkylations of dimethoxybenzene or *m*-xylene.) The initially formed *t*-butylbenzene is a liquid, whereas the product, 1,4-di-*t*-butylbenzene, which has a symmetrical structure, is a beautiful crystalline solid. The alkylation reaction probably proceeds through the carbocation under the conditions of the experiments in this chapter:

[1] Charles Friedel and James Crafts (who later became the president of MIT) discovered this reaction in 1879.

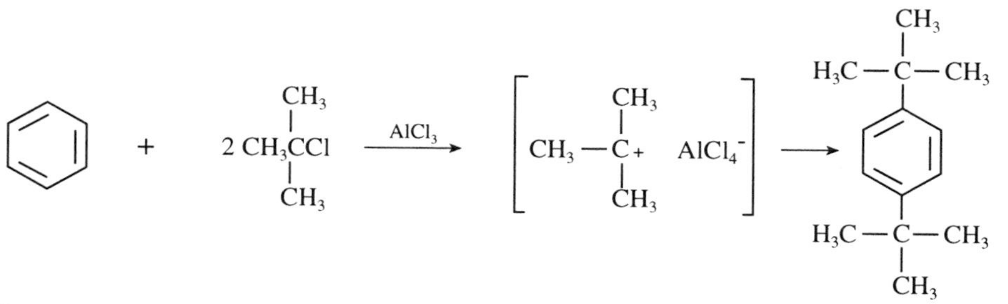

Reaction Scheme

Benzene + 2 CH₃CCl(CH₃)₂ —AlCl₃→ [(CH₃)₃C⁺ AlCl₄⁻] → 1,4-Di-*t*-butylbenzene

Benzene
MW 78.11, den. 0.88
bp 80°C

2-Chloro-2-methylpropane
(*t*-Butyl chloride)
MW 92.57, den. 0.85
bp 51°C

1,4-Di-*t*-butylbenzene
MW 190.32, mp 77–79°C
bp 167°C

The reaction is reversible. If 1,4-di-*t*-butylbenzene is allowed to react with *t*-butyl chloride and 1.3 mol of aluminum chloride at 0°C–5°C, 1,3 di-*t*-butylbenzene, 1,3,5-tri-*t*-butylbenzene, and unchanged starting material are found in the reaction mixture. Thus, the mother liquor from crystallization of 1,4-di-*t*-butylbenzene probably contains *t*-butylbenzene, the desired 1,4-di-product, the 1,3-di-isomer, and 1,3,5-tri-*t*-butylbenzene.

INCLUSION COMPLEXES: HOST-GUEST CHEMISTRY

Although the mother liquor probably contains a mixture of several components, the 1,4-di-*t*-butylbenzene can be isolated easily as an inclusion complex. Inclusion complexes are examples of host-guest chemistry. The host molecule thiourea, NH_2CSNH_2, has the interesting property of crystallizing in a helical crystal lattice that has a cylindrical hole in it. The guest molecule can reside in this hole if it is the correct size. It is not bound to the host; nuclear magnetic resonance (NMR) studies indicate the guest molecule can rotate longitudinally within the helical crystal lattice. There are often nonintegral numbers (on the average) of host molecules per guest. The inclusion complex of thiourea and 1,4-di-*t*-butylbenzene crystallizes quite nicely from a mixture of the other hydrocarbons; thus more of the product can be obtained. Because thiourea is very soluble in water, the product is recovered from the complex by shaking it with a mixture of ether and water. The complex immediately decomposes, and the product dissolves in the ether layer, from which it can be recovered.

Compare the length of the 1,4-di-*t*-butylbenzene molecule to the length of various *n*-alkanes and predict the host-guest ratio for a given alkane. You can then check your prediction experimentally. *n*-Hexane can be isolated from the mixture of isomers sold under the name *hexanes*.

Friedel-Crafts Alkylation: Synthesis of 1,4-Di-*t*-butyl-2,5-dimethoxybenzene

Adapted from Kenneth Williamson, Macroscale and Microscale Organic Experiments, 4th ed., Houghton Mifflin, Boston, MA, 2003, Ch 29, pp. 379-380 by M. J. Lusch

Reaction Equation: (Balanced)

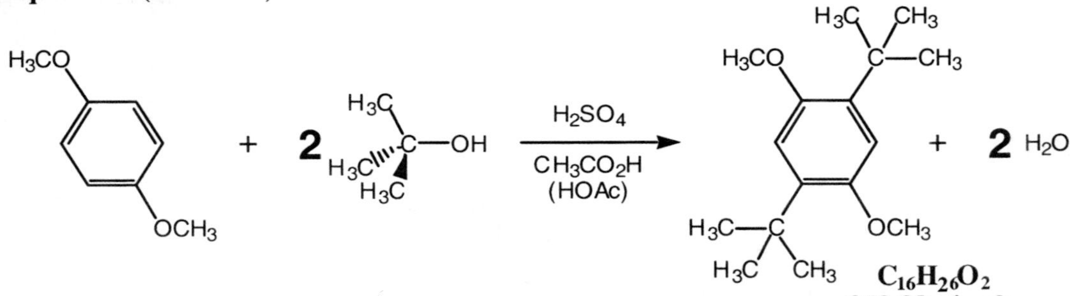

$C_{16}H_{26}O_2$
250.38 g/mole
mp = 102–104°C

Mechanism: (Specific to this Reaction)

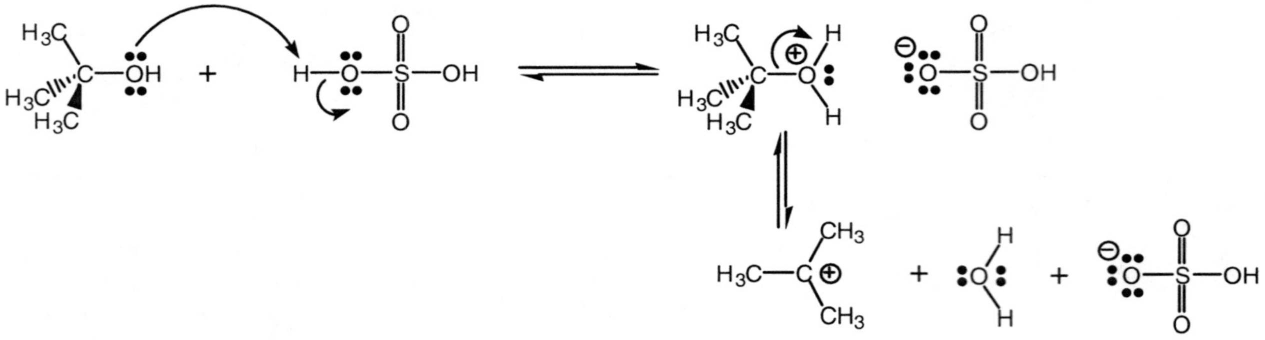

Then,

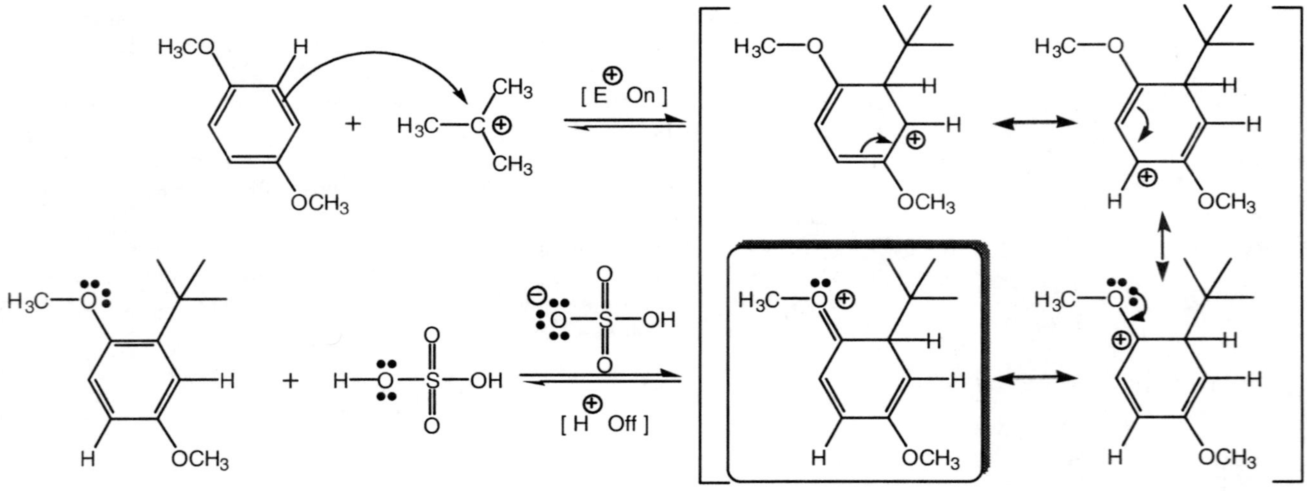

Then, Repeat with a second *tert*-Butyl Carbocation.

General: (Electron-Donating, Activating Substituent [has unshared pair of electrons next to ring])

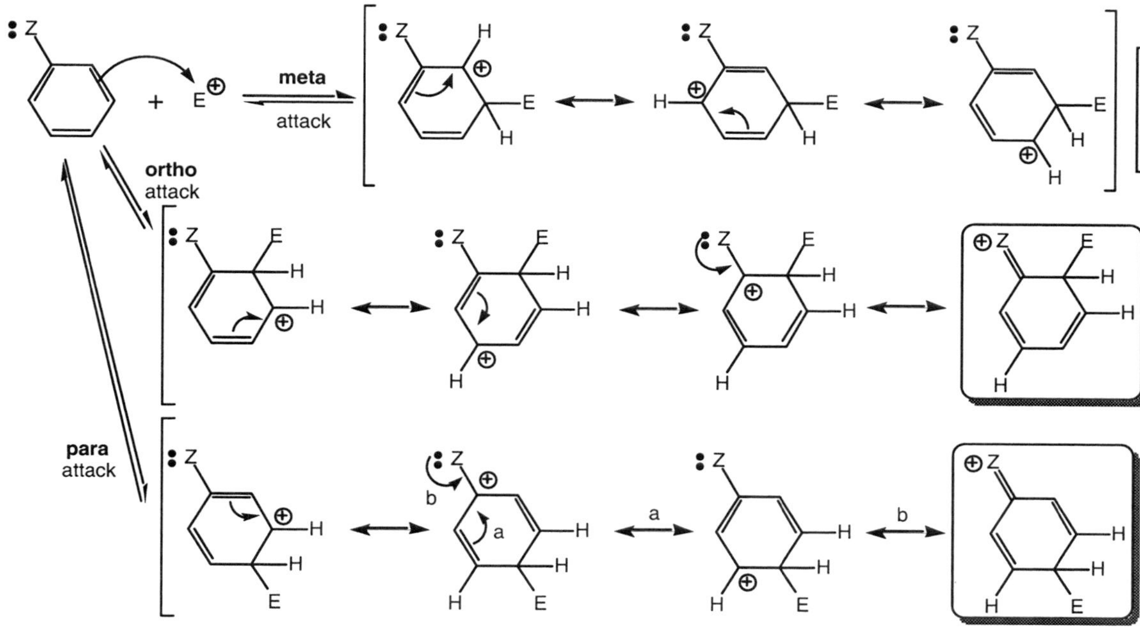

Possible Side Reactions: (Mono-Substitution Product; Formation of Di-Substitution Isomers)

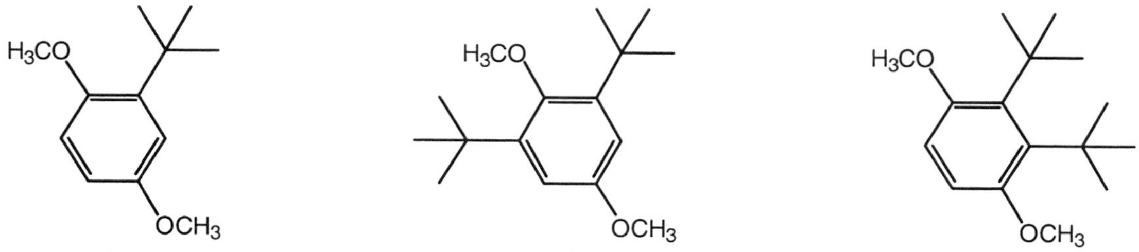

1, **2**, **3**

Procedure:

Weigh 120 mg (0.120 g) of 1,4-dimethyoxybenzene into a reaction tube and dissolve it in 0.4 mL of glacial acetic acid with gentle warming. Add 0.2 mL of t-butyl alcohol, mix thoroughly, and then cool in an ice bath. Add 0.4 mL of conc. sulfuric acid **_dropwise_** with good stirring. After each drop of acid is added, mix the solution thoroughly. After the sulfuric acid addition is complete, warm the reaction mixture to room temperature and allow to stand for 10 min. Cool the mixture in ice to complete crystallization. Slowly add 2.5 mL of water **_dropwise_** with cooling and thorough stirring after each drop. Remove the solvent with a Pasteur (glass) pipette, and wash the crystals thoroughly with more water (to remove what?). Recrystallize the crystals in the same reaction tube from methanol or a methanol-water mixture (cloud point method). Collect the product by using a Pasteur pipette to remove the mother liquor, wash with ice-cold fresh methanol, and then dry the crystals, initially under aspirator vacuum, and then on a weighing paper until the next lab. A typical yield is 80-100mg of product. Weigh, determine the melting point range, and then hand in your product in the designated container.

Waste Disposal:

Combine the aqueous reaction solution, methanol washes, and crystallization mother liquor, dilute with water and neutralize with sodium carbonate and flush down the drain. Any spilled acid should be treated with solid sodium carbonate, combined with the reaction waste, and flushed down the drain.

CHAPTER 36

Aldehydes and Ketones

When you see this icon, sign in at this book's premium web site at www.cengage.com/login to access videos, Pre-Lab Exercises and other online resources.

PRELAB EXERCISE: Outline a logical series of experiments designed to identify an unknown aldehyde or ketone with the least effort. Consider the time required to complete each identification reaction.

The carbonyl group occupies a central place in organic chemistry. Aldehydes and ketones—compounds such as formaldehyde, acetaldehyde, acetone, and 2-butanone—are very important industrial chemicals used by themselves and as starting materials for a host of other substances. For example, more than 10 billion pounds (4.5 billion kilograms) of formaldehyde-containing plastics are produced in the United States each year.

The carbonyl carbon is sp^2 hybridized, the bond angles between adjacent groups are 120°, and the four atoms R, R', C, and O lie in one plane:

$$\begin{array}{c} \ddot{\text{O}}: \\ \parallel \\ \text{C} \\ \diagup \quad \diagdown \\ \text{R} \quad\quad \text{R}' \end{array}$$

The electronegative oxygen polarizes the carbon-oxygen bond, rendering the carbon electron deficient and hence subject to nucleophilic substitution.

$$\mathrm{\underset{}{>}\!C\!=\!\ddot{O}:} \quad \longleftrightarrow \quad \mathrm{\underset{}{>}\!\overset{+}{C}\!-\!\ddot{O}\!:^{-}} \quad \text{or} \quad \mathrm{\underset{}{>}\!C\!\overset{\delta+}{=}\!\overset{\delta-}{\ddot{O}}:}$$

Geometry of the carbonyl group

Attack on the sp^2 hybridized carbon occurs via the π-electron cloud above and below the plane of the carbonyl group:

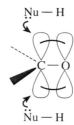

REACTIONS OF THE CARBONYL GROUP

Many reactions of carbonyl groups are acid catalyzed. The acid attacks the electronegative oxygen, which bears a partial negative charge, to create a carbocation that subsequently reacts with the nucleophile:

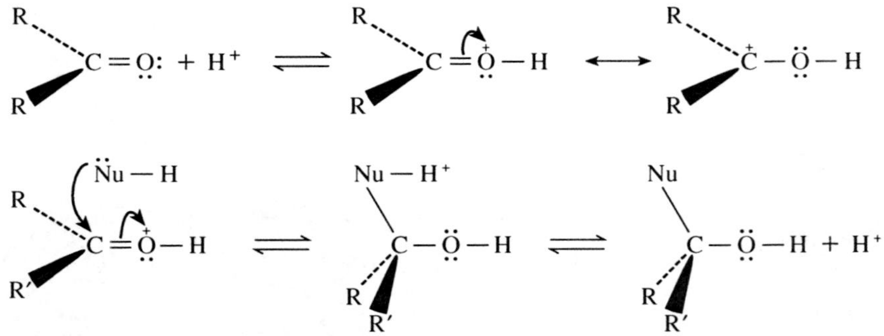

$$\underset{R'}{\overset{R}{>}}C=\ddot{O}: + H^+ \rightleftharpoons \underset{R'}{\overset{R}{>}}C=\overset{+}{\ddot{O}}-H \longleftrightarrow \underset{R'}{\overset{R}{>}}\overset{+}{C}-\ddot{O}-H$$

$$\underset{R'}{\overset{R}{>}}C=\overset{+}{\ddot{O}}-H \;\;\xrightleftharpoons[]{\text{Nu}-H}\;\; \underset{R'}{\overset{R}{>}}\underset{\text{Nu}-H^+}{C}-\ddot{O}-H \;\rightleftharpoons\; \underset{R'}{\overset{R}{>}}\underset{\text{Nu}}{C}-\ddot{O}-H + H^+$$

The strength of the nucleophile and the structure of the carbonyl compound determine whether the equilibrium lies on the side of the carbonyl compound or the tetrahedral adduct. Water, a weak nucleophile, does not usually add to the carbonyl group to form a stable compound:

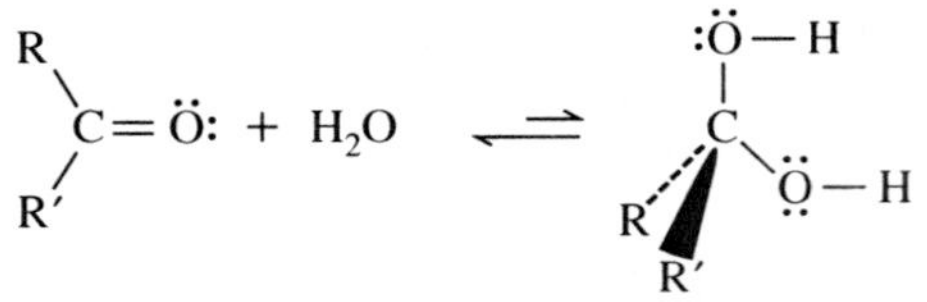

$$\underset{R'}{\overset{R}{>}}C=\ddot{O}: + H_2O \rightleftharpoons \underset{R'}{\overset{R}{>}}C\underset{\ddot{O}-H}{\overset{\ddot{O}-H}{\big<}}$$

In the special case of trichloroacetaldehyde, however, the electron-withdrawing trichloromethyl group allows a stable hydrate to form:

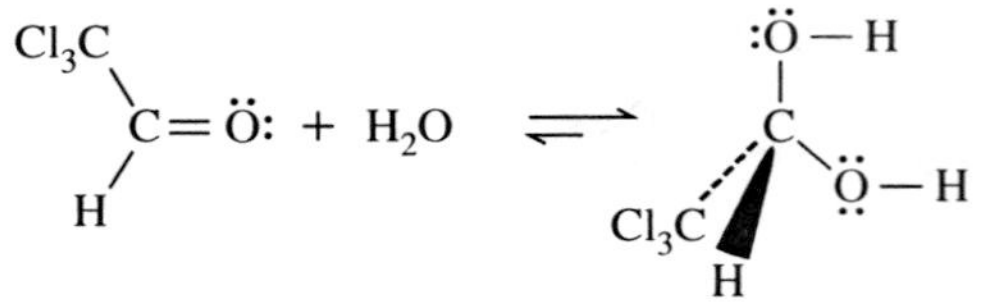

$$\underset{H}{\overset{Cl_3C}{>}}C=\ddot{O}: + H_2O \rightleftharpoons \underset{H}{\overset{Cl_3C}{>}}C\underset{\ddot{O}-H}{\overset{\ddot{O}-H}{\big<}}$$

A stable hydrate: chloral hydrate

The compound so formed, chloral hydrate, was discovered by Justus von Liebig in 1832 and was introduced as one of the first sedatives and hypnotics (sleep-inducing substances) in 1869. It is now most commonly encountered in detective fiction as a "Mickey Finn" or "knockout drops."

Reaction with an alcohol; hemiacetals

In an analogous manner, an aldehyde or ketone can react with an alcohol. The product, a hemiacetal or hemiketal, is usually not stable, but in the case of certain cyclic hemiacetals, the product can be isolated. Glucose is an example of a stable hemiacetal.

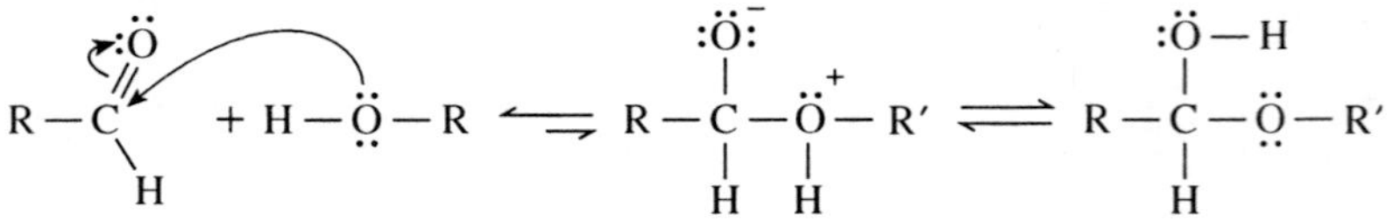

$$R-\underset{H}{\overset{\ddot{O}}{\overset{\|}{C}}} + H-\ddot{O}-R \rightleftharpoons R-\underset{H}{\overset{:\ddot{O}:^-}{\underset{|}{C}}}-\overset{+}{\ddot{O}}-R' \rightleftharpoons R-\underset{H}{\overset{:\ddot{O}-H}{\underset{|}{C}}}-\ddot{O}-R'$$

Hemiacetal
usually not stable

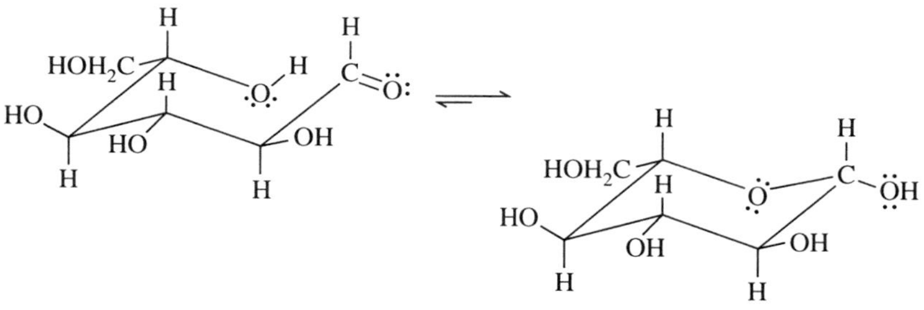

Glucose
a stable cyclic hemiacetal

BISULFITE ADDITION

The bisulfite ion is a strong nucleophile but a weak acid. It will attack the unhindered carbonyl group of an aldehyde or methyl ketone to form an addition product:

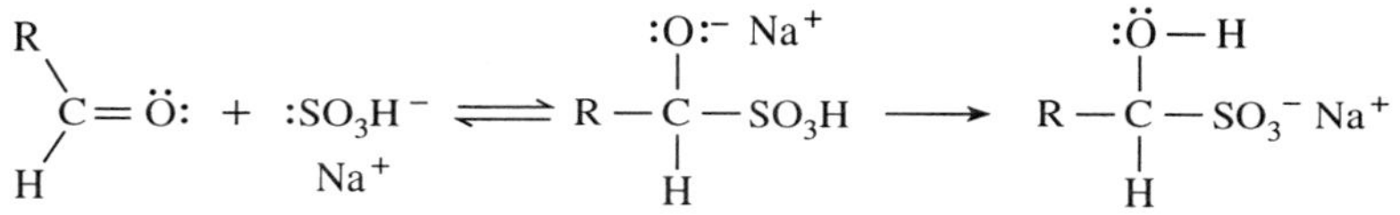

$$\underset{H}{\overset{R}{>}}C=\ddot{O}: \ + \ :SO_3H^- \ Na^+ \ \rightleftharpoons \ R-\underset{H}{\overset{:O:^- \ Na^+}{\underset{|}{C}}}-SO_3H \ \longrightarrow \ R-\underset{H}{\overset{:\ddot{O}-H}{\underset{|}{C}}}-SO_3^- \ Na^+$$

Because these bisulfite addition compounds are ionic water-soluble compounds and can be formed with a maximum 90% yield, they serve as a useful means of separating aldehydes and methyl ketones from mixtures of organic compounds. At high sodium bisulfite concentrations, these adducts crystallize and can be isolated by filtration. The aldehyde or ketone can be regenerated by adding either a strong acid or base:

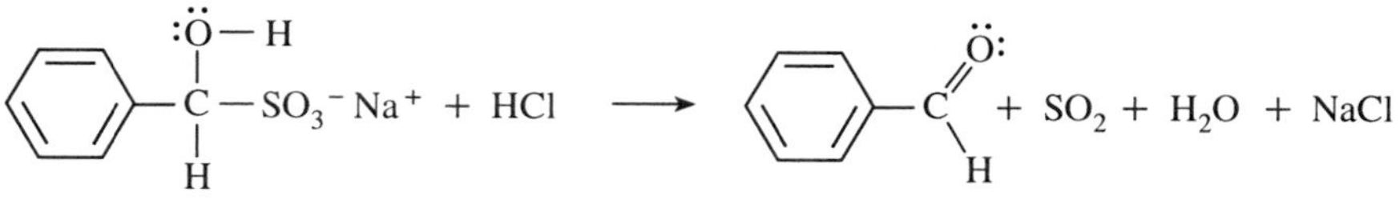

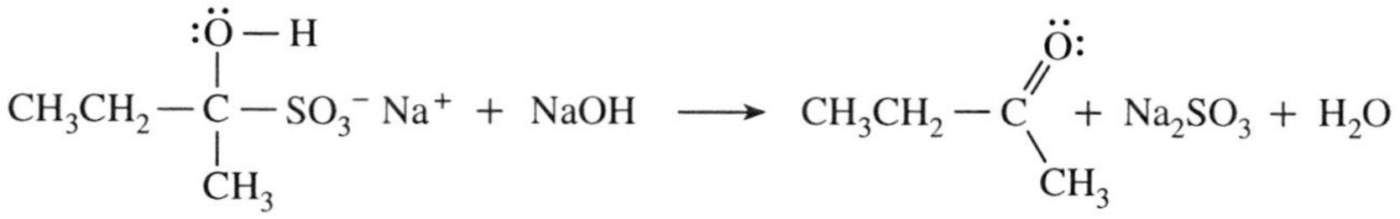

$$Ph-\underset{H}{\overset{:\ddot{O}-H}{\underset{|}{C}}}-SO_3^- \ Na^+ \ + \ HCl \ \longrightarrow \ Ph-\underset{H}{\overset{\ddot{O}:}{C}} \ + \ SO_2 \ + \ H_2O \ + \ NaCl$$

$$CH_3CH_2-\underset{CH_3}{\overset{:\ddot{O}-H}{\underset{|}{C}}}-SO_3^- \ Na^+ \ + \ NaOH \ \longrightarrow \ CH_3CH_2-\underset{CH_3}{\overset{\ddot{O}:}{C}} \ + \ Na_2SO_3 \ + \ H_2O$$

CYANIDE ADDITION

A similar reaction occurs between aldehydes and ketones and hydrogen cyanide, which, like bisulfite, is a weak acid but a strong nucleophile. The reaction is hazardous to carry out because of the toxicity of cyanide, but the cyanohydrins are useful synthetic intermediates:

Cyanohydrin formation and reactions

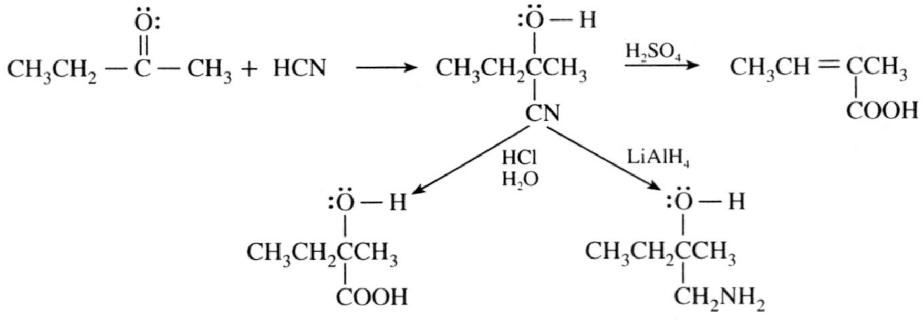

$$CH_3CH_2-\overset{\overset{\ddot{O}:}{\|}}{C}-CH_3 + HCN \longrightarrow CH_3CH_2\underset{\underset{CN}{|}}{\overset{\overset{:\ddot{O}-H}{|}}{C}}CH_3 \xrightarrow{H_2SO_4} \underset{\underset{COOH}{|}}{CH_3CH=CCH_3}$$

With HCl/H$_2$O: $CH_3CH_2\underset{\underset{COOH}{|}}{\overset{\overset{:\ddot{O}-H}{|}}{C}}CH_3$

With LiAlH$_4$: $CH_3CH_2\underset{\underset{CH_2NH_2}{|}}{\overset{\overset{:\ddot{O}-H}{|}}{C}}CH_3$

Amines are good nucleophiles and readily add to the carbonyl group:

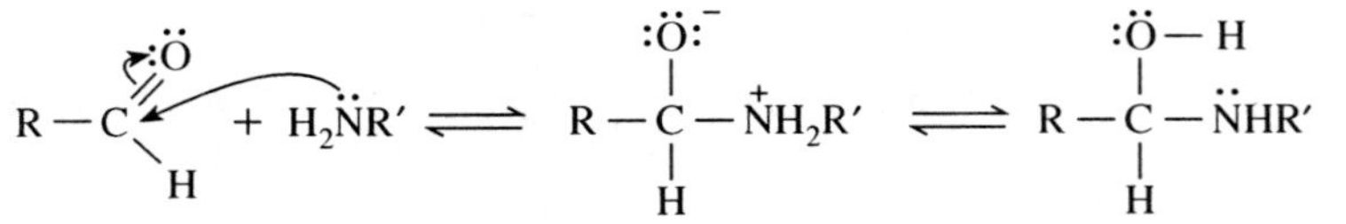

$$R-\underset{H}{\overset{\overset{:\ddot{O}}{\|}}{C}} + H_2\ddot{N}R' \rightleftharpoons R-\underset{H}{\overset{\overset{:\ddot{O}:^-}{|}}{\underset{|}{C}}}-\overset{+}{N}H_2R' \rightleftharpoons R-\underset{H}{\overset{\overset{:\ddot{O}-H}{|}}{\underset{|}{C}}}-\ddot{N}HR'$$

The reaction is strongly dependent on the pH. In acid, the amine is protonated (RN$^+$H$_3$) and is no longer a nucleophile. In strong base, there are no protons available to catalyze the reaction. But in weak acid solution (pH 4–6), the equilibrium between acid and base (**a**) is such that protons are available to protonate the carbonyl (**b**), and yet there is free amine present to react with the protonated carbonyl (**c**):

(a) $CH_3\ddot{N}H_2 + HCl \rightleftharpoons CH_3\overset{+}{N}H_3 + Cl^-$

(b) $CH_3-\underset{H}{\overset{\overset{:\ddot{O}}{\|}}{C}} + HCl \rightleftharpoons \left[CH_3-\underset{H}{\overset{\overset{:\overset{+}{O}-H}{\|}}{C}} \longleftrightarrow CH_3-\underset{H}{\overset{\overset{:\ddot{O}-H}{/}}{C^+}} \right] + Cl^-$

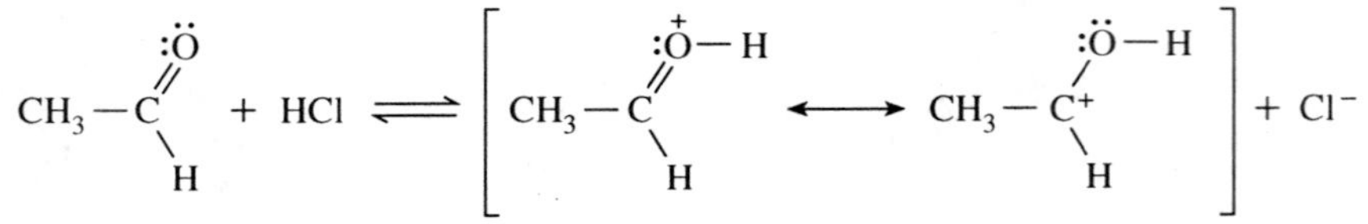

(c) $CH_3-\underset{H}{\overset{\overset{:\overset{+}{O}-H}{\|}}{C}} + CH_3\ddot{N}H_2 \rightleftharpoons CH_3-\underset{H}{\overset{\overset{:\ddot{O}-H}{|}}{\underset{|}{C}}}-\overset{+}{N}H_2CH_3 \underset{}{\overset{-H^+}{\rightleftharpoons}} CH_3-\underset{H}{\overset{\overset{:\ddot{O}-H}{|}}{\underset{|}{C}}}-\underset{H}{\ddot{N}}-CH_3$

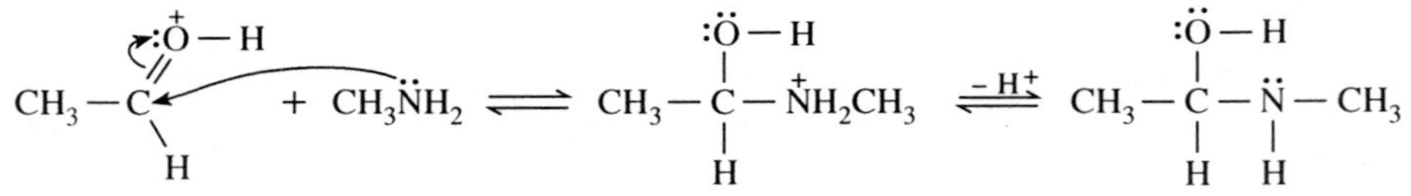

SCHIFF BASES

The intermediate hydroxyamino form of the adduct is not stable and spontaneously dehydrates under the mildly acidic conditions of the reaction to give an imine, commonly referred to as a *Schiff base*:

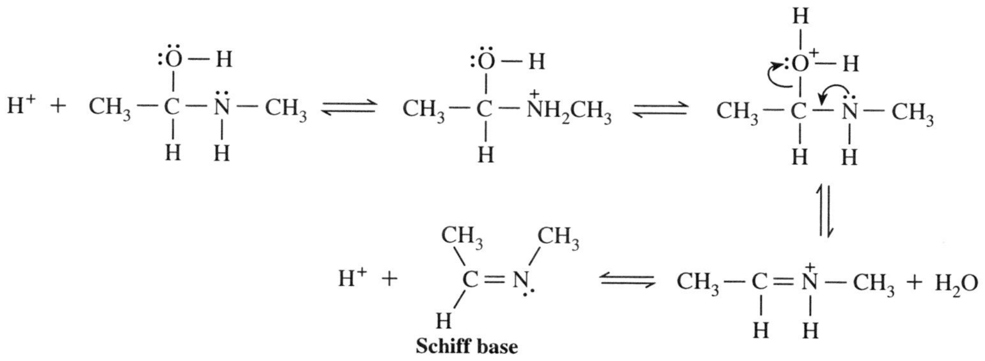

$$H^+ + CH_3-\overset{\overset{\ddot{O}-H}{|}}{\underset{\underset{H}{|}}{C}}-\overset{H}{\underset{|}{\ddot{N}}}-CH_3 \rightleftharpoons CH_3-\overset{\overset{\ddot{O}-H}{|}}{\underset{\underset{H}{|}}{C}}-\overset{+}{N}H_2CH_3 \rightleftharpoons CH_3-\overset{\overset{\overset{H}{|}}{\overset{+}{O}-H}}{\underset{\underset{H}{|}}{C}}-\overset{H}{\underset{|}{\ddot{N}}}-CH_3$$

$$H^+ + \underset{\underset{H}{|}}{\overset{\overset{CH_3}{|}}{C}}=\underset{}{\overset{CH_3}{\ddot{N}}} \rightleftharpoons CH_3-\underset{\underset{H}{|}}{\overset{|}{C}}=\overset{+}{\underset{\underset{H}{|}}{N}}-CH_3 + H_2O$$

Schiff base

Imine or Schiff base formation   The biosynthesis of most amino acids proceeds through Schiff base intermediates.

OXIMES, SEMICARBAZONES, AND 2,4-DINITROPHENYLHYDRAZONES

Three rather special amines form useful stable imines:

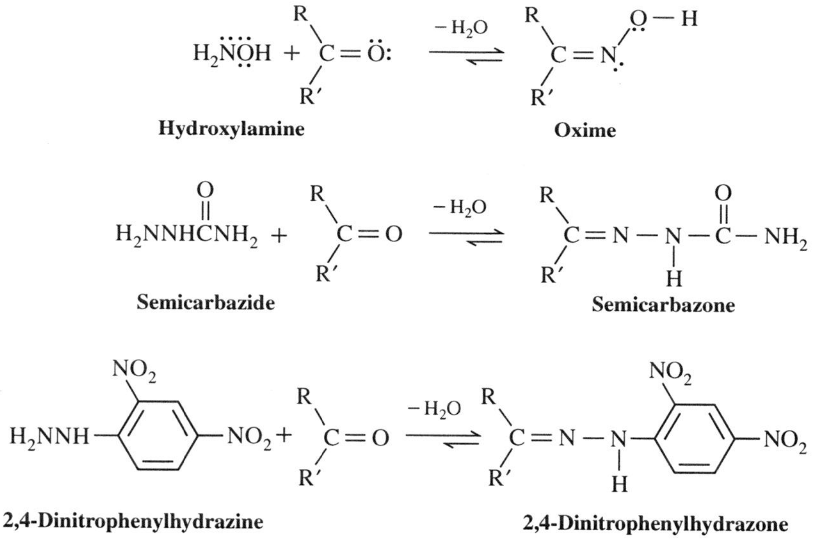

$$H_2\ddot{N}\ddot{O}H + \underset{R'}{\overset{R}{\diagdown}}C=\ddot{O}: \xrightarrow{-H_2O} \underset{R'}{\overset{R}{\diagdown}}C=N\underset{\ddot{..}}{\diagdown}\overset{O-H}{}$$

Hydroxylamine   **Oxime**

$$H_2NNH\overset{\overset{O}{\|}}{C}NH_2 + \underset{R'}{\overset{R}{\diagdown}}C=O \xrightarrow{-H_2O} \underset{R'}{\overset{R}{\diagdown}}C=N-\underset{\underset{H}{|}}{N}-\overset{\overset{O}{\|}}{C}-NH_2$$

Semicarbazide   **Semicarbazone**

$$H_2NNH-\!\!\left\langle\!\!\begin{array}{c}NO_2\\ \\ \end{array}\!\!\right\rangle\!\!-NO_2 + \underset{R'}{\overset{R}{\diagdown}}C=O \xrightarrow{-H_2O} \underset{R'}{\overset{R}{\diagdown}}C=N-\underset{\underset{H}{|}}{N}-\!\!\left\langle\!\!\begin{array}{c}NO_2\\ \\ \end{array}\!\!\right\rangle\!\!-NO_2$$

2,4-Dinitrophenylhydrazine   **2,4-Dinitrophenylhydrazone**

These imines are solids and are useful for the characterization of aldehydes and ketones. For example, IR (infrared) and NMR (nuclear magnetic resonance) spectroscopies may indicate that a certain unknown is acetaldehyde. It is difficult to determine the boiling point of a few milligrams of a liquid, but if it can be converted to a solid derivative, the melting point *can* be determined with that amount. The 2,4-dinitrophenylhydrazones are usually the derivatives of choice because they are crystalline compounds with well-defined melting or decomposition points, and they increase the molecular weight by 180. Ten milligrams of acetaldehyde will give 51 mg of the 2,4-dinitrophenylhydrazone.

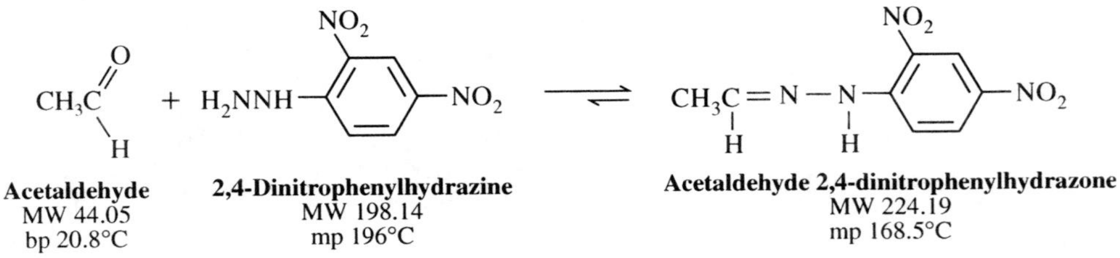

$$\text{CH}_3\overset{\displaystyle\text{O}}{\underset{\displaystyle\text{H}}{\text{C}}} + \text{H}_2\text{NNH}\!-\!\!\!\bigcirc\!\!\!-\!\text{NO}_2 \;(\text{NO}_2) \rightleftharpoons \text{CH}_3\underset{\text{H}}{\text{C}}=\text{N}-\underset{\text{H}}{\text{N}}\!-\!\!\!\bigcirc\!\!\!-\!\text{NO}_2 \;(\text{NO}_2)$$

Acetaldehyde
MW 44.05
bp 20.8°C

2,4-Dinitrophenylhydrazine
MW 198.14
mp 196°C

Acetaldehyde 2,4-dinitrophenylhydrazone
MW 224.19
mp 168.5°C

TOLLENS' REAGENT

Before the advent of NMR and IR spectroscopy and mass spectrometry, the chemist was often called on to identify aldehydes and ketones by purely chemical means. Aldehydes can be distinguished chemically from ketones by their ease of oxidation to carboxylic acids. The oxidizing agent, an ammoniacal solution of silver nitrate—Tollens' reagent—is reduced to metallic silver, which is deposited on the inside of a test tube as a silver mirror.

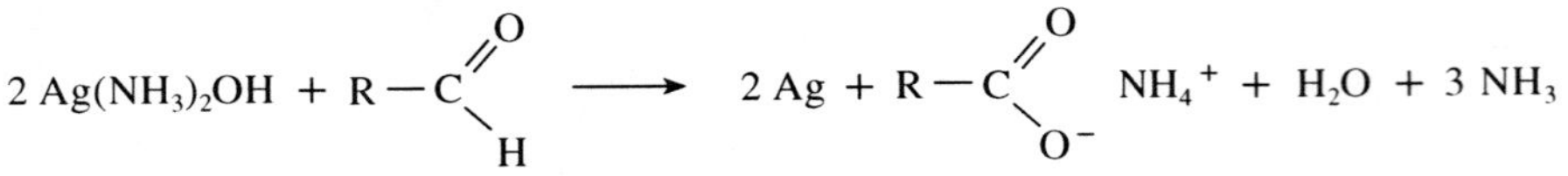

$$2\,\text{Ag}(\text{NH}_3)_2\text{OH} + \text{R}-\overset{\displaystyle\text{O}}{\underset{\displaystyle\text{H}}{\text{C}}} \longrightarrow 2\,\text{Ag} + \text{R}-\overset{\displaystyle\text{O}}{\underset{\displaystyle\text{O}^-}{\text{C}}}\;\;\text{NH}_4^+ + \text{H}_2\text{O} + 3\,\text{NH}_3$$

SCHIFF'S REAGENT

Another way to distinguish aldehydes from ketones is to use Schiff's reagent. This is a solution of the red dye Basic Fuchsin, which is rendered colorless on treatment with sulfur dioxide. In the presence of an aldehyde, the colorless solution turns magenta.

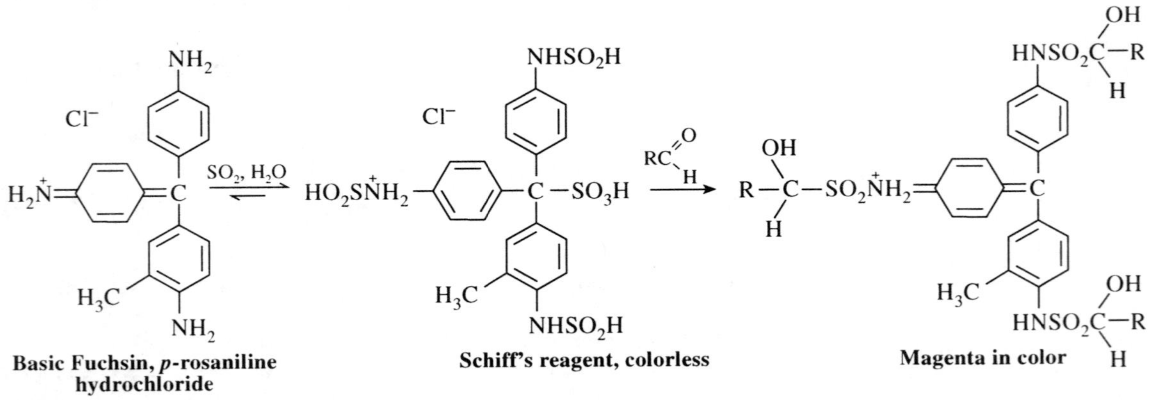

Basic Fuchsin, *p*-rosaniline hydrochloride

Schiff's reagent, colorless

Magenta in color

IODOFORM TEST

A test for methyl ketones

Methyl ketones can be distinguished from other ketones by the iodoform test. The methyl ketone is treated with iodine in a basic solution. Introduction of the first iodine atom increases the acidity of the remaining methyl protons, so halogenation

O
||
CH₃C —
Methyl ketone

stops only when the triiodo compound has been produced. The base then allows the relatively stable triiodomethyl carbanion to leave, and a subsequent proton transfer gives iodoform, a yellow crystalline solid with a melting point of 119°C–123°C. The test is also positive for fragments or compounds easily oxidized to methyl ketones, such as the fragment CH₃CHOH or the compound ethanol. Acetaldehyde also gives a positive test because it is both a methyl ketone and an aldehyde.

$$R-\overset{\overset{\ddot{O}:}{\|}}{C}-CH_3 + OH^- \rightleftharpoons R-\overset{\overset{\ddot{O}:}{\|}}{C}-\overset{-}{\ddot{C}}H_2 + H_2O$$

$$R-\overset{\overset{\ddot{O}:}{\|}}{C}-\overset{-}{\ddot{C}}H_2 + I_2 \rightarrow R-\overset{\overset{O}{\|}}{C}-CH_2I + I^- \xrightarrow{OH^-} \xrightarrow{I_2} \xrightarrow{OH^-} \xrightarrow{I_2} R-\overset{\overset{\ddot{O}:}{\|}}{C}-CI_3$$

$$R-\overset{\overset{\ddot{O}:}{\|}}{C}-CI_3 + OH^- \rightarrow R-\overset{\overset{:\ddot{O}:^-}{|}}{\underset{:OH}{C}}-CI_3 \rightarrow R-C\overset{:\ddot{O}}{\underset{:O-H}{\diagup}} + :\bar{C}I_3 \rightarrow R-C\overset{:\ddot{O}}{\underset{\ddot{O}:^-}{\diagup}} + CHI_3$$

Iodoform
mp 123°C

EXPERIMENTS

1. UNKNOWNS

You will be given an unknown that may be any of the aldehydes or ketones listed in Table 36.1. At least one derivative of the unknown is to be submitted to your instructor, but if you first do the bisulfite and iodoform characterizing tests, the results may suggest derivatives whose melting points will be particularly revealing.

You can further characterize the unknown by determining its boiling point, which is best done with a digital thermometer and a reaction tube (see Chapter 5). The boiling points of the unknowns are given on this book's web site.

In conducting the following tests, you should perform three tests simultaneously: on a compound known to give a positive test, on a compound known to give a negative test, and on the unknown. In this way you will be able to determine whether the reagents are working as they should, as well as interpret a positive or a negative test.

Carry out three tests:
 Known positive
 Known negative
 Unknown

2. 2,4-DINITROPHENYLHYDRAZONES

$$\underset{R'}{\overset{R}{\diagdown}}C=\ddot{O}: + H_2\ddot{N}-\underset{H}{\overset{}{\ddot{N}}}-\underset{NO_2}{\overset{NO_2}{\diagdown}}-NO_2 \xrightarrow{H^+} \underset{R'}{\overset{R}{\diagdown}}C=\ddot{N}\underset{H}{\overset{}{\diagdown}}\underset{NO_2}{\overset{NO_2}{\diagdown}}-NO_2$$

TABLE 36.1 Melting Points of Derivatives of Some Aldehydes and Ketones

| | | | | | | Melting Points (°C) | |
Compound[a]	Formula	n_D^{20}	MW	Water Solubility	Phenyl-hydrazone	2,4-dinitro-phenyl-hydrazone	Semi-carbazone
Acetone	CH_3COCH_3	1.3590	58.08		42	126	187
n-Butanal	$CH_3CH_2CH_2CHO$	1.3790	72.10	4 g/100 g	Oil	123	95 (106)[b]
3-Pentanone (diethyl ketone)	$CH_3CH_2COCH_2CH_3$	1.3920	86.13	4.7 g/100 g	Oil	156	138
2-Furaldehyde (furfural)	C_4H_3O-CHO	1.5260	96.08	9 g/100 g	97	212 (230)[b]	202
Benzaldehyde	C_6H_5CHO	1.5450	106.12	Insol.	158	237	222
Hexane-2,5-dione	$CH_3COCH_2CH_2COCH_3$	1.4260	114.14	∞	120[c]	257[c]	224[c]
2-Heptanone	$CH_3(CH_2)_4COCH_3$	1.4080	114.18	Insol.	Oil	89	123
3-Heptanone	$CH_3(CH_2)_3COCH_2CH_3$	1.4080	114.18	Insol.	Oil	81	101
n-Heptanal	$n-C_6H_{13}CHO$	1.4125	114.18	Insol.	Oil	108	109
Acetophenone	$C_6H_5COCH_3$	1.5325	120.66	Insol.	105	238	198
2-Octanone	$CH_3(CH_2)_5COCH_3$	1.4150	128.21	Insol.	Oil	58	122
Cinnamaldehyde	$C_6H_5CH=CHCHO$	1.6220	132.15	Insol.	168	255	215
Propiophenone	$C_6H_5COCH_2CH_3$	1.5260	134.17	Insol.	About 48	191	182

[a]Visit this book's web site for data on additional aldehydes and ketones.

[b]Both melting points have been found, depending on crystalline form of derivative.

[c]Monoderivative or diderivative.

An easily prepared derivative of aldehydes and ketones.

Video: Microscale Crystallization

To 5 mL of the stock solution[1] of 2,4-dinitrophenylhydrazine in phosphoric acid, add about 0.05 g of the compound to be tested. Five milliliters of the 0.1 M solution contains 0.5 mmol (0.0005 mol) of the reagent. If the compound to be tested has a molecular weight of 100, then 0.05 g is 0.5 mmol. Warm the reaction mixture for a few minutes in a water bath, and then let crystallization proceed. Collect the product by suction filtration (Fig. 36.1), wash the crystals with a large amount of water to remove all of the phosphoric acid, press a piece of moist litmus paper onto the crystals, and wash them with more water if they are acidic. Press the product between sheets of filter paper until it is as dry as possible and recrystallize from ethanol. Occasionally, a high molecular weight derivative will not dissolve in a reasonable quantity (10 mL) of ethanol. In that case, cool the hot suspension and isolate the crystals by suction filtration. The boiling ethanol treatment removes impurities so that an accurate melting point can be obtained on the isolated material.

An alternative procedure is applicable when the 2,4-dinitrophenylhydrazone is known to be sparingly soluble in ethanol. Measure 0.5 mmol of crystalline 2,4-dinitrophenylhydrazine into a 50-mL Erlenmeyer flask, add 15 mL of 95% ethanol, digest on a steam bath until all the solid particles are dissolved, and then add 0.5 mmol of the compound to be tested and continue warming. If there is no immediate change, add, from a Pasteur pipette, 3–4 drops of concentrated

---

[1]Dissolve 2.0 g of 2,4-dinitrophenylhydrazine in 50 mL of 85% phosphoric acid by heating, cool, add 50 mL of 95% ethanol, cool again, and clarify by suction filtration from a trace of residue.

FIG. 36.1
A Hirsch funnel filtration apparatus.

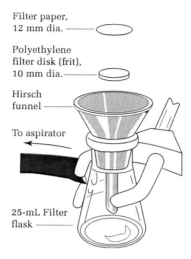

hydrochloric acid as a catalyst and note the result. Warm for a few minutes, then cool and collect the product. This procedure would be used for an aldehyde like cinnamaldehyde ($C_6H_5CH=CHCHO$).

The alternative procedure strikingly demonstrates the catalytic effect of hydrochloric acid, but it is not applicable to a substance like diethyl ketone, whose 2,4-dinitrophenylhydrazone is much too soluble to crystallize from the large volume of ethanol. The first procedure is obviously the one to use for an unknown.

Cleaning Up. The filtrate from the preparation of the 2,4-dinitrophenylhydrazone should have very little 2,4-dinitrophenylhydrazine in it, so after dilution with water and neutralization with sodium carbonate, it can be flushed down the drain. Similarly, the mother liquor from crystallization of the phenylhydrazone should have very little product in it, and so should be diluted and flushed down the drain. If solid material is detected, it should be collected by suction filtration, the filtrate flushed down the drain, and the filter paper placed in the solid hazardous waste.

3. SEMICARBAZONES

$$\underset{\substack{R\\R'}}{\overset{}{C}}=\ddot{O}: \;+\; H_3\overset{+}{N}-\underset{H}{\ddot{N}}-\overset{O}{\overset{\|}{C}}-NH_2 \;Cl^- \;+\; \underset{N}{\bigcirc} \;\longrightarrow\; \underset{R'}{\overset{R}{C}}=\ddot{N}\underset{H}{\overset{}{\ddot{N}}}-\overset{O}{\overset{\|}{C}}-NH_2 \;+\; \underset{\underset{H}{\overset{+}{N}}\;Cl^-}{\bigcirc} \;+\; H_2O$$

Semicarbazide hydrochloride   **Pyridine**   **Semicarbazone**   **Pyridine hydrochloride**

Semicarbazide (mp 96°C) is not very stable in the free form and is used as the crystalline hydrochloride (mp 173°C). Because this salt is insoluble in methanol or ethanol and does not react readily with typical carbonyl compounds in alcohol–water mixtures, pyridine, a basic reagent, is added to liberate free semicarbazide.

To 0.5 mL of the stock solution[2] of semicarbazide hydrochloride, which contains 1 mmol of the reagent, add 1 mmol of the compound to be tested and enough methanol (1 mL) to produce a clear solution; then add 10 drops of pyridine (a twofold excess) and warm the solution gently on the steam bath for a few minutes. Cool the solution slowly to room temperature. It may be necessary to scratch the inside of the test tube in order to induce crystallization. Cool the tube in ice, collect the product by suction filtration, and wash it with water followed by a small amount of cold methanol. Recrystallize the product from methanol, ethanol, or ethanol/water. The product can easily be collected on a Wilfilter.

Cleaning Up. Combine the filtrate from the reaction and the mother liquor from the crystallization, dilute with water, make very slightly acidic with dilute hydrochloric acid, and flush the mixture down the drain.

4. TOLLENS' TEST

Test for aldehydes

$$R-C(=O)H + 2\ Ag(NH_3)_2OH \longrightarrow 2\ Ag + RCOO^-NH_4^+ + H_2O + 3\ NH_3$$

Clean four or five test tubes by adding a few milliliters of 3 M sodium hydroxide solution to each and heating them in a water bath while preparing the Tollens' reagent.

To 2.0 mL of 0.03 M silver nitrate solution, add 1.0 mL of 3 M sodium hydroxide in a test tube. To the gray precipitate of silver oxide, Ag_2O, add 0.5 mL of a 2.8% aqueous ammonia solution (10 mL of concentrated ammonium hydroxide diluted to 100 mL). Stopper the tube and shake it. Repeat the process until *almost* all of the precipitate dissolves (3.0 mL of ammonia at most); then dilute the solution with water to 10 mL. Empty the test tubes of sodium hydroxide solution, rinse them, and add 1 mL of Tollens' reagent to each. Add 1 drop (no more) of the substance to be tested by allowing it to run down the inside of the inclined test tube. Set the tubes aside for a few minutes without agitating the contents. If no reaction occurs, warm the mixture briefly on a water bath. As a known aldehyde, try 1 drop of a 0.1 M solution of glucose. A more typical aldehyde to test is benzaldehyde.

At the end of the reaction, promptly destroy any excess Tollens' reagent with nitric acid because it can form an explosive fulminate on standing. Nitric acid can also be used to remove silver mirrors from test tubes.

Cleaning Up. Place all solutions used in this experiment in a beaker (unused ammonium hydroxide, sodium hydroxide solution used to clean out the tubes, Tollens' reagent from all tubes). Remove any silver mirrors from reaction tubes with a few drops of nitric acid, which is added to the beaker. Make the mixture acidic with nitric acid to destroy unreacted Tollens' reagent and then neutralize the solution with sodium carbonate, and add some sodium chloride solution to precipitate silver chloride (about 40 mg). The whole mixture can be flushed down the drain, or the silver chloride can be collected by suction filtration, and the filtrate flushed down the drain. The silver chloride would go in the nonhazardous solid waste container.

---

[2] Prepare a stock solution by dissolving 1.11 g of semicarbazide hydrochloride in 5 mL of water; 0.5 mL of this solution contains 1 mmol of reagent.

5. SCHIFF'S TEST

Very sensitive test for aldehydes.

Add 3 drops of the unknown to 2 mL of Schiff's reagent.[3] A magenta color will appear within 10 minutes if aldehydes are present. As in all of these tests, compare the colors produced by a known aldehyde, a known ketone, and the unknown compound.

Cleaning Up. Neutralize the solution with sodium carbonate and flush it down the drain. The amount of *p*-rosaniline in this mixture is negligible (1 mg).

6. IODOFORM TEST

$$R-\overset{\overset{\displaystyle \ddot{O}}{\|}}{C}-CH_3 + 3\,I_2 + 4\,OH^- \longrightarrow R-C\overset{\ddot{O}}{\underset{\ddot{O}:^-}{\diagup}} + CHI_3 + 3\,\ddot{\underset{..}{I}}:^- + 3\,H_2O$$

A methyl ketone   **Iodoform**
mp 119–123°C

Test for methyl ketones

The reagent contains iodine in potassium iodide solution[4] at a concentration such that 2 mL of solution, on reaction with excess methyl ketone, will yield 174 mg of iodoform. If the substance to be tested is water soluble, dissolve 4 drops of a liquid or an estimated 50 mg of a solid in 2 mL of water in a 20 × 150-mm test tube; add 2 mL of 3 *M* sodium hydroxide, and then slowly add 3 mL of the iodine solution. In a positive test, the brown color of the reagent disappears, and yellow iodoform separates. If the substance to be tested is insoluble in water, dissolve it in 2 mL of 1,2-dimethoxyethane, proceed as above, and at the end, dilute with 10 mL of water.

Suggested test substances are hexane-2,5-dione (water soluble), *n*-butyraldehyde (water soluble), and acetophenone (water insoluble).

Iodoform can be recognized by its odor and yellow color and, more securely, from its melting point (119°C–123°C). The substance can be isolated by suction filtration of the test suspension or by adding 0.5 mL of dichloromethane, shaking the stoppered test tube to extract the iodoform into the small lower layer, withdrawing the clear part of this layer with a Pasteur pipette, and evaporating it in a small test tube on a steam bath. The crude solid is crystallized from a methanol-water mixture (*see* Chapter 4). It can be collected on a Wilfilter.

Videos: Microscale Filtration on the Hirsch Funnel, Extraction with Dichloromethane. Photo: Use of the Wilfilter

Cleaning Up. Combine all reaction mixtures in a beaker, add a few drops of acetone to destroy any unreacted iodine in the potassium iodide reagent, remove the iodoform by suction filtration, and place the iodoform in the halogenated organic waste container. The filtrate can be flushed down the drain after neutralization (if necessary).

---

[3] Schiff's reagent is prepared by dissolving 0.1 g Basic Fuchsin (*p*-rosaniline hydrochloride) in 100 mL of water and then adding 4 mL of a saturated aqueous solution of sodium bisulfite. After 1 hour, add 2 mL of concentrated hydrochloric acid.
[4] Dissolve 25 g of iodine in a solution of 50 g of potassium iodide in 200 mL of water.

7. BISULFITE TEST

Forms with unhindered carbonyls

$$\underset{H}{\overset{R}{\diagdown}}C=\ddot{O}: \; + \; Na^+SO_3H^- \; \rightleftharpoons \; R-\underset{H}{\overset{\overset{\displaystyle :\ddot{O}H}{|}}{\underset{|}{C}}}-SO_3^-Na^+$$

Put 1 mL of the stock solution[5] into a 13 × 100-mm test tube and add 5 drops of the substance to be tested. Shake each tube during the next 10 minutes, and note the results. A positive test will result from aldehydes, unhindered cyclic ketones such as cyclohexanone, and unhindered methyl ketones.

If the bisulfite test is applied to a liquid or solid that is very sparingly soluble in water, formation of the addition product is facilitated by adding a small amount of methanol before the addition to the bisulfite solution.

Cleaning Up. Dilute the bisulfite solution or any bisulfite addition products (they will dissociate) with a large volume of water, and flush the mixture down the drain. The amount of organic material being discarded is negligible.

8. IR AND NMR SPECTROSCOPY

IR spectroscopy is extremely useful in analyzing all carbonyl-containing compounds, including aldehydes and ketones (Fig. 36.4 and Fig. 36.11). Refer to the extensive discussion in Chapter 11. In the modern laboratory, spectroscopy has almost completely supplanted the qualitative tests described in this chapter. Figures 36.4 through 36.11 present IR and NMR spectra of typical aldehydes and ketones. Compare these spectra with those of your unknown. Also compare the IR and ^{1}H NMR spectra of the hydrocarbon fluorene (Fig. 36.2 and Fig. 36.3) with those of the ketone derivative, fluorenone (Fig. 36.4 and Fig. 36.5).

FIG. 36.2

The IR spectrum of fluorene in CS_2.

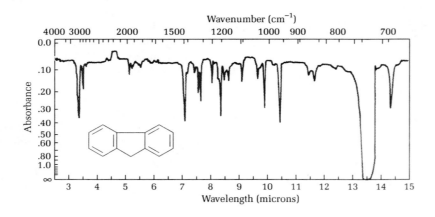

[5]Prepare a stock solution from 50 g of sodium bisulfite dissolved in 200 mL of water with brief swirling.

FIG. 36.3
The ^{1}H NMR spectrum of fluorene (250 MHz).

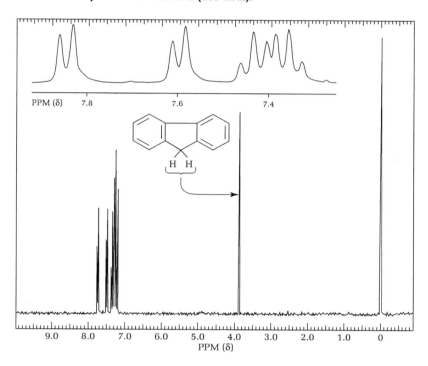

FIG. 36.4
The IR spectrum of fluorenone (KBr disk).

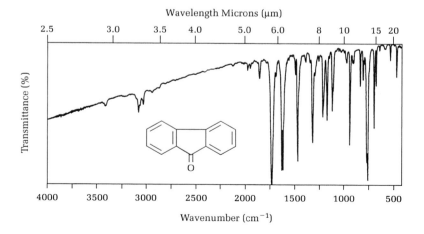

FIG. 36.5
The ¹H NMR spectrum of fluorenone (250 MHz).

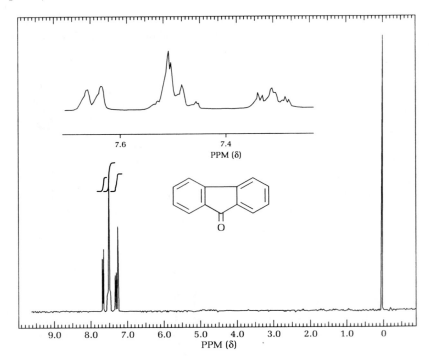

FIG. 36.6
The ¹H NMR spectrum of 2-butanone, $CH_3COCH_2CH_3$ (400 MHz).

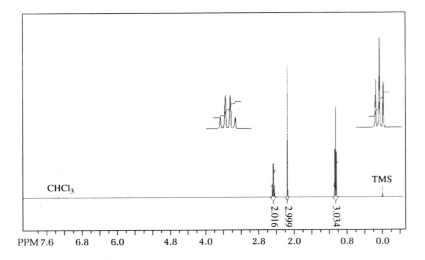

A peak at 9.6–10 ppm in the ¹H NMR spectrum is highly characteristic of aldehydes because almost no other peaks appear in this region (Fig. 36.7 and Fig. 36.10). Similarly, a sharp singlet at 2.2 ppm is very characteristic of methyl ketones; beware of contamination of the sample by acetone, which is often used to clean glassware.

FIG. 36.7
The ¹H NMR spectrum of crotonaldehyde, $CH_3CH(=)CHCHO$ (90 MHz).

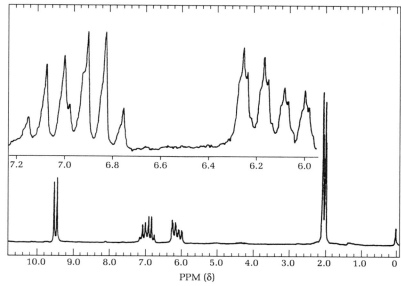

FIG. 36.8
The ¹³C NMR spectrum of 2-butanone, $CH_3COCH_2CH_3$ (100 MHz).

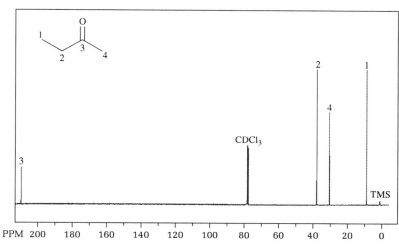

FIG. 36.9
The ¹³C NMR spectrum of crotonaldehyde (22.6 MHz).

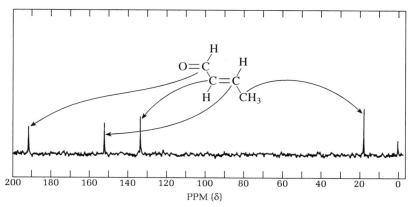

FIG. 36.10

The ¹H NMR spectrum of benzaldehyde (250 MHz).

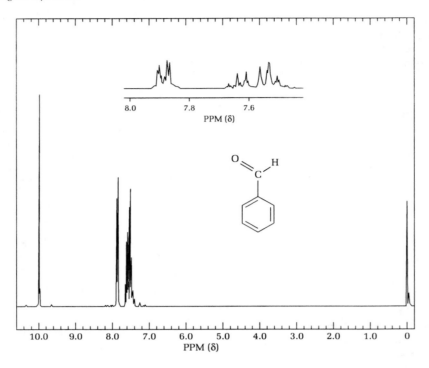

FIG. 36.11

The IR spectrum of benzaldehyde (thin film).

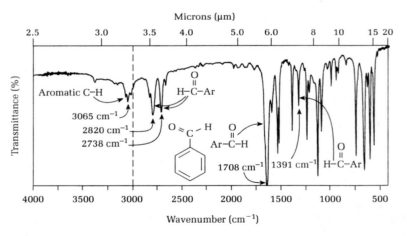

QUESTIONS

1. What is the purpose of making derivatives of unknowns?
2. Why are 2,4-dinitrophenylhydrazones better derivatives than phenylhydrazones?
3. Using chemical tests, how would you distinguish among 2-pentanone, 3-pentanone, and pentanal?

4. Draw the structure of a compound with the empirical formula C_5H_8O that gives a positive iodoform test and does not decolorize permanganate.

5. Draw the structure of a compound with the empirical formula C_5H_8O that gives a positive Tollens' test and does not react with bromine in dichloromethane.

6. Draw the structure of a compound with the empirical formula C_5H_8O that reacts with phenylhydrazine, decolorizes bromine in dichloromethane, and does not give a positive iodoform test.

7. Draw the structure of two geometric isomers with the empirical formula C_5H_8O that give a positive iodoform test.

8. What vibrations cause the peaks at about 3.6 mm (2940 cm^{-1}) in the IR spectrum of fluorene (Fig. 36.2)?

9. Locate the carbonyl peak in Figure 36.4.

10. Assign the various peaks in the ^{1}H NMR spectrum of 2-butanone to specific protons in the molecule (Fig. 36.6).

11. Assign the various peaks in the ^{1}H NMR spectrum of crotonaldehyde to specific protons in the molecule (Fig. 36.7).

CHAPTER 40

Esterification and Hydrolysis

PRELAB EXERCISE: Write the detailed mechanism for the acid-catalyzed hydrolysis of methyl benzoate.

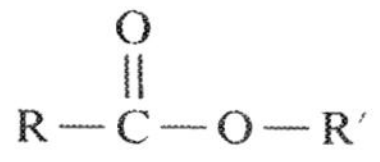

$$R-\overset{\overset{O}{\|}}{C}-O-R'$$

The ester group

Flavors and fragrances

The ester group is an important functional group that can be synthesized in a variety of ways. The low molecular weight esters have very pleasant odors, and indeed comprise the major flavor and odor components of a number of fruits. Although a natural flavor may contain nearly 100 different compounds, single esters approximate natural odors and are often used in the food industry for artificial flavors and fragrances (Table 40.1).

Esters can be prepared by reacting a carboxylic acid with an alcohol in the presence of a catalyst such as concentrated sulfuric acid, hydrogen chloride, p-toluenesulfonic acid, or the acid form of an ion exchange resin. For example, methyl acetate can be prepared as follows:

$$CH_3\overset{\overset{O}{\|}}{C}-OH + CH_3OH \underset{}{\overset{H^+}{\rightleftharpoons}} CH_3\overset{\overset{O}{\|}}{C}-OCH_3 + H_2O$$

Acetic acid   **Methanol**   **Methyl acetate**

Fischer esterification

The Fischer esterification reaction reaches equilibrium after a few hours of refluxing. The position of the equilibrium can be shifted by adding more of the acid or of the alcohol, depending on cost or availability. The mechanism of the reaction involves initial protonation of the carboxyl group, attack by the nucleophilic hydroxyl, a proton transfer, and loss of water followed by loss of the catalyzing proton to give the ester. Each of these steps is completely reversible, so this process is also, in reverse, the mechanism for the hydrolysis of an ester:

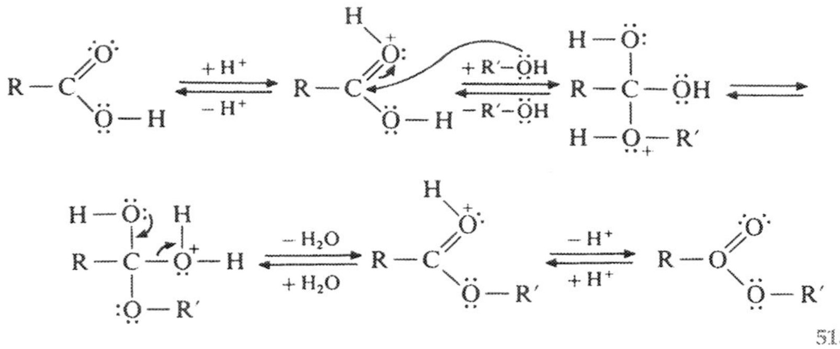

515

516   Macroscale and Microscale Organic Experiments

TABLE 40.1 Boiling Points and Fragrances of Esters

Ester		Formula	bp (°C)	Fragrance
2-Methylpropyl formate	(a)	HC(=O)—OCH$_2$CH(CH$_3$)CH$_3$	98.4	Raspberry
1-Propyl acetate	(b)	CH$_3$C(=O)—OCH$_2$CH$_2$CH$_3$	101.7	Pear
Methyl butyrate	(c)	CH$_3$CH$_2$CH$_2$C(=O)—OCH$_3$	102.3	Apple
Ethyl butyrate	(d)	CH$_3$CH$_2$CH$_2$C(=O)—OCH$_2$CH$_3$	121	Pineapple
2-Methylpropyl propionate	(e)	CH$_3$CH$_2$C(=O)—OCH$_2$CH(CH$_3$)CH$_3$	136.8	Rum
3-Methylbutyl acetate	(f)	CH$_3$C(=O)—OCH$_2$CH$_2$CH(CH$_3$)CH$_3$	142	Banana
Benzyl acetate	(g)	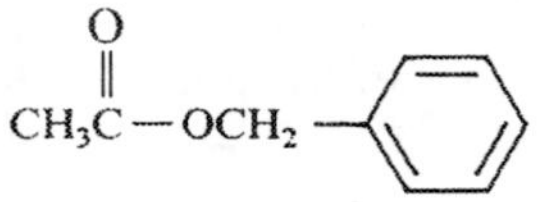 CH$_3$C(=O)—OCH$_2$—C$_6$H$_5$	213.5	Peach
Octyl acetate	(h)	CH$_3$C(=O)—OCH$_2$(CH$_2$)$_6$CH$_3$	210	Orange
Methyl salicylate	(i)	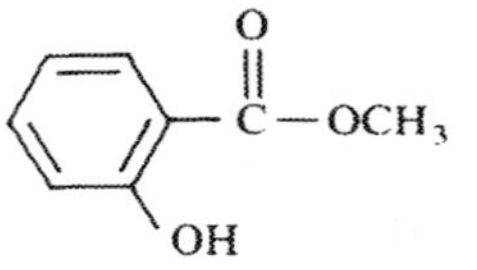 2-HO-C$_6$H$_4$-C(=O)—OCH$_3$	222	Wintergreen

Other methods are available for synthesizing esters, most of which are more expensive but readily carried out on a small scale. For example, alcohols react with anhydrides and with acid chlorides:

$$CH_3CH_2OH + CH_3C(=O)-O-C(=O)CH_3 \longrightarrow CH_3C(=O)-OCH_2CH_3 + CH_3C(=O)-OH$$
Ethanol   **Acetic anhydride**   **Ethyl acetate**   **Acetic acid**

$$CH_3CH_2CH_2OH + CH_3C(=O)-Cl \longrightarrow CH_3C(=O)-OCH_2CH_2CH_3 + HCl$$
1-Propanol   **Acetyl chloride**   ***n*-Propyl acetate**

Other ester syntheses

In the latter reaction, an organic base such as pyridine is usually added to react with the hydrogen chloride.

Other methods can also be used to synthesize the ester group. Among these are the addition of 2-methylpropene to an acid to form *t*-butyl esters, the addition of ketene to make acetates, and the reaction of a silver salt with an alkyl halide:

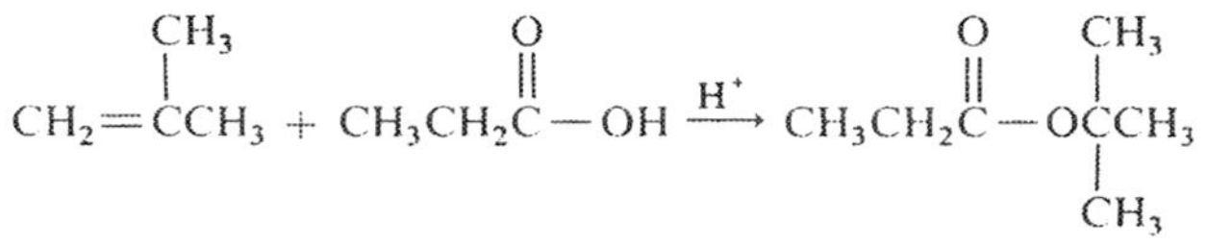

$$CH_2=\overset{\overset{\displaystyle CH_3}{|}}{C}CH_3 + CH_3CH_2\overset{\overset{\displaystyle O}{\|}}{C}-OH \xrightarrow{H^+} CH_3CH_2\overset{\overset{\displaystyle O}{\|}}{C}-O\overset{\overset{\displaystyle CH_3}{|}}{\underset{\underset{\displaystyle CH_3}{|}}{C}}CH_3$$

2-Methylpropene **Propionic acid** ***t*-Butyl propionate**
(isobutylene)

$$CH_2=C=O + HOCH_2-\text{C}_6\text{H}_5 \longrightarrow CH_3\overset{\overset{\displaystyle O}{\|}}{C}-OCH_2-\text{C}_6\text{H}_5$$

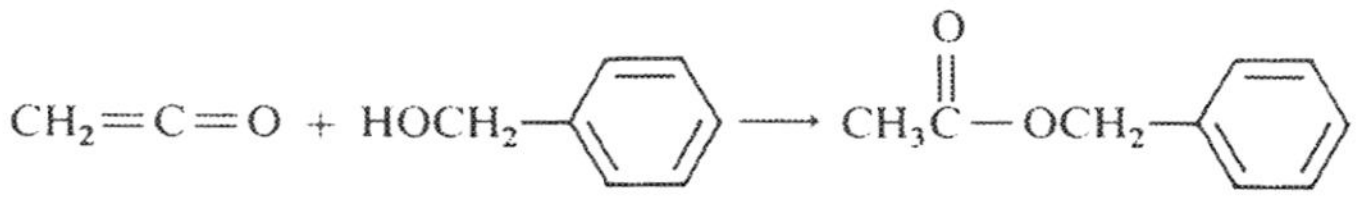

Ketene **Benzyl alcohol** **Benzyl acetate**

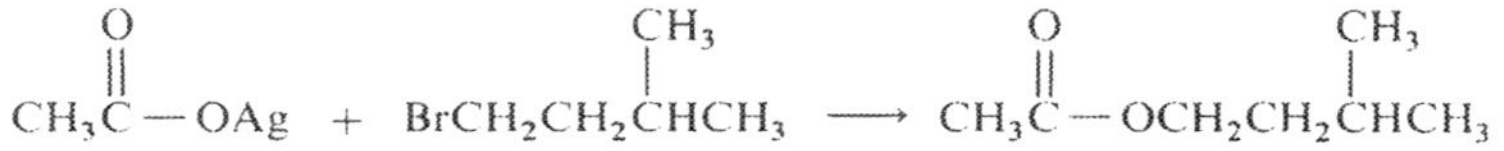

$$CH_3\overset{\overset{\displaystyle O}{\|}}{C}-OAg + BrCH_2CH_2\overset{\overset{\displaystyle CH_3}{|}}{C}HCH_3 \longrightarrow CH_3\overset{\overset{\displaystyle O}{\|}}{C}-OCH_2CH_2\overset{\overset{\displaystyle CH_3}{|}}{C}HCH_3$$

Silver acetate **1-Bromo-3-methylbutane** **3-Methylbutyl acetate**

As noted previously, Fischer esterification is an equilibrium process. Consider the reaction of acetic acid with 1-butanol to give *n*-butyl acetate:

$$CH_3\overset{\overset{\displaystyle O}{\|}}{C}-OH + HOCH_2CH_2CH_2CH_3 \underset{}{\overset{H^+}{\rightleftharpoons}} CH_3\overset{\overset{\displaystyle O}{\|}}{C}-OCH_2CH_2CH_2CH_3 + H_2O$$

The equilibrium constant is as follows:

$$K_{eq} = \frac{[n\text{-BuOAc}][H_2O]}{[n\text{-BuOH}][HOAc]}$$

For primary alcohols reacting with unhindered carboxylic acids, $K_{eq} \approx 4$. If equal quantities of 1-butanol and acetic acid are allowed to react at equilibrium, the theoretical yield of ester is only 67%. To upset the equilibrium we can, by Le Chatelier's principle, increase the concentration of either the alcohol or acid. If either one is doubled, the theoretical yield increases to 85%. When one is tripled, the yield goes to 90%. But note that in the example cited, the boiling point of the relatively nonpolar ester is only about 8°C higher than the boiling points of the polar acetic acid and 1-butanol, so a difficult separation problem exists if the product must be isolated by distillation after the starting materials are increased in concentration.

TABLE 40.2 *The Ternary Azeotrope of Boiling Point 90.7°C*

Compound	Boiling Point of Pure Compound (°C)	Percentage Composition of Azeotrope		
		Vapor Phase	Upper Layer	Lower Layer
1-Butanol	117.7	8.0	11.0	2.0
n-Butyl acetate	126.7	63.0	86.0	1.0
Water	100.0	29.0	3.0	97.0

Another way to upset the equilibrium is to remove water. This can be done by adding to the reaction mixture molecular sieves (an artificial zeolite), which preferentially adsorb water. Most other drying agents, such as anhydrous sodium sulfate or calcium chloride pellets, will not remove water at the temperatures used to make esters.

Azeotropic distillation

A third way to upset the equilibrium is to preferentially remove the water as an azeotrope (a constant-boiling mixture of water and an organic liquid). The information in Table 40.2 can be found in a chemistry handbook table of ternary (three-component) azeotropes. These data tell us that vapor that distills from a mixture of 1-butanol, *n*-butyl acetate, and water will boil at 90.7°C and that the vapor contains 8% alcohol, 63% ester, and 29% water. The vapor is homogeneous, but when it condenses, it separates into two layers. The upper layer is composed of 11% alcohol, 86% ester, and 3% water, but the lower layer consists of 97% water with only traces of alcohol and ester. If some ingenious way can be devised to remove the lower layer from the condensate and still return the upper layer to the reaction mixture, then the equilibrium can be upset, and nearly 100% of the ester can be produced in the reaction flask.

The apparatus shown in Figure 40.1, modeled after that of Ernest W. Dean and David D. Stark, achieves the desired separation of the two layers. The mixture of equimolar quantities of 1-butanol and acetic acid is placed in the flask along with an acid catalyst. Stirring reduces bumping. The vapor, the temperature of which is 90.7°C, condenses and runs down to the sidearm, which is closed with a cork. The layers separate, with the denser water layer remaining in the sidearm, and the lighter ester plus alcohol layer running down into the reaction flask. As soon as the theoretical quantity of water has collected, the reaction is over, and the product in the flask should contain an ester of high purity. The macroscale apparatus is illustrated in Figure 40.2.

Esterification using a carboxylic acid and an alcohol requires an acid catalyst. In the first experiment, the acid form of an ion-exchange resin is used. This resin, in the form of small beads, is a cross-linked polystyrene that bears sulfonic acid groups on some of the phenyl groups. It is essentially an immobilized form of *p*-toluenesulfonic acid, an organic-substituted sulfuric acid.

FIG. 40.1
A microscale Dean-Stark azeotropic esterification apparatus. A cork is used instead of a septum so that layer separation can be observed clearly.

FIG. 40.2
A macroscale azeotropic distillation apparatus, with a Dean-Stark trap where water collects.

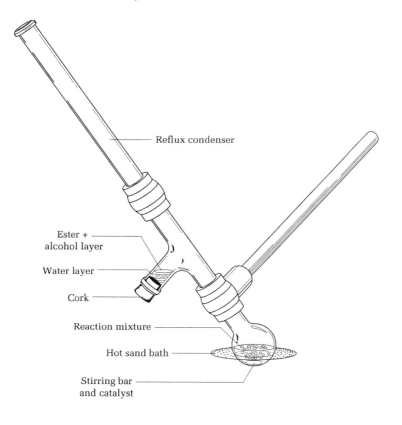
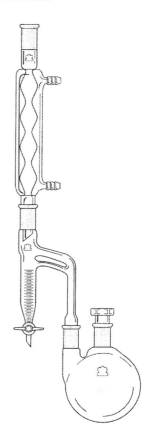

An ion-exchange catalyst

This catalyst has the distinct advantage that at the end of the reaction it can be easily removed by filtration. Immobilized catalysts of this type are becoming more and more common in organic synthesis.

If concentrated sulfuric acid were used as the catalyst, it would be necessary to dilute the reaction mixture with ether; wash the ether layer successively with water, sodium carbonate solution, and saturated sodium chloride solution; and then dry the ether layer with anhydrous calcium chloride pellets before evaporating the ether.

into a 50-mL round-bottomed flask, attach a stillhead, dry out the ordinary condenser and use it without water circulating in the jacket, and distill. The boiling point of the ester is so high (199°C) that a water-cooled condenser is liable to crack. Use a tared 25-mL Erlenmeyer flask as the receiver to collect material boiling above 190°C. A typical student yield is about 7 g. *See* Chapter 28 for the nitration of methyl benzoate, and Chapter 38 for its use in the Grignard synthesis of triphenylmethanol.

IR and nuclear magnetic resonance (NMR) spectra of benzoic acid and methyl benzoate are found at the end of the chapter (Figs. 40.6 to 40.11).

Cleaning Up. Pour the sulfuric acid layer into water, combining it with the bicarbonate layer, neutralize it with sodium carbonate, and flush the solution down the drain with excess water. The saturated sodium chloride layer can also be flushed down the drain. If the calcium chloride is free of ether and methyl benzoate, it can be placed in the nonhazardous solid waste container; otherwise it must go into the hazardous waste container. Ether goes into the organic solvents waste container, along with the pot residues from the final distillation.

5. HYDROLYSIS (SAPONIFICATION): THE PREPARATION OF SOAP

In general, the reversal of esterification is called *hydrolysis*. In the case of the hydrolysis of a fatty acid ester, it is called saponification (soap making). In this experiment, the saturated fat made from hydrogenated olive oil in Chapter 25 will be saponified to give a soap, which, in this case, will be primarily sodium stearate.

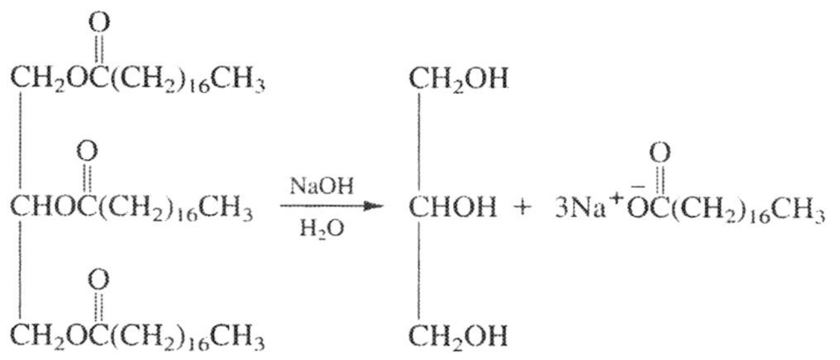

$$\begin{array}{c}
\text{CH}_2\text{OC(CH}_2)_{16}\text{CH}_3 \\
| \\
\text{O} \\
\| \\
\text{CHOC(CH}_2)_{16}\text{CH}_3 \\
| \\
\text{O} \\
\| \\
\text{CH}_2\text{OC(CH}_2)_{16}\text{CH}_3
\end{array} \xrightarrow[\text{H}_2\text{O}]{\text{NaOH}} \begin{array}{c} \text{CH}_2\text{OH} \\ | \\ \text{CHOH} \\ | \\ \text{CH}_2\text{OH} \end{array} + 3\text{Na}^+\overset{O}{\overset{\|}{\text{O}}}\text{C(CH}_2)_{16}\text{CH}_3$$

Glycerol tristearate   **Glycerol**   **Sodium stearate (soap)**

 Microscale Procedure

> IN THIS EXPERIMENT, a saturated fat is heated with sodium hydroxide in a water-ethanol mixture. The resulting soap is precipitated in a salt solution and collected on a Hirsch funnel, where it is washed free of base and salt.

Video: Microscale Filtration on the Hirsch Funnel

Place 0.18 g of the saturated triglyceride prepared in Chapter 25, Experiment 4, in a 5-mL short-necked, round-bottomed flask. Add 1.5 mL of a 50:50 water-ethanol solution that contains 0.18 g of solid sodium hydroxide (weigh this quickly). Add an air condenser and gently reflux the mixture on a sand bath for 30 minutes, taking care not to boil away the ethanol. At the end of the reaction period, some of the soap will have precipitated. Transfer the mixture to a 10-mL Erlenmeyer flask containing a solution of 0.8 g of sodium chloride in 3 mL of water. Collect the precipitated soap on a Hirsch funnel and wash it free of excess sodium hydroxide and salt using 4 mL of distilled ice water.

Test the soap by adding a very small piece (about 5–15 mg) to a centrifuge or test tube along with 3–4 mL of distilled water. Cap the tube and shake it vigorously. Note the height and stability of the bubbles. Add a crystal of magnesium chloride or calcium chloride to the tube. Shake the tube again and note the results. For comparison, do these same tests with a few grains of a detergent instead of the soap.

Macroscale Procedure

In a 100-mL round-bottomed flask, place 5 g of hydrogenated olive oil (Chapter 25) or lard or solid shortening (e.g., Crisco). Add 20 mL of ethanol and a hot solution of 5 g of sodium hydroxide in 20 mL of water. This solution, prepared in a beaker, will become very hot as the sodium hydroxide dissolves. Fit the flask with a water-cooled condenser, and reflux the mixture on a sand bath for 30 minutes. At the end of the reaction period, some of the soap may precipitate from the reaction mixture. Transfer the mixture to a 250-mL Erlenmeyer flask containing an ice-cold solution of 25 g of sodium chloride in 90 mL of distilled water. Collect the precipitated soap on a Büchner funnel and wash it free of excess sodium hydroxide and salt using no more than 100 mL of distilled ice water.

Test the soap by adding a small piece to a test tube along with about 5 mL of water. Cap the tube and shake it vigorously. Note the height and stability of the bubbles formed. Add a few crystals of magnesium chloride or calcium chloride to the tube. Shake the tube again and note the results. For comparison, do these same tests with an amount of detergent equal to that of the soap used.

QUESTIONS

For Additional Experiments, sign in at this book's premium website at www.cengage.com/login.

1. In the preparation of methyl benzoate, what is the purpose of (a) washing the organic layer with sodium bicarbonate solution? (b) washing the organic layer with saturated sodium chloride solution? (c) treating the organic layer with anhydrous calcium chloride pellets?

2. Assign the resonances in Figure 40.8 to specific protons in methyl benzoate.

3. Figures 40.9 and 40.10 each have two resonances that are very small. What do the carbons causing these peaks have in common?

CHAPTER 53

The Benzoin Condensation: Catalysis by the Cyanide Ion and Thiamine

When you see this icon, sign in at this book's premium website at www.cengage.com/login to access videos, Pre-Lab Exercises and other online resources.

> **PRELAB EXERCISE:** What purposes does sodium hydroxide serve in the thiamine-catalyzed benzoin condensation?

The reaction of 2 mol of benzaldehyde to form a new carbon-carbon bond is known as the *benzoin condensation*. It is catalyzed by two rather different catalysts—the cyanide ion and thiamine (a B-complex vitamin)—which, on close examination, seem to function in exactly the same way.

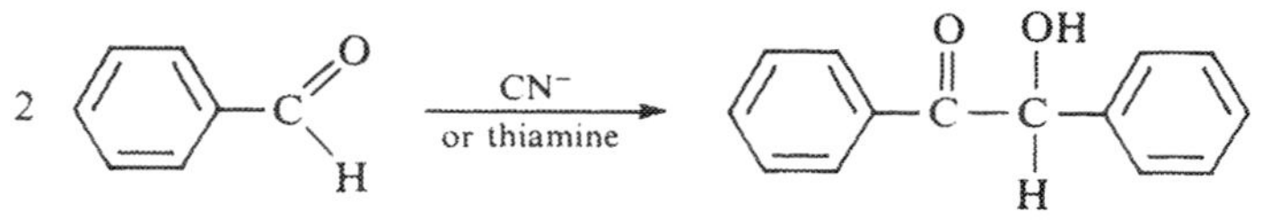

$$2 \; PhCHO \xrightarrow{CN^- \text{ or thiamine}} Ph-C(O)-CH(OH)-Ph$$

Benzaldehyde
MW 106.12
bp 178°C, den. 1.044

Benzoin
MW 212.24
mp 135°C

Consider the reaction catalyzed by the cyanide ion. The cyanide ion attacks the carbonyl oxygen to form a stable cyanohydrin, mandelonitrile, a liquid with a boiling point of 170°C that under the basic conditions of the reaction, loses a proton to give a resonance-stabilized carbanion (**A**). The carbanion attacks another molecule of benzaldehyde to give compound **B**, which undergoes a proton transfer and loses cyanide to give benzoin. Evidence for this mechanism lies in the failure of 4-nitrobenzaldehyde to undergo the reaction because the nitro group reduces the nucleophilicity of the anion in compound **A**. On the other hand, a strong electron-donating group in the 4-position of the phenyl ring makes the loss of the proton from the cyanohydrin very difficult; thus 4-dimethylaminobenzaldehyde also does not undergo the benzoin condensation with itself.

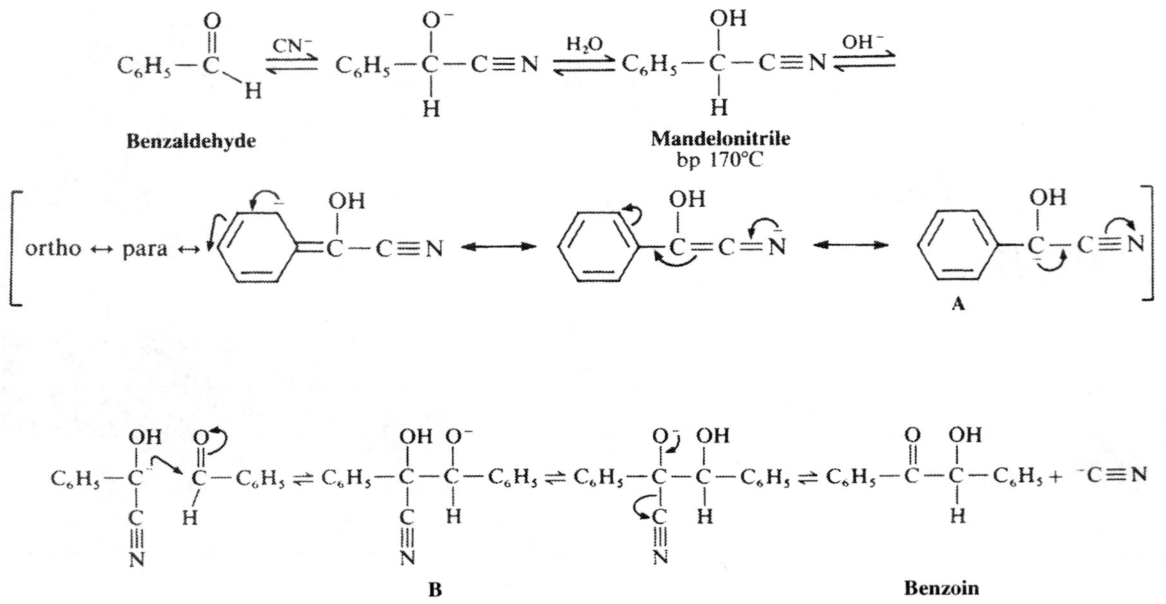

$$C_6H_5-\overset{O}{\underset{H}{\overset{\|}{C}}} \xrightarrow{CN^-} C_6H_5-\overset{O^-}{\underset{H}{\overset{|}{C}}}-C\equiv N \xrightarrow{H_2O} C_6H_5-\overset{OH}{\underset{H}{\overset{|}{C}}}-C\equiv N \xrightarrow{OH^-}$$

Benzaldehyde **Mandelonitrile** bp 170°C

$$\left[\text{ortho} \leftrightarrow \text{para} \leftrightarrow \cdots \right]$$

(resonance structures leading to A)

$$C_6H_5-\overset{OH}{\underset{\underset{N}{\overset{\|}{C}}}{\overset{|}{C}}}-\overset{O}{\overset{\|}{C}}-C_6H_5 \rightleftharpoons C_6H_5-\overset{OH}{\underset{\underset{N}{\overset{\|}{C}}}{\overset{|}{C}}}-\overset{O^-}{\underset{H}{\overset{|}{C}}}-C_6H_5 \rightleftharpoons C_6H_5-\overset{O^-}{\underset{\underset{N}{\overset{\|}{C}}}{\overset{|}{C}}}-\overset{OH}{\underset{H}{\overset{|}{C}}}-C_6H_5 \rightleftharpoons C_6H_5-\overset{O}{\overset{\|}{C}}-\overset{OH}{\underset{H}{\overset{|}{C}}}-C_6H_5 + {}^-C\equiv N$$

B **Benzoin**

The cyanide ion binds irreversibly to hemoglobin, rendering it useless as a carrier of oxygen.

Several biochemical reactions bear a close resemblance to the benzoin condensation, but are not obviously catalyzed by the highly toxic cyanide ion. Approximately 30 years ago, Ronald Breslow proposed that vitamin B_1 (also known as thiamine hydrochloride), in the form of the coenzyme thiamine pyrophosphate, can function in a manner completely analogous to the cyanide ion in promoting reactions like the benzoin condensation. The resonance-stabilized conjugate base of the thiazolium ion, thiamine, and the resonance-stabilized carbanion (C), which it forms, are the keys to the reaction. Like the cyanide ion, the thiazolium ion has just the right balance of nucleophilicity, the ability to stabilize the intermediate anion, and good leaving group qualities.

In the reactions that follow, the cyanide ion functions as a fast and efficient catalyst, although in large quantities it is highly toxic. The amount of potassium cyanide used in the first experiment (15 mg) is about eight times lower than the average fatal dose, a difference that underlines the advantage of carrying out organic experiments on a microscale.

The importance of thiamine is evident in that it is a vitamin, an essential substance that must be provided in the diet to prevent beriberi, a nervous system disease. The thiamine-catalyzed reaction is much slower, but the catalyst is completely nontoxic.

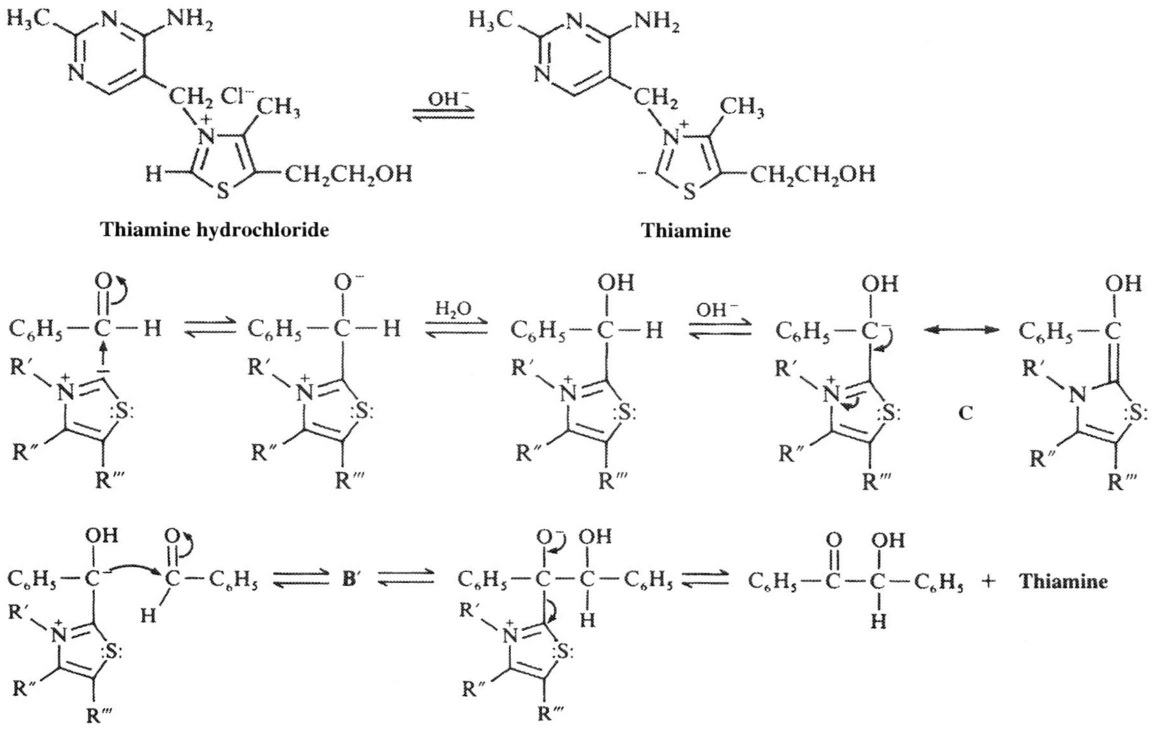

Chapter 53 ■ *The Benzoin Condensation: Catalysis by the Cyanide Ion and Thiamine* 657

Thiamine hydrochloride ⇌ (OH⁻) **Thiamine**

Thiamine-Catalyzed Benzoin Condensation of Benzaldehyde

Adapted from: Kenneth Williamson and Katherine Masters, Macroscale and Microscale Organic Experiments, 6th ed., Brooks/Cole, Cengage Learning, Belmont, CA USA, 2011. Ch 53, p. 659-60, by M. J. Lusch

Macroscale Procedure: (*room temperature* reaction)

First Lab Period: Weigh **1.30 g of thiamine hydrochloride** (catalyst) into a **50 mL** Erlenmeyer flask, add **4.0 mL of distilled water** to dissolve it, and then add **15 mL of 95% EtOH**. Swirl the flask to mix thoroughly and then cool the solution in an **ice-water** bath. Insert a thermometer into the solution and add **2.5 mL of 3M aq. NaOH** *dropwise* with swirling or stirring, keeping the temperature below 20°C. When the addition of base is complete, add **7.5 mL of benzaldehyde** in one portion and swirl to mix the solution thoroughly. Securely cork (#6) the flask (to reduce the amount of air oxidation of the benzaldehyde), place it in a small beaker to keep it from being accidentally tipped over, and allow it to stand in your equipment drawer until the next lab period (or for at least 24 hours).

Second Lab Period: At the beginning of the next lab period, cool the reaction mixture in an ice-water bath. If crystallization has not already occurred or does not occur with cooling, withdraw a drop of the solution on the tip of a glass stirring rod, and rub it against the inside surface of the flask slightly above the surface of the liquid to induce crystallization. After no more crystallization occurs (about 10 minutes in the ice bath), collect the crystallized product by suction filtration with a Büchner funnel, and wash it free of the yellow mother liquor with an ice cold 1:1 mixture of 95% EtOH and water. Draw air through the funnel by suction for 10-15 min to predry the collected product, then take a 1.0 g sample of the damp solid and recrystallize it in a 10 mL Erlenmeyer flask (from the microkit) from approximately 8.0 mL of *boiling* 95% EtOH. Collect the recrystallized product using the Hirsch suction funnel (from the microkit), wash it with ice-cold 95% EtOH, and again predry on the suction filter. Let the damp products (crude and recrystallized) dry completely in your equipment drawer until the next lab period, at which time you should determine the actual yield of the *combined* product samples and the melting point *range* of each of the two samples of the product. The product should be colorless; the usual yield is 5-6 grams.

Technical Notes

1. **Thiamine is sensitive to heat and excessively alkaline pH**; *therefore*
   - Do not allow the temperature to rise above 20°C during the addition of the NaOH solution.
   - Do not add more base than is called for.
2. Cork the reaction flask securely to reduce the amount of **air oxidation** of the benzaldehyde.
3. Removal of the **mother liquor** from solid products during filtration is extremely import to insure purity. Therefore, during suction filtration, wash the collected solid thoroughly with *ice-cold* solvent (which can also be used to remove traces of product from the reaction or recrystallization container) before allowing the product to dry.

Waste Disposal:

1. The aqueous-ethanol filtrate from the isolation of the crude product should be neutralized to pH 5–7, and then diluted with water and flushed down the drain (what does this filtrate contain besides water and EtOH?).
2. Ethanol filtrates from the recrystallization (containing dissolved benzoin) should be placed in the **Non-Halogenated Waste** containers next to the sinks.

CHAPTER 54

Nitric Acid Oxidation; Preparation of Benzil from Benzoin; and Synthesis of a Heterocycle: Diphenylquinoxaline

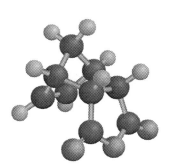

PRELAB EXERCISE: Write a detailed mechanism for the formation of 2,3-dimethylquinoxaline.

Benzoin can be oxidized to benzil, an α-diketone, very efficiently by nitric acid or by copper(II) sulfate in pyridine. When oxidized with sodium dichromate in acetic acid, the yield is lower because some of the material is converted into benzaldehyde by cleavage of the bond between two oxidized carbon atoms that is activated by both phenyl groups (a). Similarly, when hydrobenzoin is oxidized with dichromate or permanganate, it yields chiefly benzaldehyde and only a trace of benzil (b).

Ultraviolet (UV) spectroscopy is used to help characterize aromatic molecules such as benzoin. The absorption band at 247 nm in Figure 54.1 is attributable to the presence of the phenyl ketone group, in which the carbonyl group is conjugated with a benzene ring. Aliphatic αβ-unsaturated ketones, R—CH=CH—C=O, show selective absorption of UV light of comparable wavelength. See the infrared (IR) spectrum of benzoin in Figure 54.2.

FIG. 54.1

The UV spectrum of benzoin. λ_{max}^{EtOH} = 247 nm (ε = 13,200). Concentration: 12.56 mg/L = 5.92 × 10^{-5} mol/L. *See* Chapter 14 for the relationship between the extinction coefficient (ε), absorbance (*A*), and concentration (*C*).

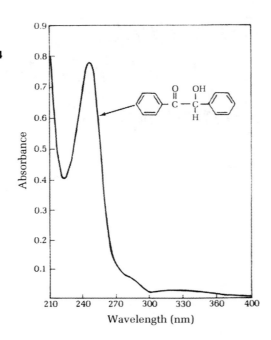

FIG. 54.2

The IR spectrum of benzoin (KBr disk).

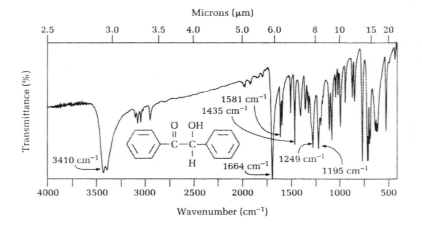

Chapter 54 ■ Nitric Acid Oxidation; Preparation of Benzil from Benzoin

EXPERIMENTS

1. NITRIC ACID OXIDATION OF BENZOIN

Benzoin
MW 212.24, mp 135°C

Benzil
MW 210.23, mp 94–95°C

CAUTION: Handle nitric acid with great care. It is highly corrosive to tissue and a very strong oxidant.

IN THIS EXPERIMENT, the hydroxyl group of benzoin is oxidized with hot nitric acid to a carbonyl group to give benzil, which is yellow in color. The product is collected on a Hirsch funnel, washed with water, and recrystallized from ethanol.

Microscale Procedure

Reaction time: 11 minutes

Heat a mixture of 100 mg of benzoin and 0.35 mL of concentrated nitric acid on a steam bath or in a small beaker of boiling water for about 11 minutes. Carry out the reaction in the hood or use an aspirator tube near the top of the tube to remove nitrogen oxides. Be sure all of the benzoin is washed down inside the tube and is oxidized. Add 2 mL of water to the reaction mixture, cool to room temperature, and stir the mixture for 1–2 minutes to coagulate the precipitated product. Remove the solvent with a Pasteur pipette and wash the solid with an additional 2 mL of water. Dissolve the solid in 0.5 mL of hot ethanol and add water dropwise to the hot solution until the solution appears to be cloudy, indicating that it is saturated. Heat to bring the product completely into solution and allow it to cool slowly to room temperature. Cool the tube in ice and isolate the product using a Wilfilter or a Hirsch funnel. Scrape the benzil onto a piece of filter paper, squeeze out excess solvent, and allow the solid to dry. Record the percent yield, the crystalline form, the color, and the melting point of the product.

Photos: *Filtration Using a Pasteur Pipette, Use of the Wilfilter*; Videos: *Filtration of Crystals, Using the Pasteur Pipette, Microscale Filtration on the Hirsch Funnel*

Cleaning Up. The aqueous filtrate should be neutralized with sodium carbonate, diluted with water, and flushed down the drain. Place the ethanol used in crystallization in the organic solvents waste container.

Macroscale Procedure

Reaction time: 11 minutes

Heat a mixture of 4 g of benzoin and 14 mL of concentrated nitric acid on a steam bath for about 11 minutes. Carry out the reaction in a hood or use an aspirator tube near the top of the flask to remove nitrogen oxides. Add 75 mL of water to the reaction mixture, cool to room temperature, and swirl for 1–2 minutes to coagulate the precipitated product. Collect and wash the yellow solid on a Hirsch funnel, pressing the solid on the filter to squeeze out the water. This crude product (the dry weight yield is 3.7–3.9 g) need not be dried, but can be crystallized at once from ethanol. Dissolve the product in 10 mL of hot ethanol, add water dropwise to the cloud point, and set aside to crystallize. Record the yield, the crystalline form,

Video: *Macroscale Crystallization*

the color, and the melting point of the purified benzil. The IR and ^{1}H NMR spectra are found at the end of the chapter (Fig. 54.5 and Fig. 54.6).

A Test for the Presence of Unoxidized Benzoin

Dissolve about 0.5 mg of crude or purified benzil in 0.5 mL of 95% ethanol or methanol, and add 1 drop of 3 M sodium hydroxide. If benzoin is present, the solution soon acquires a purplish color due to a complex of benzil with a product of autoxidation of benzoin. If no color develops in 2–3 minutes, which is an indication that the sample is free of benzoin, add a small amount of benzoin. Observe the color that develops, and note that if the test tube is stoppered and shaken vigorously, the color momentarily disappears. When the solution is allowed to stand, the color reappears.

Cleaning Up. The aqueous filtrate should be neutralized with sodium carbonate, diluted with water, and flushed down the drain. Ethanol used in crystallization should be placed in the organic solvents waste container.

55 CHAPTER

The Borohydride Reduction of a Ketone: Hydrobenzoin from Benzil

PRELAB EXERCISE: Compare the reducing capabilities of lithium aluminum hydride to those of sodium borohydride.

In 1943, H. I. Schlesinger and Herbert C. Brown discovered sodium borohydride. Brown devoted his entire scientific career to this reagent, making it and other hydrides the most useful and versatile of reducing reagents. He received a Nobel Prize in Chemistry in 1979 for his work.

Considering the extreme reactivity of most hydrides (such as sodium hydride and lithium aluminum hydride) toward water, sodium borohydride is somewhat surprisingly solid as a stabilized aqueous solution in 14 M sodium hydroxide containing 12% sodium borohydride. Unlike lithium aluminum hydride, sodium borohydride is insoluble in ether and soluble in methanol and ethanol.

Sodium borohydride is a mild and selective reducing reagent. In ethanol solution, it reduces aldehydes and ketones rapidly at 25°C and esters very slowly, and it is inert toward functional groups that are readily reduced by lithium aluminum hydride, including carboxylic acids, epoxides, lactones, nitro groups, nitriles, azides, amides, and acid chlorides.

The present experiment is a typical sodium borohydride reduction. These same conditions and isolation procedures could be applied to hundreds of other ketones and aldehydes.

EXPERIMENT

SODIUM BOROHYDRIDE REDUCTION OF BENZIL

The addition of two atoms of hydrogen to benzoin or four atoms of hydrogen to benzil gives a mixture of stereoisomeric diols, of which the predominant isomer is the nonresolvable (1R,2S)-hydrobenzoin, the *meso* isomer, accompanied by the enantiomeric (1R,2R) and (1S,2S) compounds.

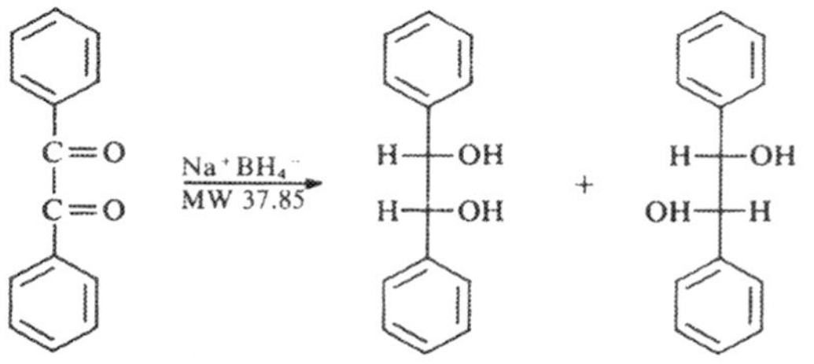

Chapter 55 ■ The Borohydride Reduction of a Ketone: Hydrobenzoin from Benzil

Benzil MW 210.22

(1R,2S)-(meso)-Hydrobenzoin mp 137°C, MW 214.25

(1R,2R) and (1S,2S)-Hydrobenzoin mp 120°C

The reaction proceeds rapidly at room temperature; the intermediate borate ester is hydrolyzed with water to give the product alcohol.

$$4\ R_2C=O + Na^+BH_4^- \longrightarrow (R_2CHO)_4B^-Na^+$$

$$(R_2CHO)_4B^-Na^+ + 2\ H_2O \longrightarrow 4\ R_2CHOH + Na^+BO_2^-$$

The procedure that follows specifies the use of benzil rather than benzoin, because you can then follow the progress of the reduction by the discharge of the yellow color of the benzil.

Chapter 54 contains the infrared (IR) and nuclear magnetic resonance (NMR) spectra for benzil (Fig. 54.5 and Fig. 54.6, respectively). The ultraviolet (UV) spectrum is shown in Figure 55.2 at the end of this chapter.

Microscale Procedure

> **IN THIS EXPERIMENT,** a solid diketone dissolved in ethanol is reduced with solid sodium borohydride to a dialcohol that crystallizes from the reaction mixture and is isolated by filtration. It is easy to follow the progress of the reaction—the diketone is yellow, and the dialcohol is colorless in solution.

Photos: Filtration Using a Pasteur Pipette, Use of the Wilfilter; Videos: Filtration of Crystals Using the Pasteur Pipette, Microscale Crystallization

In a 10 × 100-mm reaction tube, dissolve 50 mg of benzil in 0.5 mL of 95% ethanol; then cool the solution in ice to produce a fine suspension. Add to this suspension 10 mg of sodium borohydride (a large excess). The benzil dissolves, the mixture warms up, and the yellow color disappears in 2–3 minutes. After a total of 10 minutes, add 0.5 mL of water, heat the solution to the boiling point, filter the solution in case it is not clear (usually not necessary), and dilute the hot solution with hot water to the point of saturation (cloudiness)—a process that requires about 1 mL of water. (1R,2S)-Hydrobenzoin separates in lustrous thin plates (mp 136–137°C) and is best isolated by withdrawing some of the solvent using a Pasteur pipette and then using a Wilfilter (Fig. 55.1). It also can be collected on a microscale Büchner funnel. The yield is about 35 mg.

Cleaning Up. Dilute the aqueous filtrate with water, neutralize with acetic acid (to destroy the borohydride), and then flush the mixture down the drain.

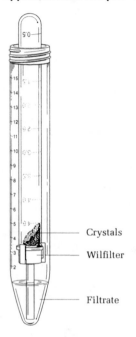

FIG. 55.1
The Wilfilter filtration apparatus. *See* Chapter 4.

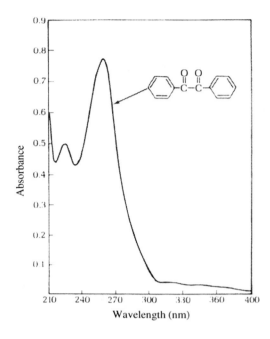

FIG. 55.2
The UV spectrum of benzil. $\lambda_{max}^{EtOH} = 260$ nm ($\varepsilon = 19{,}800$). One-centimeter cells and 95% ethanol have been employed for this spectrum.

Macroscale Procedure

In a 50-mL Erlenmeyer flask, dissolve 0.5 g of benzil in 5 mL of 95% ethanol, and cool the solution under the tap to produce a fine suspension. Then add 0.1 g of sodium borohydride (large excess). The benzil dissolves, the mixture warms up, and the yellow color disappears in 2–3 minutes. After a total of 10 minutes, add 5 mL of water, heat to the boiling point, filter in case the solution is not clear, dilute to the point of saturation with more water (10 mL), and set the solution aside to crystallize. *meso*-Hydrobenzoin separates in lustrous thin plates (mp 136–137°C); the yield is about 0.35 g. A second crop of material can be obtained by adding solid sodium chloride to the filtrate until no more of the salt will dissolve. Collect the second crop of crystals via filtration and wash the crystals with water.

Cleaning Up. Dilute the aqueous filtrate with water, neutralize it with acetic acid (to destroy the borohydride), and flush the mixture down the drain.

QUESTIONS

1. Using the *R,S* system of nomenclature, draw and name all of the isomers of hydrobenzoin.
2. Calculate the theoretical weight of sodium borohydride needed to reduce 50 mg of benzil.

Sodium Borohydride Reduction of a Ketone: Hydrobenzoin from Benzil

Adapted from: Kenneth Williamson and Katherine Masters, Macroscale and Microscale Organic Experiments, 6th ed., Brooks/Cole, Cengage Learning, Belmont, CA USA, 2011. Ch 55, p. 670, by M. J. Lusch

Macroscale Procedure:

Weigh **0.500 g of Benzil (1,2-diphenyl-1,2-ethanedione)** into a **50 mL** Erlenmeyer flask, and then dissolve it in **5 mL of 95% Ethanol (EtOH)**. Cool the resulting solution under the tap with cold water to produce a fine suspension of solid benzil. Then add **0.100 g of solid sodium borohydride (NaBH$_4$)** all at once and swirl to dissolve. The solid benzil should also dissolve, the mixture will warm up, and the yellow color of the benzil should disappear in 2–3 minutes. After a **total of 10 min**, add **15 mL of distilled water** and heat the mixture to its boiling point to dissolve all solids (although not usually necessary, hot-filter the solution if it is not clear and free of solid at this point). Finally, set the solution aside to cool slowly and crystallize *meso*-hydrobenzoin as lustrous thin plates (mp 136-137°C). Some foam formation is normal during the cooling due to decomposition of residual borohydride to give hydrogen gas.

Isolate the recrystallized product with vacuum filtration on the Hirsch funnel, wash with a 3:1 (v:v) mixture of water and EtOH (you make it up), and dry on a weighing paper in your drawer until the next lab. Determine the yield and melting point range of the dried product.

Waste Disposal:

Dilute the aqueous-ethanol filtrate from the isolation of the meso-hydrobenzoin product with an equal volume of water and then neutralize it to pH = 5–7 with glacial acetic acid (to destroy excess unreacted sodium borohydride) [**Caution!** Hydrogen gas formation!] before flushing the mixture down the drain.

58 CHAPTER

The Synthesis of an Alkyne from an Alkene; Bromination and Dehydrobromination: Stilbene and Diphenylacetylene

PRELAB EXERCISE: Calculate the theoretical quantities of thionyl chloride and sodium borohydride needed to convert benzoin to *E*-stilbene.

E-Stilbene
mp 125°C, MW 180.24
λ_{max}^{EtOH} 301 nm (ϵ = 28,500)
226 nm (ϵ = 17,700)

Heat of hydrogenation, −20.1 kcal/mol

Z-Stilbene
mp 6°C, MW 180.24
λ_{max}^{EtOH} 280 nm (ϵ = 13,500)
223 nm (ϵ = 23,500)

Heat of hydrogenation, −25.8 kcal/mol

In this chapter's experiments, benzoin, which was prepared in Chapter 53, is converted to the alkene *trans*-stilbene (*E*-stilbene), which is in turn brominated and dehydrobrominated to form diphenylacetylene, an alkyne.

One method of preparing *E*-stilbene is the reduction of benzoin with zinc amalgam in a mixture of ethanol and hydrochloric acid, presumably through an intermediate:

Chapter 58 ■ *The Synthesis of an Alkyne from an Alkene* 681

[Reaction scheme: Benzoin → (Zn/Hg, HCl, EtOH) → [intermediate diol] → (−H₂O) → *E*-Stilbene]

Benzoin

***E*-Stilbene**

The procedure that follows is quick and affords very pure hydrocarbon. It involves three steps: (1) replacing the hydroxyl group of benzoin by chlorine to form desyl chloride, (2) reducing the keto group with sodium borohydride to give what appears to be a mixture of the two diastereoisomeric chlorohydrins, and (3) eliminating the elements of hypochlorous acid with zinc and acetic acid. The last step is analogous to the debromination of an olefin dibromide.

[Reaction scheme: Benzoin (MW 212.24) → SOCl₂ Thionyl chloride MW 118.97 → Desyl chloride (mp 68°C) → NaBH₄ Sodium borohydride MW 37.85]

[Reaction scheme: Mixture of both diastereomers → Zn/HOAc → *E*-Stilbene]

The minimum energy conformations of *E*- and *Z*-stilbene are shown in Figures 58.1 and 58.2, respectively. These have been calculated using Spartan's

FIG. 58.1
The minimum energy conformation of *E*-stilbene.

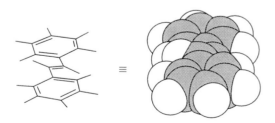

FIG. 58.2
The minimum energy conformation of *Z*-stilbene.

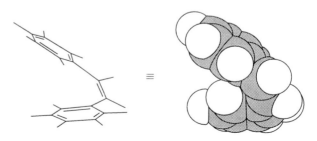

molecular mechanics routine.[1] Note that the *E*-isomer is planar, whereas the *Z*-isomer is markedly distorted from planarity.

Bromination of E-Stilbene: Synthesis of *meso*-Dibromide

Adapted from *Macroscale and Microscale Organic Experiments,* 4th Ed., K. Williamson, Houghton-Mifflin, 2003, Ch 59; pp 659-660, by M. J. Lusch and B. L. Groh

Procedure:

Weigh out **500 mg of E-stilbene** into a 50 mL Erlenmeyer flask, add **10 mL of 100% ethanol**, and get the stilbene to be *mostly* dissolved by heating on a sand bath (don't boil away the ethanol!). Weigh out **1000 mg (1.000 g) of pyridinium hydrobromide perbromide (PBB)** into a 10 mL Erlenmeyer flask (use the microkit funnel; **Caution!** Contact with stainless steel balance pans (and metal spatulas) results in corrosion of the pan – clean up any spills *immediately*). Add the solid PBB in one portion (all at once) to the warmed stilbene solution and mix by swirling; if necessary rinse the crystals of PBB reagent down the side walls of the flask with a *little* 100% ethanol; continue heating for 1-2 minutes longer. The crystalline **stilbene dibromide** separates almost at once in small plates. Cool the mixture in an ice/water bath to crystallize the maximum amount of solid, and vacuum filter to collect the dibromide product using a Hirsch funnel (**with** a filter paper). Wash the collected product with **methanol** until it is colorless or nearly so [To do this, disconnect the vacuum hose from the filter flask, add enough methanol to the filter to wet all the product, reapply the vacuum to remove the solvent]. Allow the product to air dry while pulling air though the crystals in the Hirsch funnel using the house vacuum for approximately 2 minutes. You should obtain at least 500 mg of a colorless product. Use **500 mg** of this product in the *next* part of the experiment. Save the remainder of the dibromide, allowing it to dry until the next lab period before taking the mp and turning it in to your instructor.

Waste Disposal:

(**Do this *after* you have completed the dehydrohalogenation reaction below**): Test the filtrate and washings from the original reaction mixture (ethanol/methanol) with **starch-iodide test paper** for the presence of unreacted perbromide, and if perbromide is present (likely) destroy it by reducing it to bromide ion with small amounts of **solid sodium bisulfite** (Color changes from pale yellow to white. **Starch-Iodide Test (for excess perbromide):** First wet the starch–iodide test strip with distilled water, then place a drop of the solution to be tested on the strip; a deep blue/purple/brown color indicates oxidizer (perbromide) present. Add *small amounts* of **solid sodium bisulfite** to reduce the perbromide; the test should eventually be negative (colorless). Then place the resulting solution in the *Halogenated* **Organic Waste** bottle in the first hood.

FIG. 58.5
The UV spectra of Z- and E-stilbene. Z: λ_{max}^{EtOH} = 224 nm (ε = 23,300) and 279 nm (ε = 11,100); E: λ_{max}^{EtOH} = 226 nm (ε = 18,300) and 295 nm (ε = 27,500). As in the diacetates, steric hindrance and the lack of coplanarity in these hydrocarbons cause the long-wavelength absorption of the Z-isomer to be of diminished intensity relative to the E-isomer.

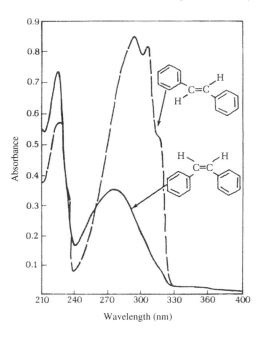

Sometimes zinc dust from a reaction like this is pyrophoric (spontaneously flammable in air) because it is so finely divided and has such a large, clean surface area able to react with air.

Cleaning Up. Combine the washings from the cotton in the trap and all aqueous layers, neutralize with sodium carbonate, remove zinc salts by vacuum filtration, and flush the filtrate down the drain with excess water. The zinc salts are placed in the nonhazardous solid waste container. Allow the ether to evaporate from the calcium chloride in the hood, and then place it in the nonhazardous solid waste container. If local regulations do not allow for the evaporation of solvents in a hood, the wet solid should be disposed in a special waste container. Ethanol mother liquor goes in the organic solvents waste container. Any zinc isolated should be spread out on a watch glass to dry and air oxidize. It should then be wetted with water, and placed in the nonhazardous solid waste container.

2. *meso*-STILBENE DIBROMIDE

> IN THIS EXPERIMENT, an acetic acid solution of stilbene, which is an alkene, is brominated with a solid bromine donor (which eliminates handling hazardous liquid bromine) to give a dibromide that crystallizes and is isolated by filtration.

E-Stilbene reacts with bromine predominantly by the usual process of *trans*-addition, and affords the optically inactive, nonresolvable *meso*-dibromide; the much lower melting enantiomeric mixture of dibromides is a very minor product of the reaction.

Pyridine + HBr + Br$_2$

↓

Pyridinium hydrobromide perbromide

In this procedure, the brominating agent will be pyridinium hydrobromide perbromide,[5] a crystalline, nonvolatile, odorless complex of high molecular weight (319.86), which dissociates, in the presence of a bromine acceptor such as an alkene, to liberate 1 mol of bromine. For microscale experiments, the perbromide is far more convenient and agreeable to measure and use than free bromine.

E-Stilbene
MW 180.24

Pyridinium hydro-
bromide perbromide
MW 319.86

(1*R*,2*S*)-*meso*-Stilbene dibromide
mp 236–237°C, MW 340.07

CAUTION: Pyridinium hydrobromide perbromide is corrosive and a lachrymator (tear producer).

Carry out the procedure in the hood.

Photo: *Use of the Wilfilter.*
Video: *Microscale Filtration on the Hirsch Funnel*

Total time required: 10 minutes

CAUTION: Pyridium hydrobromide perbromide is corrosive and a lachrymator (tear producer).

Microscale Procedure

In a reaction tube, dissolve 50 mg of *E*-stilbene in 1 mL of acetic acid by heating on a steam bath or a hot water bath; then add 100 mg of pyridinium hydrobromide perbromide. Mix by swirling; if necessary, rinse crystals of the reagent down the walls of the flask with a little acetic acid and continue the heating for an additional 1–2 minutes. The dibromide separates as small plates. Cool this mixture under the tap, collect the product on a Wilfilter (*see* Fig. 4.14 on page 73) or on a Hirsch funnel, and wash it with methanol to remove any color; the yield of colorless crystals (mp 236–237°C) is 80 mg. Use this material to prepare the diphenylacetylene after determining the percent yield and the melting point.

Cleaning Up. To the filtrate, add sodium bisulfite (until a negative test with starch-iodide paper is observed) to destroy any remaining perbromide, neutralize with sodium carbonate, and extract the pyridine released with ether, which goes in the organic solvents waste container. The aqueous layer can then be diluted with water and flushed down the drain.

Macroscale Procedure

In a 50-mL Erlenmeyer flask, dissolve 1 g of *E*-stilbene in 20 mL of acetic acid by heating on a steam bath; then add 2 g of pyridinium hydrobromide perbromide.[6] Mix by swirling; if necessary, rinse crystals of the reagent down the walls of the flask with a little acetic acid; continue the heating for an additional 1–2 minutes.

---

[5]Crystalline pyridinium hydrobromide perbromide suitable for small-scale experiments is available from the Aldrich Chemical Company. Massive crystals commercially available should be recrystallized from acetic acid (4 mL/g). Pyridinium hydrobromide perbromide can also be prepared as follows: Mix 15 mL of pyridine with 30 mL of 48% hydrobromic acid, and cool. Add 25 g of bromine gradually with swirling, cool, and collect the product (use acetic acid for rinsing and washing). Without drying the solid, crystallize it from 100 mL of acetic acid. The yield of orange needles should be 33 g (69%).
[6]*See* footnote 5.

Carry out the procedure in the hood.

Total time required: 10 minutes

The dibromide separates almost at once as small plates. Cool the mixture under the tap, collect the product, and wash it with methanol; the yield of colorless crystals (mp 236–237°C) is about 1.6 g. Use 0.5 g of this material to prepare the diphenylacetylene, and turn in the remainder to your instructor.

Cleaning Up. To the filtrate, add sodium bisulfite (until a negative test with starch-iodide paper is observed) to destroy any remaining perbromide, neutralize with sodium carbonate, and extract the pyridine released during the reaction with ether, which goes in the organic solvents waste container. The aqueous layer can then be diluted with water and flushed down the drain.

Dehydrohalogenation: Synthesis of Diphenyl Acetylene from *meso*-Stilbene Dibromide

Adapted from *Macroscale and Microscale Organic Experiments,* 4th Ed., K. Williamson, Houghton-Mifflin, 2003, Ch 59; pp 662, by M. J. Lusch and B. L. Groh

Procedure:

In a 19 X 150 mm (large) test tube, place **0.500 g (500 mg) of the *meso*-stilbene dibromide** you prepared above, **3 pellets of potassium hydroxide (approximately 250 mg)**, and **2 mL of triethylene glycol**. Clamp the tube in the vertical position using a 3-finger clamp and insert the bottom of the tube into a heated sand bath. Place a thermometer in the tube and clamp it in place using the thermometer clamp so that the thermometer is resting on the inside bottom of the test tube. Heat the mixture to 160°C (**Caution!** Vigorous bubbling and foaming can occur above 100°C as water boils off – take care to avoid bubbling out of the tube), at which point the potassium bromide byproduct begins to separate. By intermittent heating (moving the sand to or away from the test tube) keep the mixture at **160-170°C for 5 minutes**, then cool to room temperature, remove the thermometer, and add **10 mL of distilled water** (use some of this water to wash off the thermometer into the test tube). Stir the resulting mixture thoroughly with a small spatula until the water-insoluble product becomes a granular, filterable solid, and then collect this diphenyl acetylene by vacuum filtration using the Hirsch funnel [**for *this* filtration *only*, <u>do not</u> use a filter paper**, since the strongly basic solution causes cellulose filter papers to swell and clog up], washing it with a little water to remove any KOH, KBr, and triethylene glycol. The crude wet product need not be dried but may be recrystallized immediately in a 10 mL erlenmeyer flask from a minimal amount of **95% ethanol** (probably less than **1-2 mL**). Let the recrystallization solution cool <u>slowly</u> to room temperature [Slow, undisturbed cooling will favor formation of large, pure crystals]. Filter the recrystallized product by vacuum filtration on a Hirsch funnel (*with* a filter paper), wash with cold fresh ethanol to remove traces of the mother liquor, and allow it to dry in a sample envelope until the next period. Then weigh the product, determine its melting point **range** (literature mp 60-61°C), and calculate the percent yield. A typical yield is approximately 250 mg.

Waste Disposal:

Combine the filtrate from the original diluted reaction mixture (basic) and the filtrate from the recrystallization (ethanol), neutralize any base that still remains with **3M aq. HCl** (to pH 6–9), and flush the neutralized aqueous ethanol solution down the drain.

Experiment 4 - Molecular Shapes: Structural Isomerism, Conformational Analysis, and Stereoisomerism

Complete the following exercises. The report sheet (answers to questions) is due at the end of the lab period.

Molecular models are often used as visual aids to understanding the *three-dimensional* structures of organic molecules. The models used in this experiment show the correct angles between the various chemical bonds in a molecule, but do not accurately represent the relative sizes of atoms or their precise internuclear distances. Nevertheless, they do help one to appreciate the different possible arrangements of the atoms in a molecule, the different shapes that a molecule can assume, and the ways these shapes can be represented in two-dimensional formulas.

1. The Models

The molecular model set contains different colored plastic balls to represent the different kinds of atoms, such as carbon, hydrogen, and oxygen. They have holes bored in them to correspond to the number of covalent bonds that the various atoms can form in producing molecules. With atoms capable of forming more than one covalent bond, the holes are bored at angles to correspond to the normal bond angles of the atoms.

In the model kit **atoms** (balls) are typically colored as:

Carbon............black	Chlorine............green	Brown...............boron
Hydrogen..........white	Bromine............orange	Silver...............metal
Oxygen............red	Sulfur...............yellow	Phosphorus......purple
Nitrogen............blue		

Bonds (sticks):

Long Grey.........used for double bonds
Medium Grey....used for **single** bonds
Short White.......single bonds in space filling models
Purple...............for nonbonded electrons, or P=O and S=O bonds

Scale – space filling: 1.7 cm = 100 pm
– open model: 2.8 cm = 100 pm

By placing one end of a stick in a hole in one of the model atoms and the other in a hole of a second model atom, a covalent bond binding these two atoms together is represented.

2. The Carbon Atom

Procedure: Insert a stick in each hole of a model carbon atom (black), and set the model on the desk top for observation. Mentally draw imaginary lines between the ends of the sticks. Answers to the following questions are to be written on the answer sheet with the corresponding numbers at the end of each question.

1. What geometric figure is represented by these imaginary lines? **(2.1)**
2. What is the approximate angle between all possible pairs of covalent bonds in the model? **(2.2)**

Note that if one imagines a plane to include the carbon atom and two of its bonds, the remaining two bonds are symmetrically disposed on either side of that plane. The plane is therefore called a plane of symmetry. The sp³-hybridized carbon atom can be represented by the figures shown below:

The **solid bonds** lie *in the plane of the paper* (the symmetry plane), the **dashed bond** extends *away from the reader*, and the **wedge-shaped bond** extends *toward the reader*.

3. How many symmetry planes does the carbon atom model have? **(2.3)**
4. The tetrahedral carbon atom also has a threefold rotational axis of symmetry. What is meant by such an axis? **(2.4)**
5. By use of an arrow in the following figure, show the location of one such threefold axis in the carbon atom model. **(Include this on your answer sheet - 2.5)**

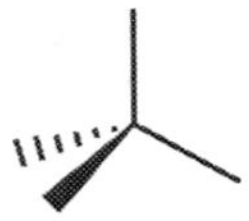

6. How many threefold rotational axes does the model have? **(2.6)**
7. The carbon atom model has an <u>additional</u> rotational axis of symmetry that is *not* threefold. Use an arrow in the following figure to show the location of one of these. **(2.7)**

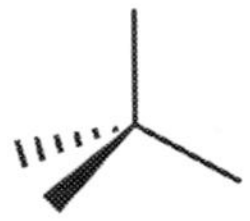

8. How many-fold is the axis you drew in (7)? **(2.8)**

3. The Methane Molecule

Procedure: Use a **black ball** (carbon) and **four white balls** (hydrogen) to construct a model of the methane molecule, CH_4. Place the model on the desk top and observe it by looking directly down on top of it. Rotate the model to place two hydrogen atoms to the right and one to the left. Take hold of the top hydrogen atom and tip the model to the right about 45° so that the methane model is resting on only two hydrogen atoms. Note that these two hydrogen atoms, which are now at the top and bottom of the carbon atom, lie in a plane beneath the carbon atom, whereas the two remaining hydrogen atoms to the left and right of the carbon atom lie in a plane above the carbon atom. Now imagine the methane model to be pressed down flat on the desk top. This imaginary flat model, which is *the projection of the methane model on a plane surface*, is the conventional structural formula for methane.

Projection formula of **methane**:

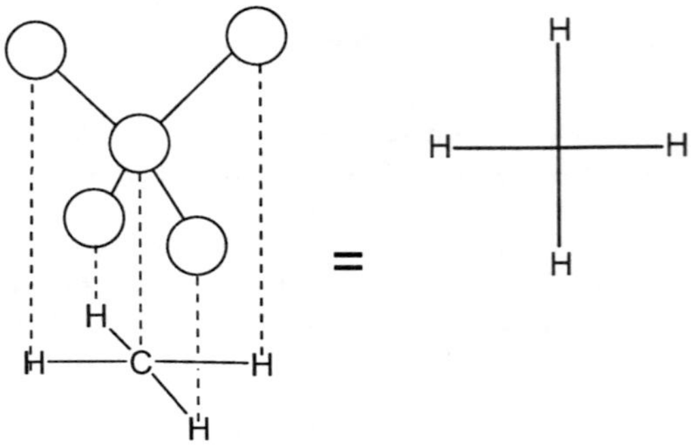

Replace *one* of the hydrogen atoms by a **chlorine atom (green)**. The resulting molecule is called **chloromethane**, or **methyl chloride**.

9. Use the numbering system in the following figure to describe *all* the symmetry planes in chloromethane. For example, there is a plane of symmetry through atoms 1,2,5. **(3.9)**

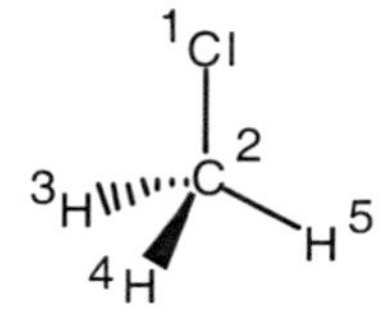

10. Use the above figure and arrows to show any **axes of symmetry** in chloromethane. **(3.10; Also state how many-fold the axes are)**

Satisfy yourself that the following plane projection formulas (these are called Fischer projections) represent the same three-dimensional structure as above.

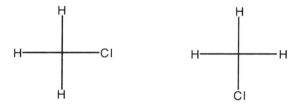

Replace a *second* hydrogen atom by a chlorine atom, to give **dichloromethane (methylene chloride), CH₂Cl₂**.

11. Use the numbering system in the following figure to describe all the symmetry planes in methylene chloride. **(3.11)**

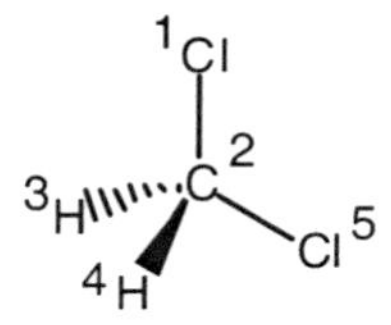

12. Use the above figure and arrows to show any symmetry axes in methylene chloride. **(3.12; Also state how many-fold the axes are)**

Satisfy yourself that the Fischer projections below represent the same three-dimensional structure as above.

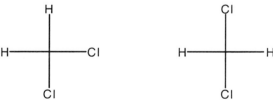

6. Rings of Carbon Atoms. Cyclohexane Conformations

Procedure: Connect six carbon atoms by means of five valence sticks. Insert sticks in all the unused holes in the carbon atom models. Note the zigzag nature of the chain when extended. Remove one valence stick from one of the terminal carbon atoms and, by twisting the atoms about their bonds, connect atom 1 to atom 6 to form a six-membered (cyclohexane) ring.

Note that the ring can be arranged in either a **chair** or a **boat** form.

With the model in the **chair** form, rest it on the table top.

1. How many hydrogens are in contact with the tabletop? **(6.1)**
2. How many hydrogens point 180° opposite these? **(6.2)**
3. Take your pencil and place it into the center of the ring perpendicular to the table. Rotate the ring around the pencil (the pencil is the axis of rotation). How may C-H bonds are parallel to the axis of rotation? **(6.3)** (These are the **axial** hydrogens.)
4. Consider the remaining hydrogens that lie roughly in a ring perpendicular to the axis through the center of the molecule. How many C-H bonds lie in this ring? **(6.4)** (These are the **equatorial** hydrogens.)
5. In the space provided on the report sheet draw the structure of cyclohexane in the chair conformation with all hydrogens attached. Label each hydrogen as either equatorial (H_e) or axial (H_a). **(6.5)**
6. Look at any C-C bond in the ring. Are the H-bonds to the carbon atoms staggered or eclipsed? **(6.6)**
7. Orient your model so that you look along an edge of the ring as shown below. Draw a Newman projection simultaneously sighting down any two parallel C-C bonds and label the axial and equatorial hydrogens. (This will require drawing two Newman projections side-by-side and connected to each other.) **(6.7)**

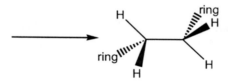

8. Replace one of the axial hydrogens with a black atom (a methyl group). Is it in an axial or equatorial position? **(6.8a)** Do a ring flip by moving one of the carbons up and moving the carbon farthest away from it down. Is the added carbon atom now axial or equatorial? **(6.8b)** Observe all axial hydrogens and do another ring flip. What is their new orientation, axial or equatorial? **(6.8c)**
9. Using your cyclohexane molecule with the added carbon, orient the ring so that the methyl group is in an axial position. On the next carbon over, place another carbon in the axial position. Are the two groups now **cis or trans**? **(6.9a)** Do a ring flip; what it the relationship between the two methyl groups now? **(6.9b)**
10. Add the hydrogen atoms to each methyl group. Flip the ring so that both methyl groups are in the equatorial positions and examine the model. Then do a ring flip and examine it again. Which conformation is more crowded, the diaxial or the diequatorial? **(6.10a)** Which conformation is more stable, that with the methyls axial or equatorial? **(6.10b)**.
11. Normally, a substituent group in the equatorial position is more stable than in the axial position. Do you agree with this statement? **(6.11)**
12. The molecule you have been working with is **trans-1,2-dimethylcyclohexane**. Now move one of the methyl groups to the other side of the ring to form **cis-1,2-dimethylcyclohexane**. By means of ring flips, examine both conformations. In what way are the conformations different? **(6.12a)** In what way are they the same? **(6.12b)**
13. Construct **1,3- and 1,4-dimethylcyclohexanes**; examine both cis and trans isomers and do ring flips on each and consider the differences in stabilities for each conformation.

a. Which **1,3-dimethylcyclohexane** is *more* stable, the *cis* or the *trans* isomer?
b. Which **1,4-dimethylcyclohexane** is *more* stable, the *cis* or the *trans* isomer?
14. Replace both methyl groups with hydrogens. With the cyclohexane molecule in the chair conformation orient the molecule with opposite ends (the two carbon atoms furthest apart) both upward. This is the **boat conformation** and should resemble the drawing below.

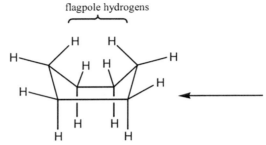

15. Sight along a C-C bond as indicated by the arrow in the diagram above. Now draw a Newman projection for this view **(6.15a)**. Compare it with the Newman projection in 6.7 above and indicate which conformation is more stable, the **boat** or the **chair (6.15b)**.
16. Now replace the flagpole hydrogens (as indicated above) with methyl groups. What happens when this is done? **(6.16)** Now twist the boat to relieve the steric strain; this is the **twist boat** conformation.
17. Review the possible conformations of the cyclohexane ring system.

7. Chiral Molecules: Enantiomers

1. Construct a molecule consisting of a tetrahedral carbon atom with four different component atoms attached: red, white, blue, and green. Each group represents a different group or atom attached to carbon. Does this molecule have a plane of symmetry? **(7.1)**
2. Molecules **without a plane of symmetry** are **chiral**. In the model you constructed in step 1, the tetrahedral carbon is the **chiral center**. (This chiral center is sometimes referred to as a **chiral carbon, stereogenic carbon or stereogenic center**.) A simple test for a chiral carbon is to look for a carbon center with **four** *different* groups attached to it; such a molecule will not have a plane of symmetry.
3. Now take the molecule you constructed and make another molecule that is the *mirror image* of it. Do either have a plane of symmetry? **(7.3a)** Are both chiral? **(7.3b)**. Now try to **superimpose** them by placing one on top of the other trying so that all matching atoms fall exactly on top of one another. Can you superimpose them on each other? **(7.3c)** *Enantiomers* are two molecules that are *nonsuperimposable* mirror images of each other. Are the two molecules you made enantiomers? **(7.3d)**
4. Molecules *with* a plane of symmetry are **achiral**. Replace the blue substituent with a second green one. The model should now have only three different groups attached to the carbon. Does the model now have a plane of symmetry? **(7.4a)** Passing a plane of symmetry though the molecule will pass through which color atoms? **(7.4b)** Construct a mirror image of the molecule. Are the two molecules superimposable or not? **(7.4c)** Are the molecules different or the same? **(7.4d)**

Each stereoisomer in a pair of enanitomers has the property of being able to **rotate monochromatic** *plane-polarized* **light**. The instrument used to measure this rotation is called a **polarimeter**. A pure solution of a single one of the enantiomers (referred to as an **optical isomer**) can rotate light in either a **clockwise** (*dextrorotatory* or *d* or *(+)*) or **counterclockwise** (*levorotatory* or *l* or *(-)*) direction. These molecules are said to be **optically active**.

8. Chiral Molecules: Meso Forms and Diasteromers

1. With your models, construct a pair of enantiomers. From each one of the models remove the same common element (e.g., the white hydrogen) and connect the two molecules together with

a single bond. You have constructed a *meso* form of a molecule, similar to *meso*-tartaric acid shown below. How many chiral carbons are in your molecule? **(8.1a)** Does this molecule have a plane of symmetry? **(8.1b)** Is the molecule chiral or achiral? **(8.1c)** Is this molecule optically active or not? **(8.1d)**

$$HO_2C-\underset{H}{\overset{OH}{C}}-\underset{H}{\overset{OH}{C}}-CO_2H$$

2. Construct a mirror image of this molecule. Are the two molecules superimposable or not? **(8.2a)** Are they optically active or not? **(8.2b)** Are these two molecules identical or different? **(8.2c)**
3. Now take one of the models you constructed in 8.2 (above), and on one of the carbon centers exchange any two colored component groups. Does the new model have a plane of symmetry **(8.3a)**? Is it chiral or achiral **(8.3b)**? How many chiral center's are present **(8.3c)**? Take this model and the model you constructed in 8.2 and see whether they are superimposable. Are the two models superimposable **(8.3d)**? Are the two models identical or different **(8.3e)**? Are the two models mirror images of each other **(8.3f)**? Here we have a pair of molecular models, each with two chiral centers that are not mirror images of each other. These two examples represent **diastereomers, stereoisomers that are not related as mirror images**.
4. Construct a mirror-image model of the molecule you created in 8.3 (above). Are the two models superimposable **(8.4a)**? What term describes the relationship of the two models **(8.4b)**?

Thus if we let these three models represent different isomers of tartaric acid, we find that there are three stereoisomers for tartaric acid - a *meso* form and a *pair* of **enantiomers**. A meso form with any *one* of the enantiomers of tartaric acid represents a *pair* of **diastereomers**. Although it may not be true for this compound because of the meso form, in general, if you have **n chiral centers**, there are a *maximum* of 2^n stereoisomers possible.

9. Drawing Stereoisomers

This section will deal with conventions for representing these *3-dimensional systems* in **2-dimensional space**.

1. Construct models of a pair of enantiomers; use tetrahedral carbon and four differently colored components for the four different groups: **red, green, blue, white**. Hold one of the models in the following way:

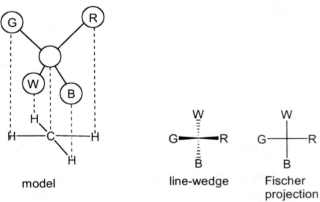

a. Grasp the blue group with your fingers and rotate the model until the green and red groups are pointing up toward you (see below). (Use the model which has the green group on the left and the red group on the right.)

b. Holding the model in this way, the blue and white groups point away from you.

c. If we use a drawing that describes a bond pointing *toward* you as a **wedge** and a bond pointing *away* from you as a **dashed-line**, the model can be drawn as shown below.

Dashed
Line-Wedge

Fischer
Projection

If this model were compressed into 2-dimensional space, we would get the projection shown on the *right* above. This is termed a **Fischer projection** and is named after a pioneer in stereochemistry, Emil Fischer. The Fischer projection has the following requirements:
   (a) the **center** of the cross represents the *chiral* carbon and *is in the plane of the paper*;
   (b) the **horizontal line** of the cross represents those *bonds projecting out front from the plane of the paper (toward you)*;
   (c) the **vertical line** of the cross represents *bonds projecting behind the plane of the paper (away from you)*.

2. Using the enantiomer of the model created in 9.1 above, **draw both the dashed-line-wedge and Fischer projection (9.2)**.

3. Take the model from 9.2 and **rotate it by 180° about a horizontal axis through the chiral carbon** (turn upside down). Draw its Fischer projection **(9.3a)**. Does this keep the requirements of the Fischer projection **(9.3b)**? Is the projection representative of the same compound or of a different compound (i.e., the enantiomer, diastereomers, meso, etc.) **(9.3c)**?

In general, if you have a Fischer projection and **rotate it *in* the plane of the paper by 180°** the resulting projection is of the same system. Test this assumption by taking the Fischer projection in 9.1, rotating it in the plane of the paper by 180°, and comparing it to the drawing you did in section 9.1c. However, rotating a Fischer projection 180° *out* of the plane of the paper **(flipping it over, either horizontally or vertically)** is not permitted.

4. Again, take the model you made in section 9.1, exchange the red and the green components. Does this exchange give you the enantiomer **(9.4a)**? Now exchange the blue and the white components. Does this exchange return you to the original model **(9.4b)**?

In general, for a given chiral center, whether we use the dashed-line-wedge or the Fischer projection, an **odd-numbered** *exchange of groups* leads to the **mirror image** of that center; an **even-numbered** *exchange of groups* leads back to the **original system**.

5. Test the above by starting with the Fischer projection given below and carrying out the operations directed in a, b, and c below; use the space provided on the Report Sheet **(9.5a-c)** for the answers.

a. Exchange R and G; draw the Fischer projection you obtain; label this new projection as either the same as the starting model of 9.1 or the enantiomer. **(9.5a)**

b. Using the new Fischer projection from above, exchange B and W; draw the Fischer projection you now have. **(9.5b)**
c. Now rotate the last Fischer projection you obtained by 180° (in the plane of the page, the axis of rotation is perpendicular to the page). Draw the Fischer projection you now have; label this as either the same as the starting model or the enantiomer. **(9.5c)**

6. Let us examine models with two chiral centers by using **tartaric acid** as the example, **HOOC-CH(OH)-CH(OH)-COOH**; use your colored components to represent the various groups. Hold your models so that each stereoisomer is oriented as 9.6a (below). In the space provided on the Report Sheet **(9.6)**, draw each of the corresponding Fischer projections. **Circle** the Fischer projection that shows a plane of symmetry. **Underline** all the Fischer projections that would be optically active.

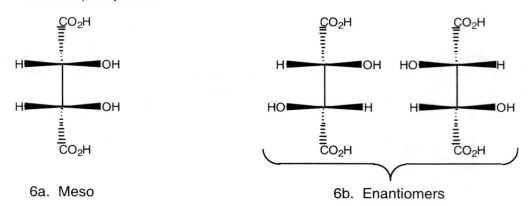

6a. Meso   6b. Enantiomers

7. Use the Fischer projection of meso-tartaric acid (6a, above) and carry out even and odd exchanges of the groups; follow these exchanges with a model. Does an odd exchange lead to an enantiomer, a diastereomer, or to a system identical to the meso form **(9.7a)**? Does an even exchange lead to an enantiomer, a diastereomer, or to a system identical to the meso form **(9.7b)**?

4. The Conformations of Ethane (Optional)

Procedure:

Join two carbon atoms with a bond, then attach six hydrogens to the remaining valences to construct a model of ethane, C_2H_6. Sight down the carbon-carbon bond, and slowly rotate one carbon atom with respect to the other until a C–H bond on one carbon atom bisects the H-C-H angle on the other carbon atom. This is called the **staggered** conformation of the molecule, which can be represented by several conventions (Newman, "sawhorse," dashed line-wedge), as shown below:

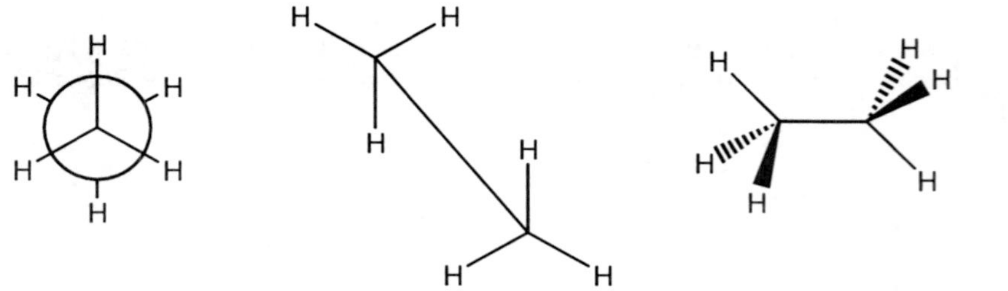

Newman projection   "sawhorse" formula   dashed line-wedge

1. Use the numbering system in the following figure to describe all the planes of symmetry in the staggered conformation of ethane **(4.1)**.

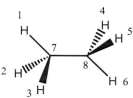

2. Use the above figure and arrows to show any rotational symmetry axes in the staggered conformation of ethane. **(4.2; Also state how many-fold the axes are)**

3. This conformation of ethane also has a **center of symmetry**. A **center of symmetry** is defined as a point within an object such that a straight line drawn from any part or element of the object to the center and extended an equal distance on the other side encounters an equal part or element. Where is the center of symmetry located in the staggered conformation of ethane? **(4.3)**

Rotate one carbon atom 60° with respect to the other, to construct the **eclipsed conformation** of ethane, as shown below.

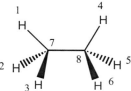

4. Draw the eclipsed conformation of ethane as a **Newman projection**. **(4.4)**

5. Use the numbering system in the following figure to describe all the **symmetry planes** in the eclipsed conformation of ethane. **(4.5)**

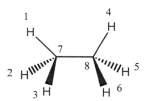

6. Use the above figure and arrows to show any rotational symmetry axes in the eclipsed conformation of ethane. **(4.6; Also state how many-fold the axes are)**

7. Does the eclipsed conformation of ethane have a center of symmetry? If so, where is it? **(4.7)**

8. In which conformation, staggered or eclipsed, does a hydrogen atom on one carbon atom come closer to a hydrogen atom on the other carbon atom **(4.8a)**. If nonbonded atoms repel each other when they come close to one another, which conformation of ethane would you expect to be the less stable? **(4.8b)**

9. What type of motion converts this conformation to the more stable conformation? **(4.9)**

10. Plot the energy of the ethane molecule (vertical axis) against the angle of rotation of the C--C bond (horizontal axis), starting in the staggered conformation as 0°. Go through one full revolution (360°). **(4.10)**

The S_N1 Reaction: Synthesis of 2-Chloro-2-methylpropane (*tert*-butyl chloride)

From Joy of Organic Chemistry by Berton C. Weberg and John E. McCarty
Mankato State University, Mankato
Edited by B. L. Groh and M. J. Lusch

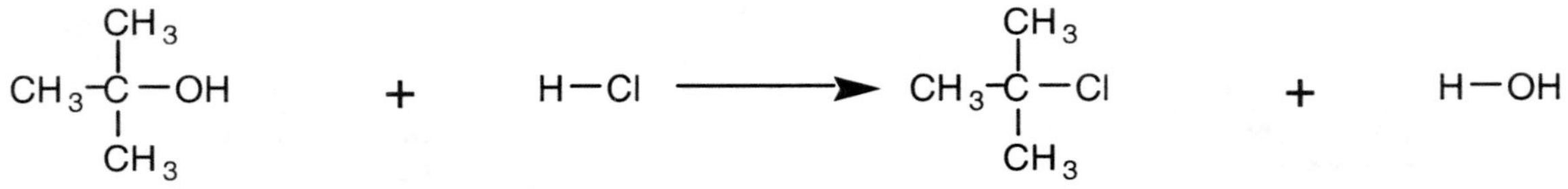

$$(CH_3)_3C{-}OH \; + \; H{-}Cl \longrightarrow (CH_3)_3C{-}Cl \; + \; H{-}OH$$

Theory. Conversion of tertiary alcohols to the corresponding halides is easily effected at room temperature by reaction with hydrohalogen acids. Preparation of the halides of secondary alcohols by this procedure is much slower, and the formation of the halides of primary alcohols usually requires prolonged heating. The addition of the anhydrous zinc halide[1] to the hydrohalogen acid accelerates the rate of the reaction so that the halides of secondary alcohols may be prepared conveniently at room temperature. If concentrated sulfuric acid is added to the hydrohalogen acid, the halide derived from primary alcohols may be formed after a few hours of heating.

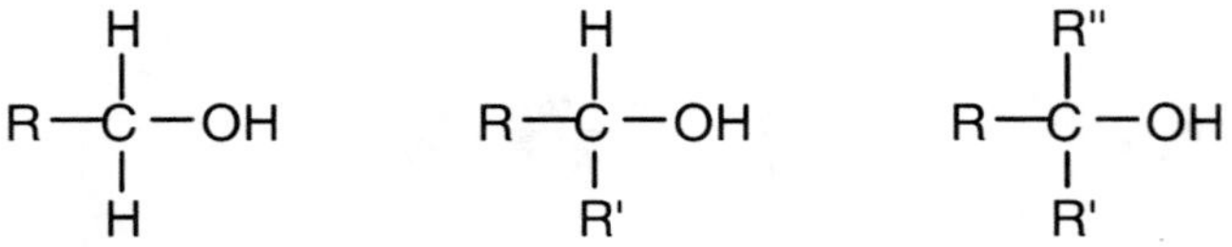

$$R{-}CH_2{-}OH \quad\quad R{-}CHR'{-}OH \quad\quad R{-}CR'R''{-}OH$$

Primary   **Secondary**   **Tertiary**

For example:

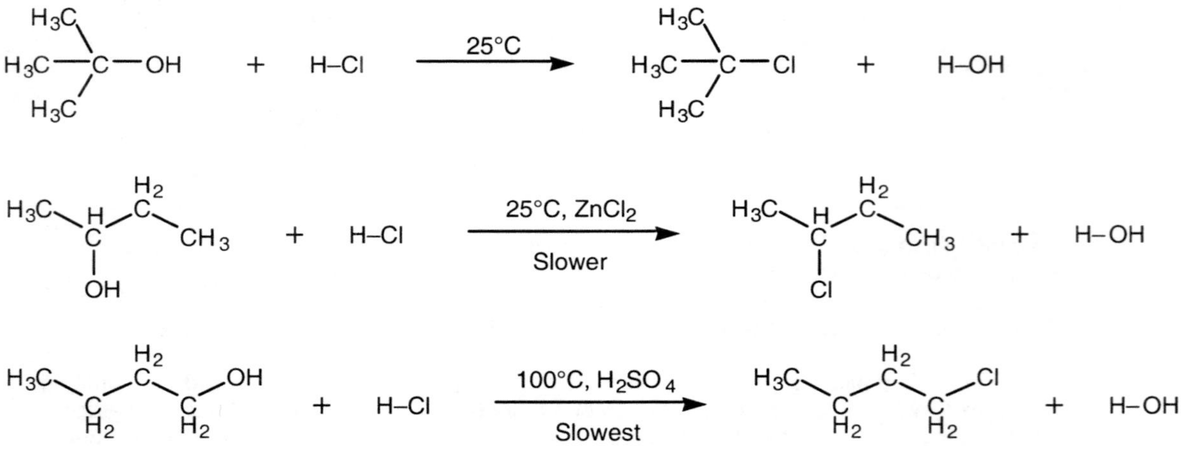

$$(CH_3)_3C{-}OH \; + \; H{-}Cl \xrightarrow{25°C} (CH_3)_3C{-}Cl \; + \; H{-}OH$$

$$(CH_3)_2CH{-}OH \; + \; H{-}Cl \xrightarrow[\text{Slower}]{25°C,\; ZnCl_2} (CH_3)_2CH{-}Cl \; + \; H{-}OH$$

$$CH_3CH_2CH_2{-}OH \; + \; H{-}Cl \xrightarrow[\text{Slowest}]{100°C,\; H_2SO_4} CH_3CH_2CH_2{-}Cl \; + \; H{-}OH$$

The reaction with hydriodic acid is faster than the corresponding reaction with hydrobromic acid, which is faster in turn than that with hydrochloric acid. This order is due primarily to the nucleophilic character of the halide ion, which is a function of its ionic radius. Primary, secondary, and tertiary alcohols all react with the phosphorus halides, and this reaction is often more convenient experimentally with primary and secondary alcohols.

Two different substitution mechanisms, S_N1 or S_N2, can be involved in the conversion of alcohols to the corresponding halides with the hydrohalogen acids. Under the conditions of this preparation (tertiary alcohol and highly polar medium), the S_N1 mechanism, involving the formation of the intermediate carbocation, is operating.

Review: *Extraction* (Chap. 7, pp. 131-40), and Simple Distillation (Chap. 5.4 Macroscale)

Procedure. In a **125 mL Erlenmeyer flask** (in a hood), place 40g (**35 mL**) of **concentrated (11.6M) hydrochloric acid**. Add **7.8 g of *tert*-butyl alcohol (2-methyl-2-propanol)** to the acid and swirl the solution. *Tert*-butyl alcohol is miscible with hydrochloric acid and results in a clear solution. *Tert*-butyl chloride, however, is immiscible with hydrochloric acid, and the solution rapidly becomes cloudy. *Loosely* stopper the reaction flask with a cork and swirl the flask frequently for a thirty-minute period. During this time set up a separatory funnel and ring stand, and a distillation apparatus using the 19/22 glassware kit (blue box in the common drawer; see diagram on the previous page).

After a total of **30 minutes** of reaction time (or more, doesn't hurt), pour the reaction mixture into a **125 mL separatory funnel** (using a narrow-necked funnel to aid in the transfer) and remove the aqueous hydrochloric acid layer (bottom). Wash the *top* layer with **5 mL of 5% or saturated aqueous sodium bicarbonate solution** (*caution, foaming!*) and again remove the bottom layer. Finally, wash the organic layer with **5 mL of water**, and once again remove the bottom aqueous layer. Transfer the crude *tert*-butyl chloride to a *dry* **50 mL Erlenmeyer flask** (cork stoppered) and dry the liquid with a small amount of **anhydrous calcium chloride**.

Prepare the distillation assembly (see diagram above), with a <u>**25 mL**</u> 1-neck round bottom (R.B.) flask as the **distilling flask** to hold the crude product, a **50 mL** single-neck R.B. flask as the **receiving flask** (weighed before assembling), and a heating mantle with sand bath. The entire distillation assembly must be <u>dry</u>. Transfer the dry *tert*-butyl chloride to the distilling flask with a long-stemmed funnel having a small piece of cotton as a plug in the neck of the funnel to hold back the calcium chloride. Record the weight of the 50 mL receiver (<u>with a glass stopper</u>) to be used to hold the main fraction. Add a boiling chip to the distillation flask, and distill the product, retaining the fraction boiling from **48-51°C**. Record the weight and the *actual* boiling range. Calculate percent of theoretical yield.

Clean up: The saturated sodium bicarbonate solution used earlier to wash the crude product should have been drained into the aqueous HCl solution originally removed from the product layer. Then, in the hood or in the sink, cautiously add (foaming) enough **solid sodium carbonate** to neutralize the acid present. When neutral (litmus), dilute the solution with water and flush it down the drain. This operation is best carried out in a **250 mL or larger** beaker or flask.

References:

1. The mixture of anhydrous zinc chloride with concentrated hydrochloric acid is a convenient reagent for distinguishing between primary, secondary, and tertiary alcohols containing four to six carbon atoms. It is named the **Lucas Reagent** after H. J. Lucas, former Professor of Chemistry at the California Institute of Technology.

S_N2 Reaction: Synthesis of 1-Chlorotetradecane from 1-Tetradecanol

Brian Groh, Jason Pendleton, Duane Anderson, and Chad Kratochwill
Department of Chemistry and Geology
Minnesota State University
Mankato, MN 56001

Introduction: This experiment was developed by undergraduate students at MSU,M as a research project with the intent of creating a new experiment that would afford the product of an S_N1 reaction in better yield and using greener chemistry than the typical synthesis of *n*-butyl bromide from *n*-butanol (see http://www.epa.gov/greenchemistry/ or search "green chemistry" on the web). Of interest in this reaction is lack of a solvent as a reaction medium and the combination of two solid reagents that react to form a liquid product. This latter point makes it easy to observe the progress of the reaction. The reaction is quite clean and proceeds in very good yield making it easy to isolate the product. Final purification of the product is carried out by rapid column chromatography with analysis of the isolated product accomplished by gas chromatography.

Reaction Equation: (Balanced)

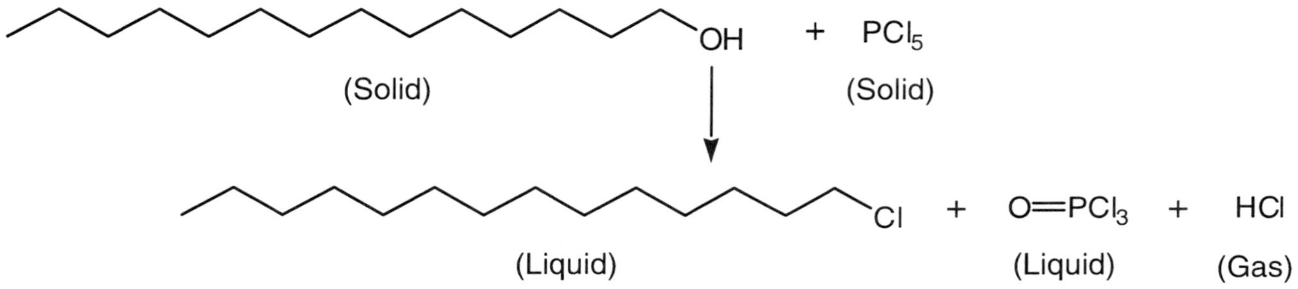

CH$_3$(CH$_2$)$_{12}$CH$_2$OH (Solid) + PCl$_5$ (Solid) → CH$_3$(CH$_2$)$_{12}$CH$_2$Cl (Liquid) + O=PCl$_3$ (Liquid) + HCl (Gas)

Generalized Reaction:

R—CH$_2$—OH + PCl$_5$ ⟶ R—CH$_2$—Cl + O=PCl$_3$ + HCl

Mechanism:

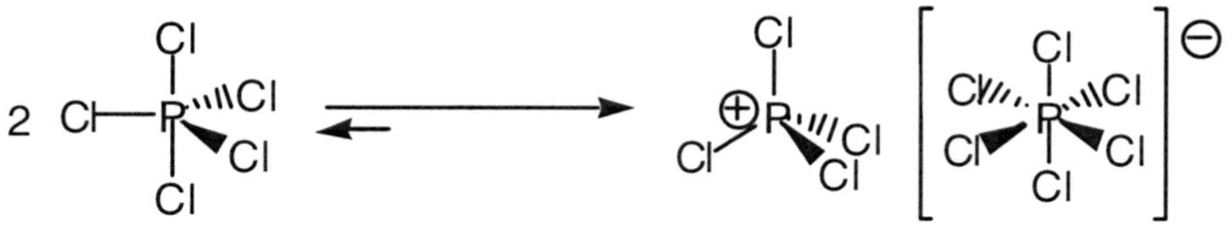

2 PCl$_5$ ⇌ [PCl$_4$]$^+$ [PCl$_6$]$^-$

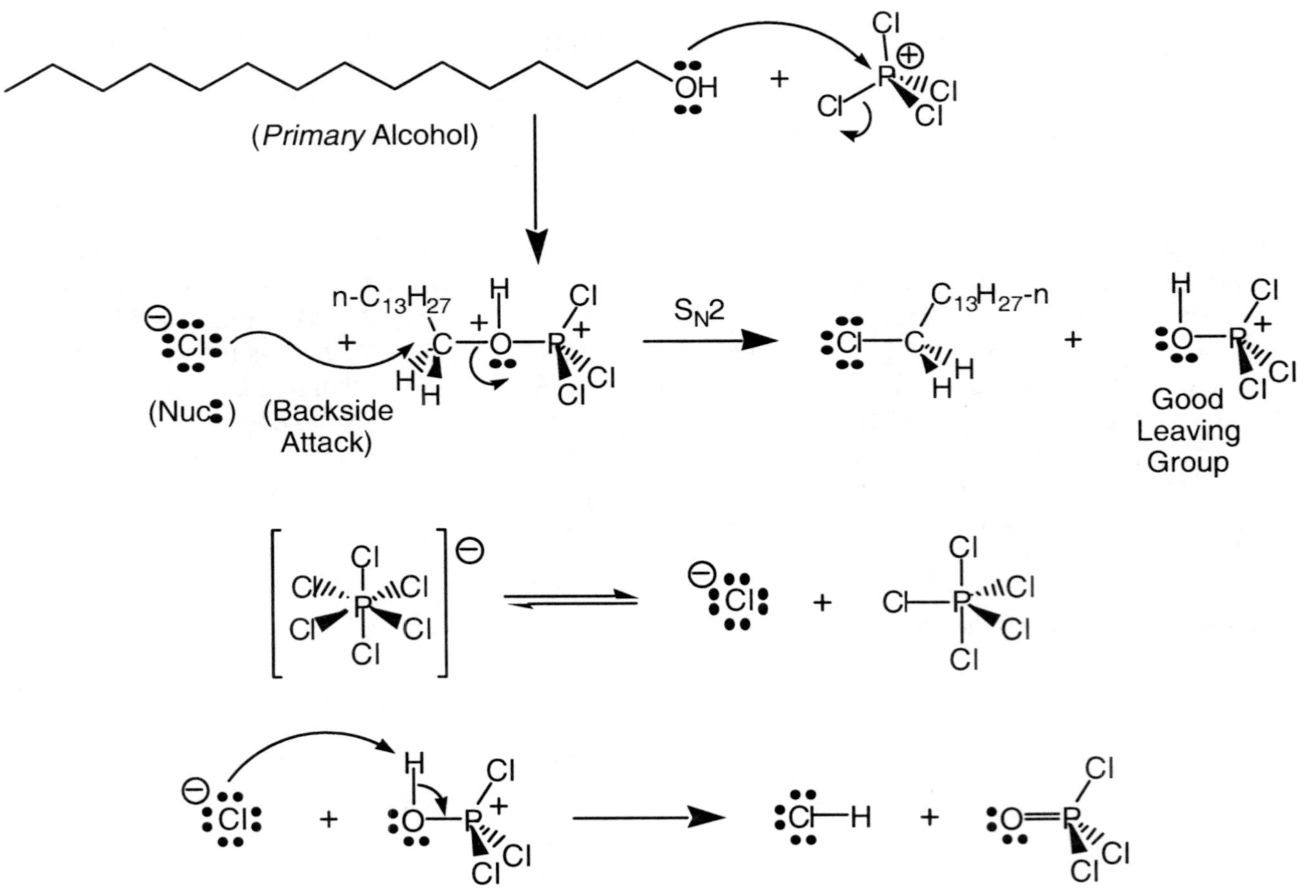

Workup: $O=PCl_3 + 3\ H_2O \longrightarrow O=P(OH)_3\ (H_3PO_4) + 3\ HCl$
Water Soluble (Neutralize)

Procedure: Weigh out **0.534 grams** of **phosphorus pentachloride (PCl$_5$)** into an 18 x 150 mm (large) test tube and *immediately* stopper with a cork. [**Caution:** PCl$_5$ reacts quickly and energetically with water to generate phosphoric acid and HCl gas; therefore the water vapor in the air will react with and *destroy* the PCl$_5$. You can minimize this *hydrolysis* (cleavage by water) by quickly weighing out the PCl$_5$ and keeping it sealed away from moisture as much as possible. **Keep the Reagent Bottle *Tightly* Closed When It Is NOT in Use!**] Using a clean spatula, weigh out **0.500 grams of 1-tetradecanol** into a 16 x 100 mm (small) test tube (using the small funnel from the microkit will facilitate this transfer). In the hood, quickly add the 1-tetradecanol *into* the PCl$_5$ in the large test tube and stir the mixture of solids with a glass stirring rod for **3–5 min**, cooling as necessary in an ice–water bath to moderate the vigorousness of this reaction (**exothermic** = heat given off!); during this time the mixture will liquefy. *Loosely* cork the test tube once again and place the test tube in a water bath (125 mL of water in a 250 mL beaker) heated on a digital stirring hotplate set at 99°C (use to be demonstrated) for the next **40 minutes**; agitate the mixture briefly 4–5 times during this heating period by "finger flicking" or swirling the tube. After heating for 40 min the reaction mixture should be clear with no solids remaining (if not, see your instructor).

Cool the reaction mixture in an ice–water bath and while in the ice bath add **2 mL of distilled water** and mix well with a glass rod. Remove the test tube from the ice bath, warm it to room temperature, and then transfer the entire mixture into a **centrifuge tube** with a disposable glass pipette. Allow the layers to separate and remove the aqueous layer with the glass pipette [you may have to test, by adding a few drops of water, to determine which layer is the aqueous one]. Extract the organic product by adding **2 mL of pentane** and **2 mL of saturated aqueous sodium chloride solution** to the centrifuge tube and mixing well by drawing up and expelling the mixture quickly with a pipette. Allow the layers to separate and again remove the aqueous layer with a glass pipette. [Note: Should a precipitate of sodium chloride form, dissolve it by adding 1–2 mL of additional water.] Transfer the organic layer with a glass pipette into a dry 16 x 100 mm (small) test tube and add a small amount ("just enough") of **anhydrous sodium sulfate** to remove the last traces of water. *Stopper* the test tube with a cork and allow the mixture to stand for 15 minutes with occasional *mixing*. When dry, transfer the solution away from the drying agent with a glass pipette into a second glass pipette packed with a small piece of **cotton** and a **2 cm high column of** *silica gel*, which will function as a chromatography column. Allow the solution of product in pentane to drain through the column of silica gel into a **weighed** 10 mL Erlenmeyer flask. Pass an additional *two* **1 mL portions of pentane (2 x 1 mL)** through the column to wash the last traces of product from the silica gel, also collecting these elutions of pentane in the weighed Erlenmeyer flask. When no more pentane will drain through the column by gravity, attach a hose to a compressed *air* line and connect it to the top of the pipette column, and then expel the remaining liquid from the column with a gentle stream of compressed air. (This will also dry the column and reduce the fire hazard of disposing of the used column in the broken (waste) glass containers.)

In the hood, evaporate the pentane solvent by blowing a slow stream of air from one of the hoses at the side of each hood into the flask until there no longer seems to be cooling of the flask due to the evaporation of the solvent. Take the sample back to your bench and connect the flask to the vacuum (suction) to remove the last traces of the pentane solvent. [To do this, set up a vacuum trap connected via vacuum tubing (red, thick wall) between the side-arm of the trap and the vacuum outlet on the bench top, and connect the product flask to the vacuum using the rubber thermometer adapter from the microkit, through which is inserted a short length of glass tubing that is connected with vacuum tubing to the adapter out on the trap.] Warm the flask with your hand and keep it under vacuum until it no longer bubbles and is no longer cooling from evaporation. Typically 1–2 minutes under *good* vacuum is sufficient. Weigh the flask and contents to determine the amount of 1-chlorotetradecane you have obtained, and calculate your percent yield.

Submit your *entire* **sample** of product (should be less than 1 mL) in one of the small autosampler vials for **gas chromatographic (GC) analysis** of its purity. A standard GC chromatogram will be posted showing the retention times of the 1-chlorotetradecane and the 1-tetradecanol starting material for comparison purposes.

Waste Disposal:

1. After separation of the **aqueous layer** from crude product, combine with the **saturated NaCl wash solution**, and neutralize this acidic mixture slowly and carefully with **solid Na_2CO_3**, and then discard into the sink (wash with a lot of water).
2. **Drying agent (Na_2SO_4)** – neutral salt: dry, then dissolve in water and flush down drain.
3. Dispose of the used pipette chromatography column (including silica gel) in the waste glass container.

Saponification: Synthesis of Salicylic Acid from Methyl Salicylate
(Oil of Wintergreen)

Adapted from: Joy of Organic Chemistry by Berton C. Weberg and John E. McCarty
Mankato State University, Mankato
by M. J. Lusch

Reaction Equation:

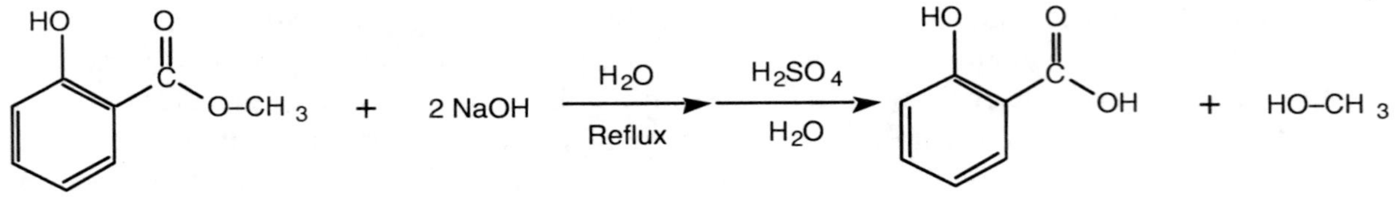

Mechanism: (General – Nucleophilic Acyl <u>Substitution</u>)

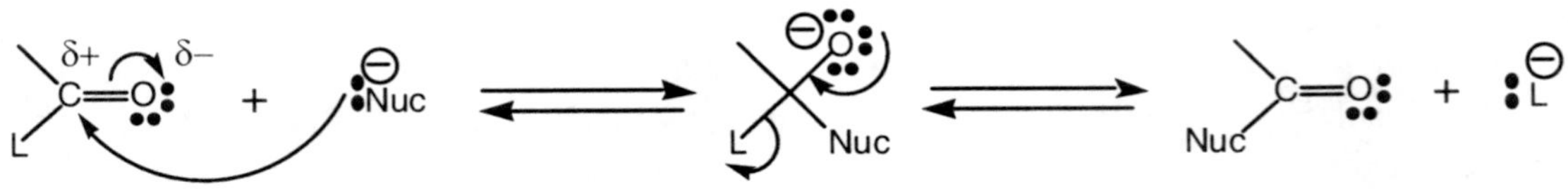

Mechanism: (General – Nucleophilic Acyl <u>Substitution</u>)

A16

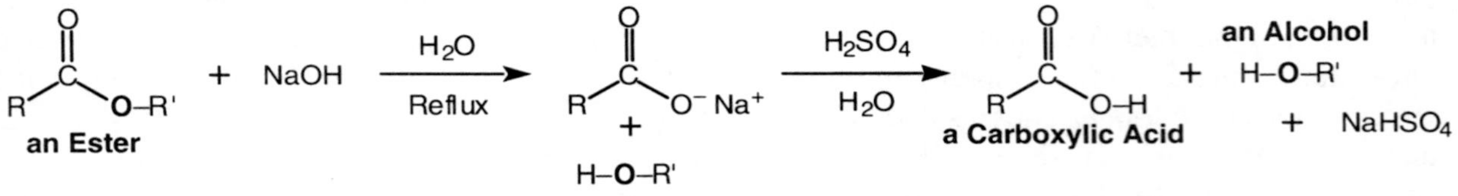

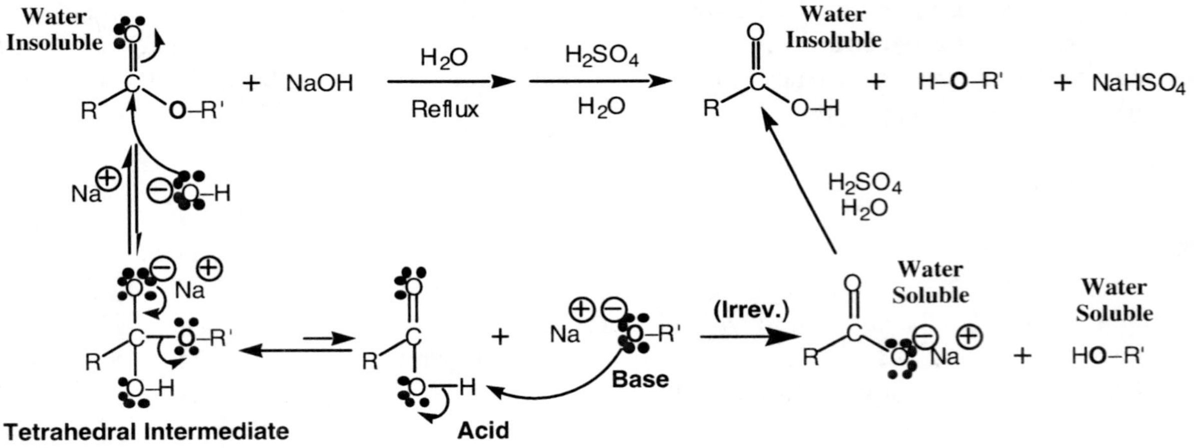

Mechanism: (Specific for Methyl Salicylate – Nucleophilic Acyl Substitution)

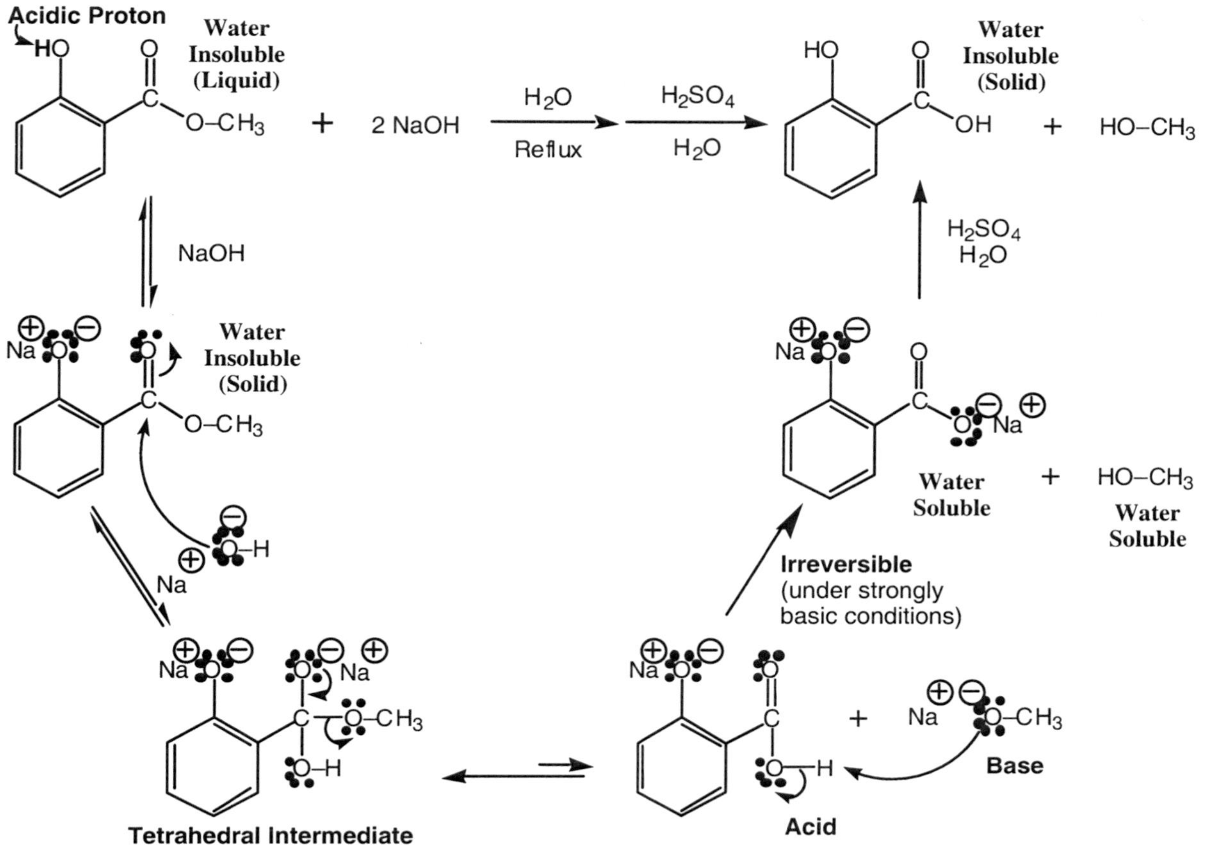

Procedure: Synthesis of Salicylic Acid by Basic Hydrolysis of Methyl Salicylate

Weigh **8.00 g of solid sodium hydroxide (pellets)** [**Caution!** Corrosive, Hygroscopic (water absorbing)] into a 100 mL round bottom flask, add **40 mL of water**, and swirl until all of the sodium hydroxide is dissolved (heat generated; no cooling necessary). Add a *teflon* boiling chip or two, and then add, in one portion, **4.00 g of liquid methyl salicylate (oil of wintergreen)** [note formation of a white solid]. Place a Liebig (fat) condenser vertically into the opening of the flask (lubricate the ground glass joint with stopcock grease), and, with water running through the condenser, heat the reaction mixture at a gentle boil (reflux) with a small electric heating mantle (no sand necessary – 100 mL flask fits snugly) until the white solid dissolves and the mixture becomes homogeneous (only one phase is present), and then for **5 minutes longer** (requires **approximately 10–15 minutes of reflux in total**).

Cool the reaction mixture in an ice/water bath to ice temperature and then transfer the reaction solution into a **250 mL beaker** (rinse flask with a small amount of water and add to the beaker). Slowly, with cooling and stirring, add **40 mL of 6N (3M) aqueous sulfuric acid** in portions of roughly 5 mL at a time, neutralizing the sodium hydroxide and making the reaction mixture acidic. As the last of the sulfuric acid is added and the solution becomes strongly acidic, a white crystalline solid (salicylic acid) will separate from the solution. Cool this mixture in an ice/water bath and remove the solid by vacuum filtration on a 5.5 cm Buchner funnel, then wash it thoroughly with cold water to remove as much sulfuric acid as possible. Recrystallize this damp salicylic acid from a minimum amount of boiling (heating mantle) pure water (up to approx. 125 mL or slightly more; cover the beaker with a watch glass while heating). After cooling (slowly at first and eventually in an ice bath) recover the recrystallized salicylic acid by vacuum filtration, wash with ice-cold water, and dry the resulting crystals until next week in your lab drawer. When dry, determine the weight of product obtained and determine its melting point. Pure salicylic acid melts at 159°C.

Technical Notes

1. Use **Caution!** when weighing out the solid NaOH: It is extremely corrosive to skin and eyes; it is also Hygroscopic (water absorbing) - don't leave spilled pellets out in the air for any length of time, and keep bottles capped when not in use.
2. Set up the reflux using an electrical heating mantle.
3. Add boiling chips before adding the methyl salicylate, and before beginning the heating for reflux.
4. For this experiment, put a 100 mL beaker inverted loosely over the opening at the top of the condenser (doesn't seal the top of the condenser, leaves it still open to the atmosphere) - to prevent spray of hot conc. NaOH solution should the boiling reaction mixture "bump".
5. Heat the mixture under gentle reflux until only **one** clear phase (no solids) is present, and then for 5 min longer.
6. Cool the reaction flask and its contents thoroughly in an ice-water bath, then pour reaction **solution** into a 250 mL beaker for the neutralization step (can rinse R.B. flask with a small amount of water) (100 mL R.B. flask will be too full otherwise, and it is difficult to remove precipitated solid from it).
7. During the filtration to collect the crude salicylic acid, you can use small amounts of cold water to rinse out the beaker to recover as much of the salicylic acid as possible.
8. Use the same 250 mL beaker as used in the acidification step for the recrystallization of the salicylic acid; don't even have to clean it because any solid residue is the same salicylic acid you are recrystallizing. Stir the solution continuously to speed up the dissolution of the salicylic acid.
9. Cover the beaker with a watch glass (supplied by instructor) while heating/boiling - will diminish the irritating vapor/odor given off (phenolic odor?), and will also diminish evaporation of the water so a skim of salicylic acid doesn't form on the surface of the boiling solution.
10. Leave the purified salicylic acid in the top of the Buchner funnel for drying until next week (easier to remove dry solid than wet solid from funnel).

During Lab and Post-lab Information: (To be completed during and after the experiment.)

1. You should include K, M, N, and O, (pp. 18-19), with the following provisions:
   M - Calculate percent yield based on the theoretical yield you calculated in your prelab.
   N - See the above mechanisms for Nucleophilic Acyl Substitution.
   O - Possibly something about the potential reasons (mechanical and chemical) for less than a perfect 100% yield.
2. Turn in your purified salicylic acid next week, after you have allowed it to dry, weighed it to determine the % yield, and taken its mp (as an indication of its purity). Place your product in the large bottle provided by the instructor for "**Student Prep Salicylic Acid.**"

Waste Disposal:

1. Slowly and carefully <u>neutralize</u> the acidic filtrate from the isolation of the <u>crude</u> salicylic acid (H_2SO_4, $NaHSO_4$, methanol, and traces of salicylic acid) with **solid Na_2CO_3**, and then discard into the sink (wash down with a lot of water).
2. The filtrate from the recrystallization (traces of salicylic acid) can be flushed down the drain directly.
3. **Collect** Student Prep **Salicylic Acid** after calculation of % Yield and melting point determination.

Thin Layer Chromatography - Analysis of Analgesics

Reference: Chapter 8, Williamson and Masters, 6th ed., pp. 164-77; Exp. 1 Analgesics

Procedures:

Preparation of TLC Plates:

- The *dull* side of the plate is the silica gel side (active). The shiny side is the plastic backing.
- Handle TLC plates ***only by the edges!!*** (Finger prints contains fats and oils, which will contaminate the silica gel surface of the TLC plate.)
- Draw a *light* pencil line on the silica gel side of the plate, about **1 cm from one end** of the TLC plate. Place 4 pencil dots along this line to mark the location for each of the samples to be spotted, 7 mm from each edge of the plate and 7 mm from each other (TLC plates are 3.5 cm x 6.7 cm [35 mm x 67 mm]).
- Make sure that no sample spot is *less than* 5 mm from the *edge* of the plate.

Spotting of Samples:

- Spotting solutions should be approximately 1% w/v (weight/volume) in a spotting solvent.
- You will need to spot and develop the 6 *known* analgesic materials: **acetaminophen, aspirin, caffeine, ibuprofen, ketoprofen, and naproxen**, and then determine the R_f of each substance.
- You will be able to spot only 4 samples per TLC plate. I would suggest 4 standards on one plate and 4 standards on a second plate, being sure to have at least *one* standard substance present on *both* plates to allow the most accurate side-by-side comparison.
- Make small sample spots, preferably no larger than 1 mm in diameter. You may wish to practice spotting on one end of a sample plate. You goal is to apply a small, concentrated spot (somewhat dark when visualized using the UV lamp) on the TLC plate. Too much sample will result in a large spot that may streak when developed. Check the spot size using the UV lamp (short UV). You should see a small dark spot for each compound at the baseline.
- Standard sample solutions of **acetaminophen, aspirin, caffeine, ibuprofen, ketoprofen, and naproxen** in 100% ethanol will be located in the lab (see instructor), as will the UV lamps. Please spot your *standard* plates there. The standard solutions will be in screw cap vials. The spotting capillaries will be in or next to the vials, a different spotting capillary for each different standard. Leave the capillaries with the correct sample vial after you finish with them; take care not to mix up the spotting capillaries.

Chromatographic Chambers:

- Each student should have a chromatographic chamber at their lab station. This consists of a tall beaker in which has been placed a piece of filter paper, and a small watch glass to be used as a cover for the chamber.

Elution Solvents:

- The primary elution solvent to be used, 2:1 hexanes/ethyl acetate (EtOAc) with 1.7% acetic acid (HOAc), will be located at your bench.
- Place enough elution solvent in the chamber to wet the filter paper completely and give approximately 0.25 inch of liquid on the bottom of the chamber (approximately 4 mL; need not be accurately measured).
- Be sure that the sample spots lie *above* the level of the solvent in the developing chamber.

- If you should need a more polar solvent (1% acetic acid in ethyl acetate; do not use this solvent for the other tests) there will be some sample chambers in the back of the room for that use. You may re-develop your current TLC plate in the more polar solvent if you handle it carefully.

Chromatogram Development:

- Place the TLC plate in the chamber (use a pair of tweezers to hold the plate) so that it is not touching the sides of the chamber except at the top of the plate (stand it in the *middle* of the bottom of the chamber if possible), and preferably leaning *away from* the filter paper.
- Allow the solvent to travel up the plate by capillary action.
- Develop the plate until the solvent reaches the top, or nearly so (you must remove your plate as soon as the solvent reaches the top or you will get inaccurate R_f's).
- Mark the *leading* edge of the solvent as soon as you remove the sample plate from the developing chamber. If you allow the solvent to reach the top of the plate, use the top as the solvent front.

Sample Visualization:

- Allow the TLC plate to dry *completely* (preferably in the *hoods*) before trying to visualize the sample spots by UV light or staining.
- **NEVER LOOK DIRECTLY INTO A UV LAMP; IT MAY DAMAGE YOUR EYESIGHT!!!!**
- Using the **short** wavelength UV (254 nm), sample spots should appear as dark spots against the fluorescent green background of the plates.
- Once the plate has been visualized, trace *lightly* around the spots with a *soft* lead pencil, to accurately reflect both the *size* and the *position* of the spot.
- Determine the R_f values for *each* component spot in the sample. Place a pencil mark at the **center** of each spot. Measure from the point of origin of the chromatography (the ***baseline***; **not** the bottom of the plate) to the *center* of each component spot, and from the baseline to the ***solvent front*** (measure in millimeters, to an accuracy of 0.5 mm, for best results):

$$R_f = \frac{\text{Distance (mm) a spot travels}}{\text{Distance (mm) the solvent travels}} \quad (0 \leq R_f \leq 1.0)$$

- You will need to trace your TLC plates *to scale* into your notebook and finally into a lab report.
- *Measure R_f's using the <u>original</u> TLC plates, not the drawings in your notes. Use the <u>center</u> of the spot for the R_f calculation.*
- Set up a table in your notebook with the <u>name, structure, and R_f calculation</u> for each known analgesic compound. You will have to look up the **structures** of each compound.
- You will find that at least one spot will barely move. *After* circling and measuring the Rf's of this and the other spots on the plate, take the plate and re-develop it in ethyl acetate containing 1% acetic acid. Look to see how far the spot moved and whether it consists of one or more spots. Circle the new spot(s) with a *<u>dashed</u> line, ignoring the other spots that moved to the <u>top</u> of the plate in this more polar solvent.* You should also calculate the R_f of the material on the lower portion of the redeveloped plate. (You should measure from the center of the spot *after* the *first* development to its destination after the *second* development, *not* the origin of the *first* spotting.) You may need to repeat the TLC of the spots of the compounds that barely move. If you do so you only need to develop the plate once using the more polar solvent (ethyl acetate/1% acetic acid), and then calculate the R_f's for this more polar solvent.

Unknown Sample Preparation:

- One to two unknown solid analgesics should be located at each balance. You need only a tiny amount (**3-5 mg; weigh!**) of each unknown in *separate* CLEAN reaction tubes or small test tubes. Add 0.5 mL of 100% ethanol to each unknown solid and allow as much of the unknown to dissolve with stirring or agitation as is soluble (the binder material will not dissolve). Allow the insoluble material to settle or centrifuge the mixture before spotting the solution. You will need only one spotting capillary for each of your unknown samples.

Identification of Unknowns:

- You will determine the **brand identity** (Table 1) of **2 (two)** over-the-counter analgesic unknowns (A-H) *of your choosing* by comparing the R_f's of their component(s) with the R_f's of the standard samples.
- You will need *accurate* measurements of R_f's, or in some cases **side-by-side comparisons** (run a fresh TLC with the *unknown* sample spotted side-by-side with *authentic* samples of suspected *known* analgesics, to determine conclusively the identity of some of the unknowns.
- Careful measurements and technique will allow you to distinguish all unknowns.

Table 1. Over-the-Counter Analgesics

Brand\ingredient(s):	Acetaminophen	Aspirin	Caffeine	Ibuprofen	Naproxen	Ketoprofen
Advil				200 mg		
Aleve					500 mg	
Anacin		400 mg	32 mg			
Bayer		325 mg				
Excedrin	250 mg	250 mg	65 mg			
No-Doz			200 mg			
Orudis KT						12.5 mg
Tylenol	500 mg					

Waste Disposal:

1. Waste from the experiment can go into the *non-halogenated organic* waste containers.
2. Empty the solvent from TLC developing chamber into the *non-halogenated organic* waste containers, and leave the chamber uncovered when you are finished with the experiment so that residual solvent will evaporate. *Do not* **throw away the filter paper from the chamber!** Once dry it can be reused.

Aldol Condensation Reaction: Synthesis and Identification of an Aldol Reaction Product

Reference: This experiment was adapted with permission by B. L. Groh and M. J. Lusch from an experiment designed by John M. Ferguson, Department of Chemistry, University of Central Oklahoma, Edmond, OK 73034.

Further reading: Chapter 37, pp. 484-485, Williamson and Masters, *Macro-scale and Microscale Organic Experiments*, 6th ed., Brooks/Cole Cengage Learning, Belmont, CA, 2011.

Reaction Equation:

General:

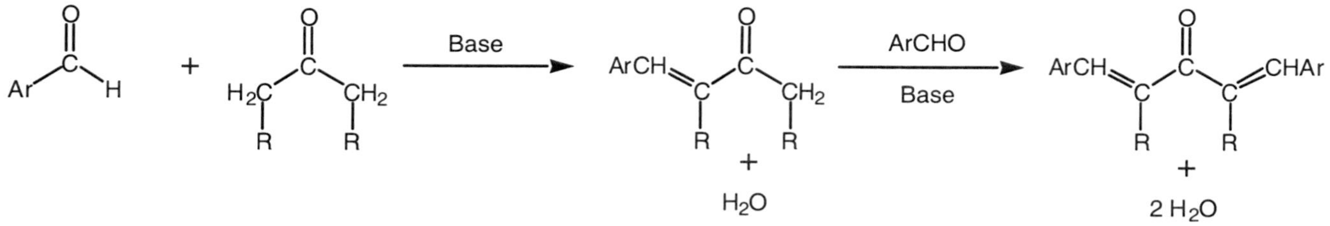

Mechanism:

General:

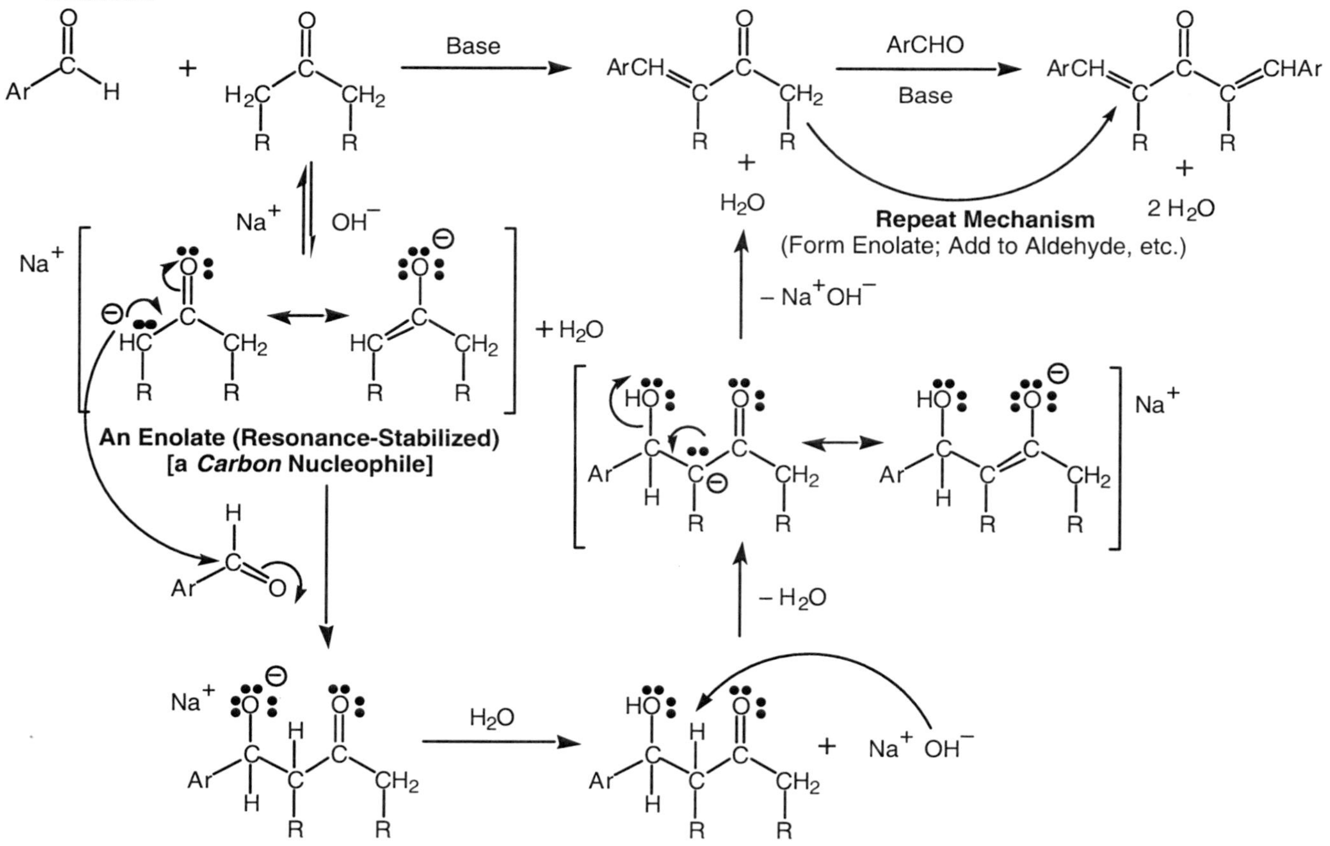

Procedure:

You are to select *one* **ketone** [acetone (*reagent grade*, not the acetone used for washing glassware), **cyclohexanone, or cyclopentanone**] as your ketone, and any *one* **of the aldehydes** available in lab [**benzaldehyde, p-methylbenzaldehyde, p-isopropylbenzaldehyde, p-meth-oxybenzaldehyde, p-ethoxy-benzaldehyde, piperonal, p-N,N-dimethylaminobenzaldehyde, or trans-cinnamaldehyde**]. Then carry out the **aldol reaction** under the conditions described below, **purify** the crude product by recrystallization from an appropriate recrystallization solvent, assess its purity by **TLC analysis** and melting point range determination, and eventual-ly **predict** the structure of the Aldol product, and finally verify the structure of the product based upon IR and NMR data. **Only the Aldol Reaction and the recrystallization (and possibly the TLC analysis) will be done in lab this week.** The purified and dried product will be submitted *next* week for the NMR spectrum to be run by the instructor. The IR spectrum will eventually be run by you at a convenient time in the next few weeks.

Liquid Aldehydes - About 3–4 mL of the aldehyde is added to a 50 mL round bottomed flask containing a solution of 1.0 mL of the selected ketone, 5 mL of 6M aq. NaOH, and 10 mL of 95% ethanol.

Solid Aldehydes - About 4.0 g of the aldehyde is dissolved in 15 mL of 95% ethanol. It is usually necessary to warm the ethanol to completely dissolve the solid. The aldehyde solution is added to a 50 mL round bottomed flask containing a solution of 1.0 mL of ketone, 5 mL of 6M aq. NaOH, and 10 mL of ethanol.

The reaction mixture containing either solid or liquid aldehyde is agitated to mix and then allowed to stand. The formation of a precipitate is usually quite rapid. If no precipitate is observed after 15 minutes, the flask should be fitted with a reflux condenser and warmed in a *sand bath* for 20 minutes. To ensure best possible reaction you may wish to reflux the reaction from the start rather than waiting 15 minutes to see whether the reaction proceeds.

[**Note:** *Acetone* **tends to undergo side reactions when** *heated* **in the presence of base. When using acetone it may be best to allow sufficient time at room temperature for the reaction to proceed on its own. If you are using the combination of the reactants** *acetone and benzaldehyde*, **the reaction should** not **be heated, but rather should be allowed to react only at room temperature to minimize side reactions. If no product has crystallized within an hour consult your instructor.**]

Once product has been formed the reaction mixture is chilled with an ice/water bath and the product isolated by vacuum filtration on a Hirsch or Buchner funnel. The crude product is then washed sequentially with ice-cold 10–15 mL portions of **95% ethanol, 4% acetic acid in ethanol** (what is being removed by this wash solution?), and finally with **95% ethanol** again. [Remember, when washing the product, disconnect the vacuum, add solvent, carefully mix the solid in the funnel with solvent (so you don't make a hole in the filter paper), and reconnect the vacuum. Do this each time you wash with a solvent when you isolate a product by filtration.]

Recrystallization - The damp crude product is then recrystallized from an appropriate solvent. *You* will have to determine the best solvent from which to recrystallize your product. You can do this by placing equal samples (ca. **10 mg** each) of the aldol product into small test tubes and adding equal amounts, but not more than **0.5 mL**, of the different solvents to each tube. If the sample dissolves *at room temperature*, the material is *too soluble* in the solvent to use it for recrystallization. If the sample does not dissolve at room temperature, try heating it over a steam bath. If the sample does *not dissolve when heated*, then it is *too insoluble* for that solvent to be used for recrystallization. **If it dissolves** *when heated*, **allow it to cool slowly to room temperature. If it crystallizes well without cooling in ice it** *may* **be a suitable solvent for recrystallization.**

The solvents from which to choose are the following: **100% ethanol** (if your sample is very soluble in 100% ethanol, try 95% ethanol), **ethyl acetate, aceton**e, **toluene,** or **2-butanone**.

After isolating the recrystallized product, assess its purity by TLC analysis (see below). The melting point *range* of the purified and dried product will be determined the next week, and the IR and NMR spectra will be obtained at a later time.

TLC Analysis of Recrystallized Aldol Product (see pp. 156-160 in Williamson for general information on TLC) - Assess the purity of your sample (need not be dry yet) by thin layer chromatography (TLC) on **silica gel** *before* submitting it for spectral analysis: Dissolve a small sample of your product in CH_2Cl_2. Make a small spot approximately 8–10 mm from the bottom of a TLC plate. Develop the plate in the **1.8% 1-propanol in CH$_2$Cl$_2$ solvent** provided. You should see a *single* distinct spot without a tail at R$_f$ between 0.3 and 0.8. If not, you will need to use a **more polar** solvent – **5% 1-propanol in CH$_2$Cl$_2$**. Carefully sketch the plate in your notebook and calculate the R$_f$ directly from the plate (R$_f$ is determined by measuring from the origin spot to the center of the sample spot and dividing that by the distance from the origin spot to the solvent front when you remove it from the developing chamber. The product(s) should be easily observed using a UV lamp. *NEVER LOOK DIRECTLY AT THE UV LIGHT!!*

IR Analysis - Samples for IR analysis can be conveniently prepared by dissolving a small amount (5 mg) of product in a few drops of CH$_2$Cl$_2$. Apply 1-2 drops of the solution onto the face of a NaCl plate (IR plate), then evaporate the solvent completely using a stream of air from the bench air lines. This will deposit a thin crystalline sample that is suitable for analysis. Place this single salt plate in a sample holder and take the IR spectrum. The baseline of the IR spectrum will most likely be tilted and will have to be corrected. Instructions for this procedure (base line correction) are next to the IR spectrometer. Your instructor will help you with this procedure if you wish. Always save your spectrum to the hard drive under a unique name so you may always return to the original spectrum. (A good way to save the spectrum is to use your initials and aldol; e.g. *BGAldol*).

Interpret the data from the IR using the bands above 1500cm^{-1} and *label the position of all peaks on the spectrum between 4000-1500^{-1}cm*.

^{1}H NMR Analysis - To request an NMR spectrum, you will need to submit a form with your name, aldehyde and ketone reagents used, recrystallization solvent used, and mp range along with your sample. The ^{1}H NMR spectrum of your compound will be taken and returned to you along with your sample. You will find your spectrum based upon the aldehyde and ketone pair you used.

Interpret the NMR data. To do this you may have to measure the integrals if they are not already measured for you. This is done by measuring the height of the integral (in mm) from the left side of a peak (flat line) to the right side of the peak (again to the flat line to the right of the peak, but higher). See your lecture text or instructor for more information. It is acceptable for integral ratios to be off by as much as 10% of the height of the line. Determine the integral ratios of the peaks. You now have the simplest ratios between different hydrogens.

Label all peaks with the corresponding fragment of your molecule. This is most easily done by labeling the signals in your spectrum from *right* to *left* just under the baseline with **a, b, c, d,** etc. Then draw your product on the NMR spectrum and circle all H's of a particular resonance on the structure, labeling the circled H's with the alphabetical label of the signal to which they correspond.

Waste Disposal:

The reaction mixture filtrate should be disposed of in the ***basic* organic waste**. The filtrate from recrystallization should be disposed of in the non-halogenated organic waste container.

GENERAL PROCEDURES FOR IDENTIFYING UNKNOWNS

1. **Purify and Obtain Physical Constants.**
2. **Perform (1) Solubility and (2) Other Test-Tube Tests, to Obtain As Much Information as Possible Concerning the Structure of the Unknown.**
3. **Compile a Short (5-20) List of Possible Compounds.**
4. **Prepare at least Two Derivatives of the Unknown.**
5. **Decide Upon the Identity of the Unknown and Submit Report.**

EXPLANATION

Important: Treat the Unknown Sample Received as Though It Will Be the Only Quantity of the Material Available to You.

1. **Purify and Obtain Physical Constants:**

For Solids - Determine the Melting Point Range of the Unknown as received; if the mp range is less than 2-3°C, no further purification will be necessary; if the mp range is greater than 3°C, recrystallize a small sample (0.5-1.0 gm) of the unknown from a suitable (to be determined by you) solvent, and determine the mp range of the purified material. Repeat the recrystallization from the same or a different solvent until the mp range becomes less than 2-3°C (preferably less than 1°C). For the limited number of solids that sublime readily, sublimation may be preferable to recrystallization as a purification technique.

For Liquids - Distill (Microscale) the Unknown as received to purify and obtain a Boiling Point Range; if necessary, determine the bp of the purified unknown more accurately using the Micro Boiling Point Technique (Procedure Available). The Refractive Index of the liquid Unknown may also be determined IF the substances under consideration have refractive indices that differ by at least 0.0020 units (decide after Step 3 above).

2. **Perform Test-Tube Tests to Obtain As Much Information as Possible Concerning the Structure of the Unknown:**

First, perform *solubility tests* on your unknown to get some information about potential functional groups in the molecule (Williamson, 4th Ed., pp 766-768, esp. **Figure 70.2**, pg 767).

Second, perform the **Beilstein copper wire test for halogens** (Cl, Br, and I only; F does not show up; Williamson, 4th Ed., pg 766), and *functional group* classification tests (for alcohols, phenols, aldehydes/ketones, carbon-carbon double/triple bonds, and, if necessary, esters, amides, and nitriles) (Williamson, 4th Ed., pp 769-778 *and* Handouts).

Third, perform other refining test tube tests: Lucas and Jones Reagents for 1°, 2°, and 3° alcohols, Jones and $KMnO_4$ tests for Aldehyde vs Ketones, Iodoform Test for Alcohols and Ketones, Hinsberg Tests for 1°, 2°, and 3° amines, and other more specialized tests if necessary.

Before performing *any* test on an Unknown compound, always run the test (especially if you have never run the test before) on a compound known to give a positive result, and also on a *blank* (all reagents except the compound to be tested) or on a compound known to give a negative result. In this way you will know whether the reagents are good (if the positive result works properly), *and* you will have a basis for comparison with the result from the Unknown, and will be able to decide whether it gives a positive, negative, or inconclusive result.

Use as many tests as are available to you for a given functional group and attempt to obtain as much corroborating information as possible in order to guard against misleading or erroneous results. Try not to base your entire case on a single test result. Keep in mind that some compounds do not behave as we generally would expect them to behave and therefore give inconclusive or misleading test results.

Take care to avoid too narrow an interpretation or concluding too much from the results of a test (for instance, formation of a muddy brown precipitate and the disappearance of purple color when a compound reacts with cold $KMnO_4$ may mean that the compound is an alkene, but it could also be an alkyne, or it might also be an aldehyde, or any one of a number of more exotic but easily oxidized organic compounds).

3. Compile a Short List of Possible Compounds (5-15):

Routinely consider all compounds in the available tables that have melting points (for solids) or boiling points (for liquids) that are ± 5°C on either side of the melting point range or boiling point range *you* determined for your Unknown. (In some cases ± 10°C may be suggested by the instructor.) Exclude from this short list those compounds that have been eliminated on the basis of the results of the test tube tests (*but,* be prepared to reconsider them if subsequent information just doesn't fit the other compounds).

4. Prepare at least Two Derivatives of the Unknown (Handouts):

Decide on the basis of the possible compounds in the short list which of the available derivatives would *best serve* to distinguish between them. The greater the differences in melting points for a given type of derivative, the more easily and certainly will the possibilities be distinguished. It would be desirable to have derivative melting points at least 5°C and preferably 10°C or more apart, in order to avoid mistakes due to impurities in the derivative you have produced (keep in mind that impurities lower and broaden the melting point range of a solid). It is also desirable to choose derivatives that have melting points well above room temperature (above 60-70°C) if possible, in order to avoid impurity-induced lowering of the melting point to the extent that it becomes difficult to obtain a solid derivative at all.

Prepare *as many* derivatives as necessary to *unambiguously* determine the identity of the Unknown.

5. Decide Upon the Identity of the Unknown and Submit Report:

Report forms will be made available on which you report each piece of data you obtained for your Unknown, and what if anything you concluded from it. You will be graded *not only* on the correct identification of the Unknown, but also upon the *completeness and validity* of the data you collect and the *thoroughness and logic* of your conclusions as presented in your report.

It is therefore extremely important that you keep an *accurate* and *thorough* laboratory notebook during the course of your investigations, recording each piece of information as you obtain it - physical properties, test results and the conclusions to be drawn from them, compounds being considered and their structures, procedures for preparing and purifying derivatives (including identity and quantity of recrystallization solvent), and the melting point ranges of the resulting solids. A test result that was not recorded is no more useful than a test that was not even run. Also be careful to distinguish between an *observation* (what you see happening or not happening) and the *conclusion* being drawn from that observation (what does the result mean about the Unknown).

Tests for Presence of C=C (double) or C≡C (triple) Bonds (NOT Aromatic unsaturation)[1]

[1 mL ≈ 20-25 drops]

- Subject <u>both</u> **your unknown** <u>and</u> a **known alkene** (e.g., cyclohex<u>ene</u>) to the bromine/CCl$_4$ and aqueous KMnO$_4$ tests, side-by-side, and do a blank test as well (all reagents except the unknown or alkene. The <u>known alkene</u> will tell you what a **positive** test will look like (and also that the test reagents are good), and the <u>blank</u> tells you what a **negative** test (<u>no reaction</u>) looks like by comparison.

- **Bromine in CCl$_4$:**

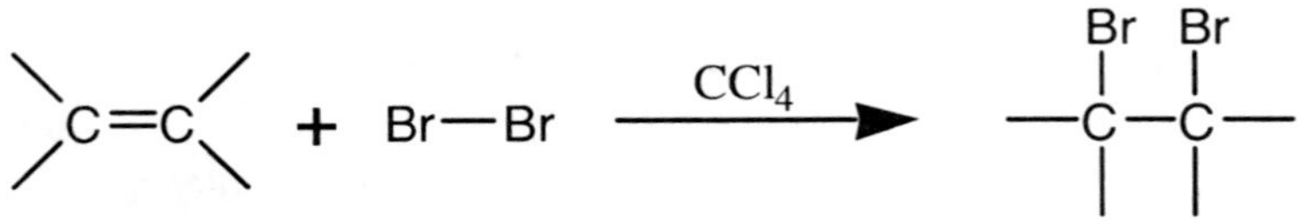

$$\underset{\text{Colorless}}{\diagdown\text{C}=\text{C}\diagup} + \underset{\text{Red-Orange}}{\text{Br}-\text{Br}} \xrightarrow{\underset{\text{Colorless}}{\text{CCl}_4}} \underset{\text{Colorless}}{-\underset{|}{\overset{|\text{Br}}{\text{C}}}-\underset{|}{\overset{|\text{Br}}{\text{C}}}-}$$

<u>Alkenes (& Alkynes)</u>– React Rapidly to Decolorize the Red-Orange Bromine Solution.

<u>Alkanes</u> – *Don't* React within 30 sec (No Decolorization of the Red-Orange Bromine Solution, except <u>slowly</u> in light over 10-15 min, with HBr evolution.)
- do bromine test with **1 mL (approx. 20 drops) 5% Br$_2$ in CCl$_4$ (Caution! Bromine corrosive**; use only in hood) and enough drops (<u>count</u> how many) of cyclohexene to decolorize the bromine; repeat with the **same** number of drops of cyclohex<u>ane</u> added to 1 mL 5% Br$_2$ in CCl$_4$ in each of 2 additional test tubes, one kept in the dark and one in the light (no <u>rapid</u> decolorization should occur; no reaction <u>at all</u> in the dark, <u>slow</u> reaction with HBr evolution (fuming) in the light).

Aqueous Potassium Permanganate:

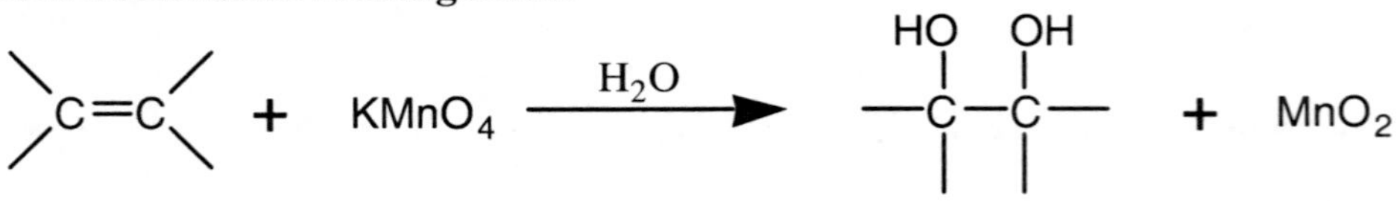

$$\underset{\text{Colorless}}{\diagdown\text{C}=\text{C}\diagup} + \underset{\text{Royal Purple}}{\text{KMnO}_4} \xrightarrow{\text{H}_2\text{O}} \underset{\text{Colorless}}{-\underset{|}{\overset{|\text{HO}}{\text{C}}}-\underset{|}{\overset{|\text{OH}}{\text{C}}}-} + \underset{\text{Muddy Brown Ppt}}{\text{MnO}_2}$$

<u>Alkenes (& Alkynes)</u> – React with agitation over a few minutes to change the Royal Purple Solution to a Muddy Brown.

---

[1] *From Joy of Organic Chemistry by Berton C. Weberg and John E. McCarty, Mankato State University, Mankato. Edited by M. J. Lusch*

Alkanes – *Don't* **React, Royal Purple Solution remains Unchanged**

- Do permanganate test by placing **1 mL (20 drops) of 1-2% aqueous permanganate solution** (royal purple) in a test tube and adding **2-3 drops** of unknown alkene to give a distint muddy brown precipitate. Compare the result with **2-3 drops** of cyclohex<u>ene</u> and to the permanganate solution without any organic compound added.

Waste Disposal:

(a) Br_2 in CCl_4 test solutions → Halogenated Organic Waste.

(b) Permanganate test solutions → Add Sodium Bisulfite to decolorize, Flush down the drain.

(c) Conc. Sulfuric Acid test solutions → Aqueous Acidic Waste.

Test for Phenols

Adapted from Kenneth Williamson, Macroscale and Microscale Organic Experiments, 4th ed., Houghton Mifflin, Boston, MA 2003, Ch. 70, pp. 752, by B.L. Groh

Iron (III) Chloride Test.

Dissolve 15 mg of the unknown compound in 0.5 mL of water or water-alcohol mixture, and add 1 or 2 drops of 1% iron (III) chloride solution. A red, blue, green or purple color is a positive test.

Waste Disposal:

Since the quantity of material is extremely small, the test solution can be diluted with water and flushed down the drain. Alternatively, it can be placed in the aqueous acidic or basic waste container.

Unknown Alcohols: Functional Group Characterization[1]

1. Ceric Ammonium Nitrate (in 2M aq. Nitric Acid) $Ce^{IV}(NH_4)_2(NO_3)_2$

Used to detect the presence of <u>alcohol</u> hydroxyl groups: primary, secondary, or tertiary alcohols give a positive test. The alcohol forms a red-colored complex with the initially yellow-orange ceric ion solution. Phenols do not give the characteristic red color, but may be oxidized to give brown or black products.

General procedure:

Place **1 mL** of the **ceric ammonium nitrate solution** in a small test tube, and add **3–5 drops** of your unknown. Mix thoroughly. A rapid change in color from yellow-orange to red or red-brown indicates the presence of an alcohol (primary, secondary, or tertiary). Use **1-butanol, cyclohexanol,** and **tert-butyl alcohol** for comparison as *known* examples of primary, secondary, and tertiary alcohols, respectively. If your unknown is *very insoluble* in water, it may be necessary to dissolve it in a water–soluble organic solvent first: place **5 drops** of your unknown in a small test tube and add **1 mL of acetonitrile (MeCN)** to dissolve the unknown, and then add **1 mL** of the **ceric ammonium nitrate solution**.

2. Jones Oxidation (Chromium Trioxide in conc. Sulfuric Acid = Chromic Acid)

Used to detect primary and secondary hydroxyl groups: oxidizes primary alcohols to carboxylic acids and secondary alcohols to ketones, respectively. *Aldehydes* **also** give a positive Jones Oxidation test (become oxidized to carboxylic acid). Tertiary alcohols and ketones cannot be further oxidized and give negative results.

1° ROH:

$$R-CH_2OH + Cr_2O_7^{2-} \xrightarrow{H^+} R-\overset{O}{\underset{\|}{C}}-OH + Cr^{3+}$$

2° ROH:

$$R_2CHOH + Cr_2O_7^{2-} \xrightarrow{H^+} R-\overset{O}{\underset{\|}{C}}-R + Cr^{3+}$$

(orange-yellow solution)   (greenish blue solution)

General procedure: Do this test in the fume hood!

Place **1 mL of acetone** and **2 drops of liquid alcohol** or **10 mg of solid alcohol** in a test tube. Mix the contents thoroughly, and then add **2 drops of the chromium trioxide/sulfuric acid reagent (Jones reagent)**. Mix thoroughly and note the result <u>within 2 sec</u>. The orange-yellow dichromate color should change to **greenish-blue** upon oxidation of primary and secondary alcohols (**and aldehydes**). Tertiary alcohols (and ketones) give no visible reaction in 2 sec, the orange color of the solution remaining unchanged. Disregard any changes taking place

---

[1] *From Joy of Organic Chemistry by Berton C. Weberg and John E. McCarty, Mankato State University, Mankato. Edited by M. J. Lusch*

after 2 seconds. Run a control test on the acetone solvent (a blank-no reaction) and compare the result. Use **1-butanol**, **cyclohexanol**, and **tert-butyl alcohol** for comparison as **known** examples of primary, secondary, and tertiary alcohols, respectively.

3. Lucas Test (Zinc Chloride in conc. HCl)

Used to detect secondary and tertiary alcohol hydroxyl groups: conversion of an alcohol to an alkyl halide **via a S_N1 reaction**. The more stable the carbocation intermediate, the faster the reaction.

2° ROH: $\quad R_2CHOH + HCl \xrightarrow{ZnCl_2} R_2CHCl$

3° ROH: $\quad R_3COH + HCl \xrightarrow{ZnCl_2} R_3CCl$

General procedure:

Before performing the test with the Lucas reagent, test the solubility of your alcohol in concentrated HCl by adding **5 drops of alcohol** to **1 mL of conc. HCl**. *The Lucas test will not work for alcohols which are **insoluble** in concentrated hydrochloric acid.*

Place **1 mL of the Lucas reagent** in a test tube and add **5 drops of your unknown alcohol**. Mix thoroughly. The formation of insoluble alkyl chloride is typically detected as a milkiness and sometimes even as a separate upper layer. **Tertiary alcohols react immediately, secondary alcohols usually react within 5-30 minutes**, and **primary alcohols do not react in less than 1 hr**, if ever (because the carbocation does not form). Use **1-butanol**, **cyclohexanol**, and **tert-butyl alcohol** for comparison as **known** examples of primary, secondary, and tertiary alcohols, respectively.

4. Iodoform Test

Used to detect **methyl carbinols** and **methyl ketones**. Use the iodoform test *only* if you have evidence that your compound contains an alcohol or a ketone. Ethanol is the only primary alcohol, and acetaldehyde (ethanal) the only aldehyde, that gives a positive iodoform test (R = H).

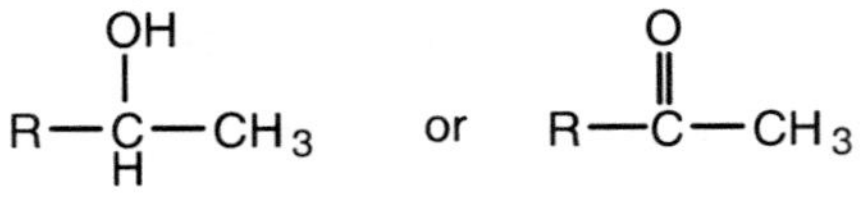

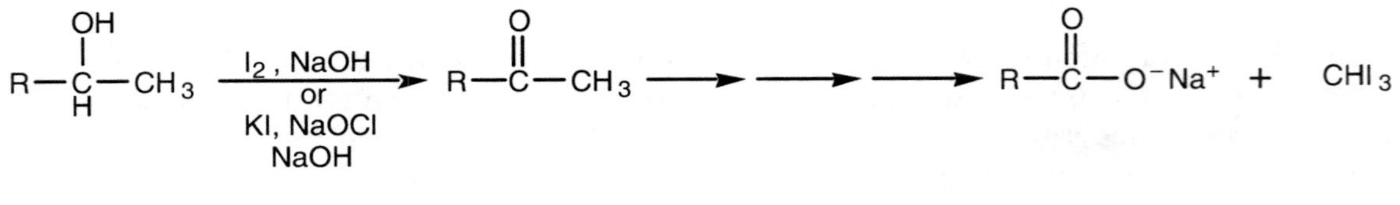

$$R-\underset{H}{\overset{OH}{C}}-CH_3 \quad \text{or} \quad R-\overset{O}{\overset{\|}{C}}-CH_3$$

$$R-\underset{H}{\overset{OH}{C}}-CH_3 \xrightarrow[\substack{\text{or} \\ KI, NaOCl \\ NaOH}]{I_2, NaOH} R-\overset{O}{\overset{\|}{C}}-CH_3 \longrightarrow \longrightarrow R-\overset{O}{\overset{\|}{C}}-O^-Na^+ + CHI_3$$

General procedure:

Place **5 drops of liquid unknown or 0.1 g of solid unknown** in a *large* test tube and dissolve in **5 mL acetonitrile (MeCN)**. To this add **1 mL of 10% aq. NaOH**, and then add, portion-wise *with stirring*, **iodine/potassium iodide (I_2/KI) solution** (requires a *large* amount; e.g., 10-20 mL) until there is a definite excess of iodine (brown color). (The iodine will be decolorized by the sodium hydroxide (forming NaOI), but this fact *alone* does not indicate a positive test.) When the color of iodine is no longer discharged after **two minutes** of stirring, add a **few drops of 10% NaOH** to discharge the brown color, and then dilute the reaction mixture with **water** to fill the test tube. The formation of a **lemon yellow precipitate (CHI_3)** with the distinct odor of iodoform (mp = 119-21°C) is an indication of the presence of a **methyl carbonyl or methyl carbinol group**. Test **acetophenone (methyl phenyl ketone)** to see what a **positive** test looks like, and **cyclohexanol** or **cyclohexanone** to see what a **negative** test (no iodoform) looks like.

Unknown Alcohols: Derivative Preparation[1]

1. Aryl (Phenyl and α-Naphthyl) Urethane Derivatives

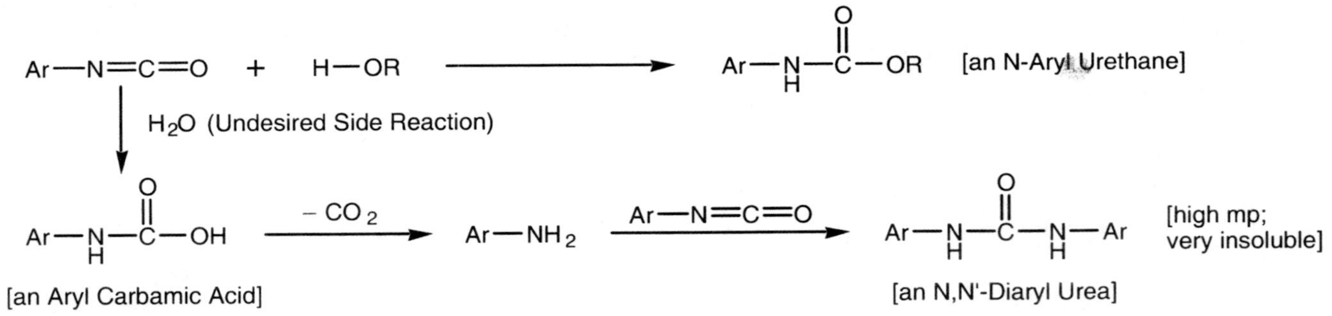

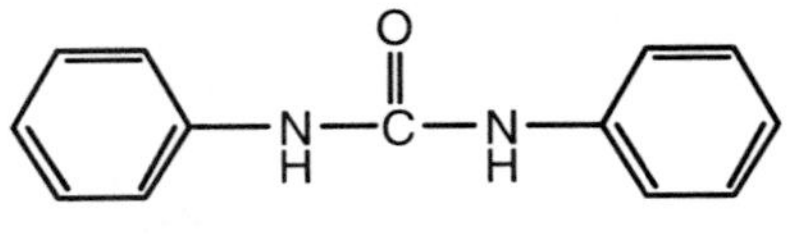

Ar—N=C=O + H—OR ⟶ Ar—NH—C(=O)—OR [an N-Aryl Urethane]

↓ H_2O (Undesired Side Reaction)

Ar—NH—C(=O)—OH —$-CO_2$→ Ar—NH_2 —Ar—N=C=O→ Ar—NH—C(=O)—NH—Ar [high mp; very insoluble]

[an Aryl Carbamic Acid]   [an N,N'-Diaryl Urea]

N,N'-Diphenyl Urea, mp 238°C

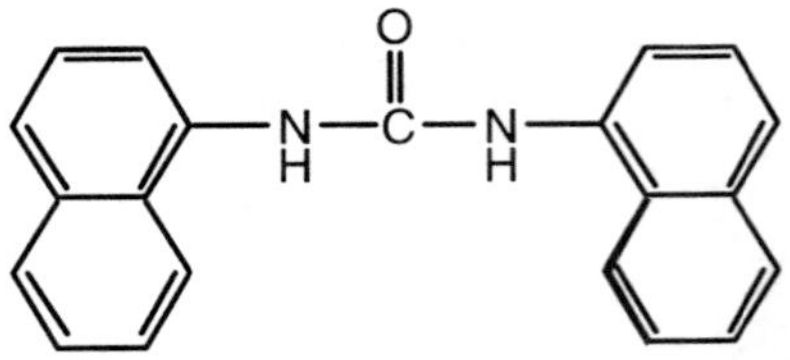

N,N'-Di-1-Naphthyl Urea, mp 296°C

General Procedure:

CAUTION!! All isocyanates are powerful <u>lacrymators</u> and are chemically reactive with water, alcohols, phenols, amines, and other active hydrogen compounds. Measure out isocyanates <u>in the hood only</u>, and ***tightly*** replace the bottle caps immediately after use.

Place **1.0 mL (about 25 drops)** of ***phenyl*** isocyanate (to prepare a *phenyl* urethane) or 1- (or α-) *naphthyl* isocyanate (to prepare an *α-naphthyl* urethane) in a **dry** test tube, and **add 1.0 mL (about 25 drops) of liquid alcohol or 1.0 g of solid alcohol.** Loosely stopper the test tube and heat on a steam bath for 10 minutes or longer. Cool the mixture in an ice bath and if necessary scratch the sides of the test tube with a glass rod to induce crystallization.

The crude urethane thus formed must be purified and separated from (the sometimes *large* amount of) insoluble N,N'-diaryl urea byproduct by adding **3-5 mL of hexanes** solvent (bp 68°C, flammable), pulverizing any solid, and heating the hexanes to boiling (boiling stick) while thoroughly mixing it with the solid (this dissolves any desired *urethane* that is present, but <u>not</u> the undesired, insoluble *urea* byproduct. The insoluble urea must then be removed by **rapidly hot filtering** the mixture into another test tube to remove insoluble solids (**using a disposable 3 mL plastic pipet filter; procedure to be demonstrated**). If there appears to be a solid in the resulting hexanes filtrate, warm the solution on a steam bath to see if the solid will redissolve. **If the** resulting filtrate is clear *when hot*, cooling in an ice bath will usually give colorless crystals of the **desired urethane derivative**. These are filtered off (using a small scale Hirsch funnel and the side-arm test tube or microkit filter flask, instead of the large Buchner funnel and 250 mL filtering flask), washed with **ice cold, fresh hexanes**, and dried.

---

[1] *From Joy of Organic Chemistry by Berton C. Weberg and John E. McCarty, Mankato State University, Mankato. Edited by M. J. Lusch*

2. Nitrobenzoate Ester Derivatives (3,5-Dinitrobenzoates & p-Nitrobenzoates)

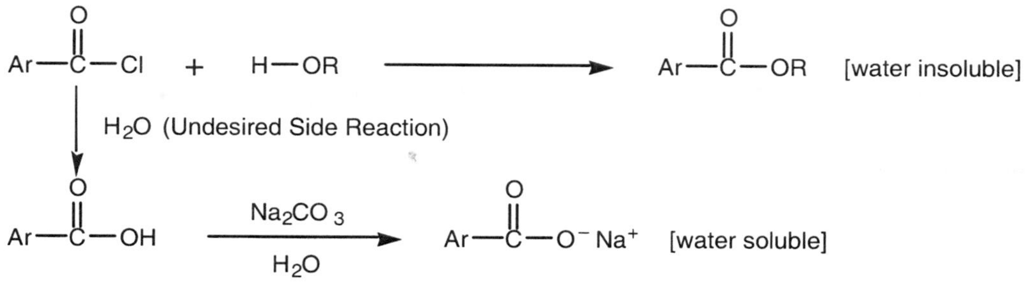

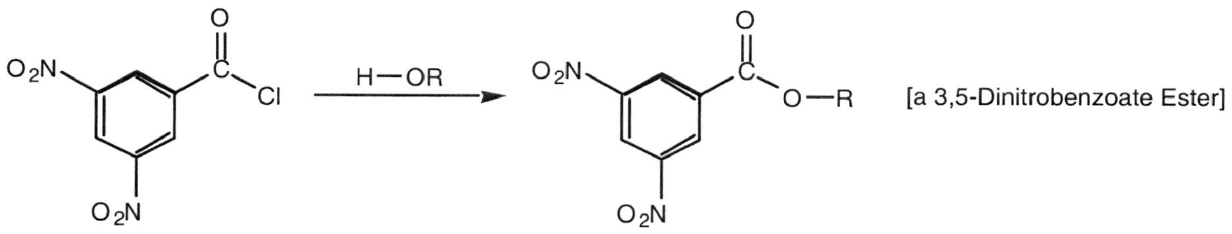

$$Ar-C(=O)-Cl + H-OR \longrightarrow Ar-C(=O)-OR \quad \text{[water insoluble]}$$

$\downarrow H_2O$ (Undesired Side Reaction)

$$Ar-C(=O)-OH \xrightarrow{Na_2CO_3,\ H_2O} Ar-C(=O)-O^-Na^+ \quad \text{[water soluble]}$$

[3,5-Dinitrobenzoyl Chloride] + H—OR → [a 3,5-Dinitrobenzoate Ester]

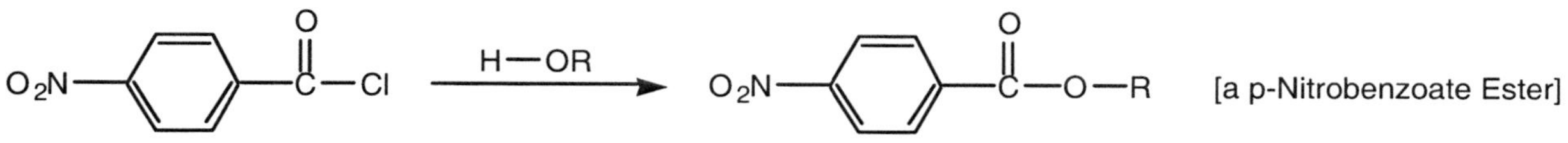

[p-Nitrobenzoyl Chloride] + H—OR → [a p-Nitrobenzoate Ester]

General Procedure:

CAUTION!! The nitrobenzoyl chlorides employed in these procedures (*especially* the **3,5-dinitrobenzoyl chloride, which is usually kept in a desiccator**) are reactive with water **and** *atmospheric moisture*; they must be weighed out quickly and only immediately before use, and the bottles *tightly* capped *after* use, to prevent deterioration of the reagents due to unwarranted exposure to atmospheric moisture.

Place **1.0 g of 3,5-dinitrobenzoyl chloride** (to make a 3,5-dinitrobenzoate) **or p-nitrobenzoyl chloride** (to make a p-nitrobenzoate) in a *large* **dry** test tube and add **1.0 mL (about 25 drops) of liquid alcohol or 1.0 g of solid alcohol**. Loosely stopper the test tube and heat on a steam bath for 10 minutes or more. If the odor of alcohol is still prominent, add another 0.5 g of the acid chloride and heat for an additional 10 minutes.

Cool the reaction mixture and attempt to induce solidification of the product by scratching with a glass rod. **Even if the product refuses to crystallize**, however, add **10 mL of 5% (saturated) sodium carbonate solution** and thoroughly mix it with the reaction products.

In the *most* favorable cases, byproduct nitrobenzoic acid and unreacted acid chloride will dissolve in the basic solution (caution: foaming; CO_2 evolution), and a ***water-insoluble,* solid nitrobenzoate ester** will separate. This can then be filtered off, washed with water, and recrystallized from 95% ethanol or water/ethanol mixtures (sometimes hexanes or hexanes/ethyl acetate mixtures work better).

In the *less* favorable cases, in which the water-insoluble nitrobenzoate ester will not solidify in the aqueous mixture, decant the aqueous layer from the liquid nitrobenzoate and attempt to dissolve the nitrobenzoate in a minimum amount of boiling recrystallization solvent. Solid nitrobenzoate can sometimes be obtained by cooling such solutions. If this proves successful, the resulting solid can be filtered, washed, and dried as usual.

Unknown Aldehydes and Ketones: Functional Group Characterization[1]

Experimental Procedures:

A. Test Tube Tests – Reactions which *Characterize* the Carbonyl Compound

1. Test for the *Presence of Aldehyde/Ketone Carbonyls*: 2,4-Dinitrophenylhydrazone (2,4-DNP) Formation. Few other functional groups react in any way (alcohols, phenols, carboxylic acids and esters do not react at room temperature; amines may form insoluble, high melting (>250°C) salts in the acidic reagent, but these are usually not very highly colored).

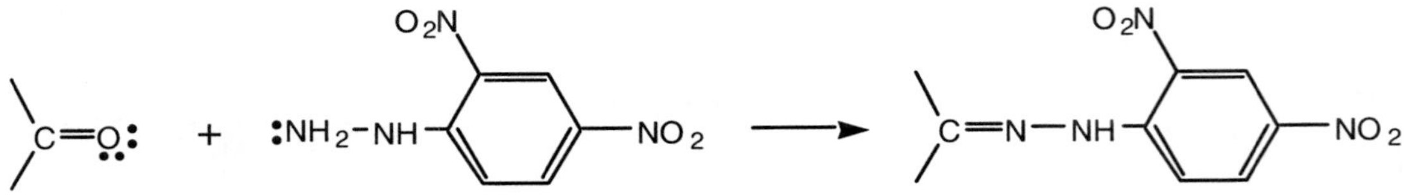

$$\text{C=O} + \text{NH}_2\text{-NH-}\underset{\text{2,4-Dinitrophenyl Hydrazine}}{\underset{\text{(2,4-DNPH)}}{\text{Ar(NO}_2)_2}} \longrightarrow \text{C=N-NH-}\underset{\text{A 2,4-Dinitrophenyl Hydrazone}}{\underset{\text{(A 2,4-DNP)}}{\text{Ar(NO}_2)_2}}$$

Experimental:

Place **2 mL of 2,4-Dinitrophenyl Hydrazine (2,4-DNPH) Reagent** solution (already prepared: 2,4-DNPH + conc. H_2SO_4 dissolved in ethanol/water) in a large test tube and add **2 drops of liquid compound to be tested**, or **0.1 g of solid carbonyl compound dissolved in 1 mL of methanol**. Thoroughly mix the contents and notice the almost immediate **formation of a colored (yellow, orange, or red) precipitate** of the 2,4-Dinitrophenyl Hydrazone (2,4-DNP) derivative. **Save this precipitate! It can be used as one of the derivatives you need to prepare.** (See Below)

2. Distinguishing between Aldehydes and Ketones (by Oxidation)

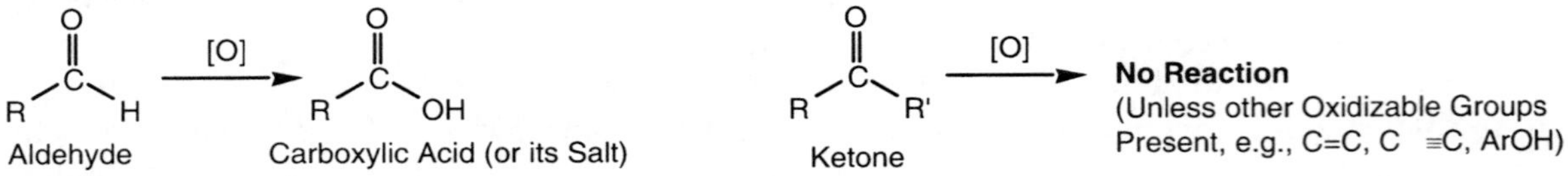

Aldehyde —[O]→ Carboxylic Acid (or its Salt)

Ketone —[O]→ **No Reaction** (Unless other Oxidizable Groups Present, e.g., C=C, C≡C, ArOH)

(a) Permanganate Test (Aldehydes React, Ketones Don't)

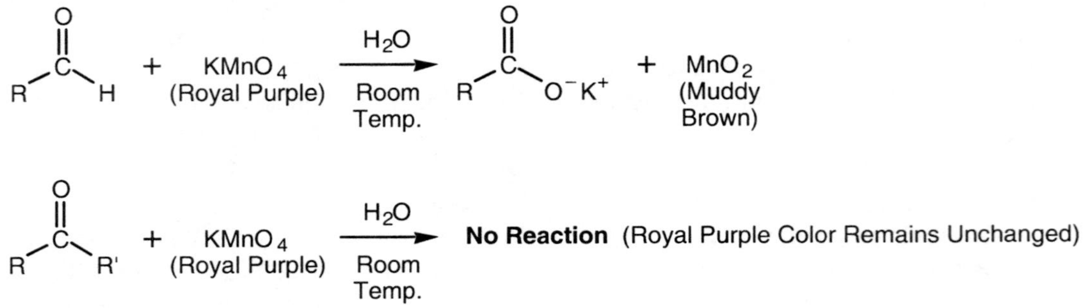

R-CHO + $KMnO_4$ (Royal Purple) —H_2O, Room Temp.→ R-COO⁻K⁺ + MnO_2 (Muddy Brown)

R-CO-R' + $KMnO_4$ (Royal Purple) —H_2O, Room Temp.→ **No Reaction** (Royal Purple Color Remains Unchanged)

---

[1] *From Joy of Organic Chemistry by Berton C. Weberg and John E. McCarty, Mankato State University, Mankato. Edited by M. J. Lusch*

Experimental:

To **2 mL of water** in a test tube, add **one (1) drop of 2% aqueous Potassium Permanganate solution** and **two (2) drops of the compound to be tested**. *Agitate* the contents of the test tube to thoroughly mix the contents and look for a change in the royal purple color of the permanganate to a dull reddish brown or a muddy brown precipitate. Test the following known compounds: **benzaldehyde, acetone, cyclopentanone**.

2. **Iodoform Test** – Test for the Presence or Absence of a *Methyl Carbinol* or a *Methyl Carbonyl* Compound (via Enolate Iodination/Cleavage)

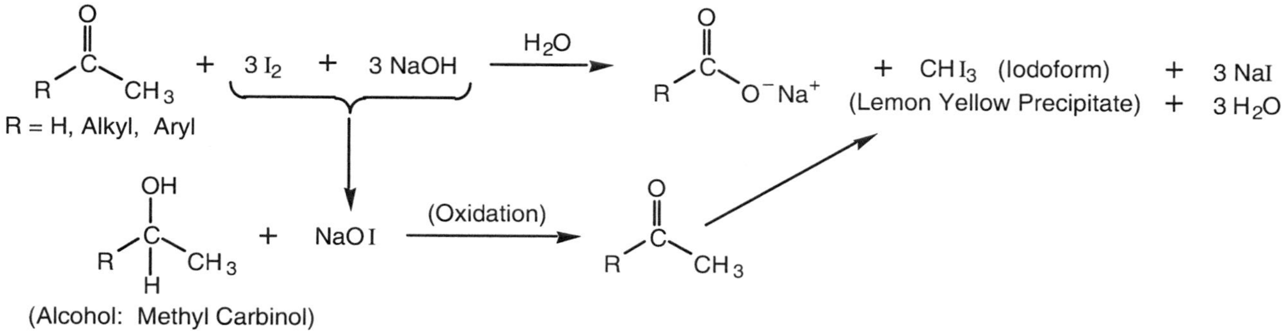

$$R-CO-CH_3 \ (R = H, Alkyl, Aryl) + 3\,I_2 + 3\,NaOH \xrightarrow{H_2O} R-CO-O^-Na^+ + CHI_3 \text{ (Iodoform, Lemon Yellow Precipitate)} + 3\,NaI + 3\,H_2O$$

$$R-CH(OH)-CH_3 \text{ (Alcohol: Methyl Carbinol)} + NaOI \xrightarrow{\text{(Oxidation)}} R-CO-CH_3$$

Experimental:

Place **5 drops of liquid unknown or 0.1 g of solid unknown** in a *large* test tube and dissolve in **5 mL acetonitrile (MeCN)**. To this add **1 mL of 10% aq. NaOH**, and then add, portion-wise *with stirring*, **iodine/potassium iodide (I$_2$/KI) solution** (requires a *large* amount; e.g., 10-20 mL) until there is a definite excess of iodine (brown color). (The iodine will be decolorized by the sodium hydroxide (forming NaOI), but this fact *alone* does not indicate a positive test.) When the color of iodine is <u>no longer discharged after two minutes</u> of stirring, add a **few drops of 10% NaOH** to discharge the brown color, and then dilute the reaction mixture with **water** to fill the test tube. The formation of a **lemon yellow precipitate (CHI$_3$)** with the distinct odor of iodoform (mp = 119-21°C) is an indication of the presence of a **methyl carbonyl or methyl carbinol group**. Test **acetophenone (methyl phenyl ketone)** to see what a **positive** test looks like, and **cyclohexanol** or **cyclohexanone** to see what a **negative** test (no iodoform) looks like.

Reactions of Enolate Anions –

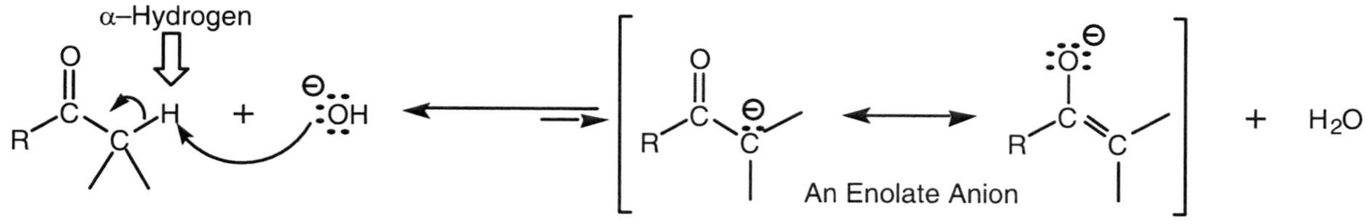

$$R-CO-CH(R')-H + {}^-OH \rightleftharpoons \left[R-CO-C^-(R') \leftrightarrow R-C(O^-)=C(R') \right] + H_2O$$

(α–Hydrogen; An Enolate Anion)

Mechanism:

Unknown Aldehydes and Ketones: Derivative Formation[1]

By Addition to the Carbonyl Group, Followed by Elimination of Water –

1. 2,4-Dinitrophenylhydrazones (2,4-DNP)

2,4-Dinitrophenyl Hydrazine (2,4-DNPH)

A 2,4-Dinitrophenyl Hydrazone (A 2,4-DNP)

Experimental: Place **2 mL of 2,4-Dinitrophenyl Hydrazine (2,4-DNPH) Reagent** solution (already prepared: 2,4-DNPH + conc. H_2SO_4 dissolved in ethanol/water) in a large test tube and add **2 drops of liquid compound to be tested**, or **0.1 g of solid carbonyl compound dissolved in 1 mL of methanol**. Thoroughly mix the contents and notice the almost immediate formation of a colored (yellow, orange, or red) precipitate of the 2,4-Dinitrophenyl Hydrazone (2,4-DNP) derivative. Filter off the solid 2,4-DNP derivative, wash thoroughly with 1:1 95% Ethanol/Water (to remove sulfuric acid and excess 2,4-DNPH), then wash with pure water (to remove sulfuric acid). Recrystallize from 95% Ethanol or some other appropriate solvent.

This reaction can also be used as a <u>test</u> for aldehydes and ketones. Few other functional groups react in any way (alcohols, phenols, carboxylic acids and esters do not react at room temperature; amines may form insoluble, high melting (>250°C) salts in the acidic reagent, but these are usually not very highly colored).

2. Semicarbazones

Semicarbazide Hydrochloride

A Semicarbaz<u>one</u>

Experimental: Dissolve **500 mg (0.50 g) of semicarbazide hydrochloride** and **1.0 g of sodium acetate trihydrate in 3 mL of water** in a *large* test tube. Add **500 mg (0.50 g) or 0.5 mL (12-13 drops) of the carbonyl compound to be derivatized**, warm in a steam bath and **add sufficient 95% ethanol** to dissolve the carbonyl compound and make the solution homogeneous *when hot*. Heat the solution on the steam bath for **5 minutes** or longer (the ethanol may boil), then cool slowly and eventually in an ice/water bath. If crystals do not form, even after scratching the sides of the test tube with a glass rod, reheat on the steam bath for another 5 minutes and repeat the cooling. Addition of more carbonyl compound (0.25 – 0.50 g) during this heating may also be effective in generating a solid derivative, especially if the carbonyl compound is volatile and may be lost to evaporation during the heating. Once solid derivative has been formed, filter off the solid and recrystallize from water or ethanol-water.

---

[1] *From Joy of Organic Chemistry by Berton C. Weberg and John E. McCarty, Mankato State University, Mankato. Edited by M. J. Lusch*

3. Oximes

$$\ce{>C=O} + :NH_2-OH \cdot HCl \longrightarrow \ce{>C=N-OH}$$
$$\text{Hydroxylamine Hydrochloride} \qquad \text{An Oxime}$$

Experimental: Dissolve **500 mg (0.50 g) of hydroxylamine hydrochloride** in **3 mL of water** in a large test tube. Add **1.0 mL of 6 N aq. NaOH** and **200 mg (0.20 g) of solid or 0.20 mL (4-6 drops) of liquid carbonyl compound to be derivatized**. Warm the mixture in a steam bath and **add sufficient 95% ethanol** to dissolve the carbonyl compound and make the solution homogeneous <u>when hot</u>. Heat the solution on a steam bath for 10 minutes or longer (the ethanol may boil), then cool, slowly at first and eventually in an ice bath to crystallize the oxime derivative. Scratching the sides of the test tube with a glass rod may be necessary to initiate crystallization. Once solid derivative has been formed, filter off the solid and recrystallize from water or ethanol-water.

4. Phenylhydrazones and 4-Nitrophenylhydrazones

Experimental: Dissolve **0.50 g (500 mg) of solid or 0.5 mL of liquid** unknown aldehyde or ketone in **10 mL of 95% ethanol**.

    For **Phenylhydrazones** add *either* (a) **0.50 g of Phenylhydrazine Hydrochloride (solid)** and **0.50 g of sodium acetate trihydrate**, *or* (b) **0.5 mL of Phenylhydrazine (liquid)** and **a drop of glacial acetic acid**. The resulting mixture is heated to boiling on a steam bath and enough **water is added** to make the solution homogeneous (no solids) without the formation of milkiness (ethanol may be added to remove the milkiness if it forms). The solution is heated on the steam bath for **5 minutes** or longer, and then cooled slowly to room temperature and then to ice temperature. The resulting phenylhydrazone solid that crystallizes is filtered off, washed with water or aqueous ethanol, and then recrystallized from 95% ethanol, water, or an ethanol-water mixture.

    For **4-Nitrophenylhydrazones** add **0.50 g of 4-Nitrophenylhydrazine (solid)** and **a drop of glacial acetic acid**. The solution is heated on the steam bath for **5 minutes** or longer, and then cooled slowly to room temperature and then to ice temperature. The resulting 4-nitrophenylhydrazone solid that crystallizes is filtered off, washed with 95% ethanol, and then recrystallized from 95% ethanol, water, or an ethanol-water mixture.

Waste Disposal:

1. Mostly <u>Aqueous</u> Waste (with small amounts of organics):
    (a) ACIDIC Aqueous Solution in ACIDIC Aqueous Waste Bottle;
    (b) BASIC Aqueous Solutions in BASIC Aqueous Waste Bottle.
2. Mostly <u>Organic</u> Waste (with small amounts of water and inorganics):
    (a) NON–Halogenated Organic Solutions in NON–Halogenated Organic Waste Bottles;
    (b) HALOGENATED Organic Solutions in HALOGENATED Organic Waste Bottle.

3:40 - 4:10pm

Unknown Carboxylic Acids
Revised Procedures for Preparation of Derivatives[1]

I. Preparation of Amide Derivatives (Primary Amide, Anilide, and p-Toluidide.

All Amide Derivatives require (a) the preparation of the Acid Chloride of the acid, followed by (b) reaction of the acid chloride with an amine (Ammonia, Aniline, or p-Toluidine).

1. Synthesis of an Acid Chloride

$$R-\underset{\underset{O}{\|}}{C}-OH + SOCl_2 \xrightarrow{\text{reflux}} R-\underset{\underset{O}{\|}}{C}-Cl + HCl\uparrow + SO_2\uparrow$$

Do this reaction in the fume hood! Thionyl chloride (bp 79°C) and the byproducts (SO_2 and HCl) are noxious and irritating to the eyes, nose and throat (lachrymator).

[If you intend to make *both* an amide *and* an anilide (or toluidide) derivative, do the acid chloride synthesis **once** on **twice** the scale specified below, and physically divide the resulting solution <u>by volume</u> into two equal portions for reaction with the necessary amine.]

Place **1.0 g of the unknown acid** into a **25 mL round bottom flask**, add **3.0 mL of thionyl chloride ($SOCl_2$)** and heat at reflux **in the hood** for **30 minutes**. Cool the resulting solution and protect from atmospheric moisture (stopper with a glass stopper using stopcock grease or tightly stopper with a cork in a test tube). Use the acid chloride as is for the preparation of any of the amide derivatives described below.

2. Preparation of Primary Amides (Reaction with Ammonia from Ammonium Hydroxide)

$$R-\underset{\underset{O}{\|}}{C}-Cl + NH_3 \xrightarrow[0-5\ °C]{H_2O} R-\underset{\underset{O}{\|}}{C}-NH_2 + NH_4Cl$$
(H_2O insoluble)

$\downarrow H_2O$ (undesirable side reaction)

$R-\underset{\underset{O}{\|}}{C}-OH \xrightarrow{NH_3} RCO_2^-NH_4^+$
(H_2O soluble)

<u>General procedure:</u>
Cooling is Critical to the Success of this Preparation! Cool the acid chloride solution to ice temperature (about 5°C) and **add it** <u>cautiously</u> and <u>dropwise</u> to an **ice cold** solution of **15 mL of *concentrated* ammonium hydroxide (NH_4OH) solution (30% w/w)**. <u>Vigorously</u> agitate during the addition and <u>keep the mixture cold</u>. When the addition is complete there should be a water-insoluble solid in the aqueous solution (This should be the primary amide derivative.)
Filter off this resulting solid amide, wash with **5% (Saturated) aqueous Sodium Bicarbonate solution** (to react with and dissolve any carboxylic acid or acid chloride), and then with water. Recrystallize this crude amide from **95 % ethanol, or an ethanol /water mixture**. Alternative recrystallization solvents: toluene, hexanes, or ethyl acetate /hexanes mixtures.

---

[1] *From Joy of Organic Chemistry by Berton C. Weberg and John E. McCarty, Mankato State University, Mankato, MN. Edited by M.J. Lusch*

2. Preparation of Anilide (Reaction of Acid Chloride with Aniline)

$$R-\overset{O}{\underset{\|}{C}}-Cl + C_6H_5-NH_2 \text{ (excess)} \xrightarrow{0-5\,°C} R-\overset{O}{\underset{\|}{C}}-\underset{H}{N}-C_6H_5 \text{ (H}_2\text{O insoluble)}$$

$$\longrightarrow C_6H_5-\overset{+}{N}H_3Cl^- \text{ (H}_2\text{O soluble)}$$

General procedure:
Cool the acid chloride solution to ice temperature (about 5°C) and add it *dropwise* (slowly) with stirring and cooling, to **10.0 mL of liquid aniline** in a small beaker. Solid anilide and aniline hydrochloride will form, so that the resulting mixture may become an entirely solid, peanut butter-like mass. After mixing well, if the acid chloride odor is still present, add more liquid aniline dropwise with stirring and cooling until the odor of acid chloride can no longer be detected (**Don't** add more than an additional **10 mL of aniline**; a moderate excess of aniline will not adversely effect the product and will be removed in the next step, but *too much* apparently prevents the anilide product from precipitating from the aqueous solution).

When all the acid chloride has been reacted, add **30–50 mL of 6M HCl (or more as necessary)** and stir the mixture well, pulverizing the solid to a fine, lump-free powder (excess aniline is converted to aniline hydrochloride, which is soluble in the aqueous acid).

Collect the solid derivative by vacuum filtration and wash the collected solid with **15–30 mL of water**. Then resuspend the solid in **15 mL of 5 % NaOH** *in the same beaker as used previously for the reaction* (no need to clean first) and stir thoroughly. Filter off the solid again and wash it thoroughly with cold water.

Recrystallization should be attempted from **95 % ethanol or an ethanol/ water mixture**. In some cases, for which ethanol is too good a solvent or causes **oiling out** of the derivative, toluene or hexanes may be a better recrystallization solvent.

3. Preparation of p-Toluidide (Reaction of Acid Chloride with p-Toluidine)

$$R-\overset{O}{\underset{\|}{C}}-Cl + CH_3-C_6H_4-NH_2 \text{ (excess)} \xrightarrow{0-5\,°C} R-\overset{O}{\underset{\|}{C}}-\underset{H}{N}-C_6H_4-CH_3 \text{ (H}_2\text{O insoluble)}$$

General procedure:
Follow the same procedure as for the anilide derivative, substituting **solid p-toluidine** (about 5 g) in place of liquid aniline. Stir well in order to get complete reaction with the solid p-toluidine reagent. Work-up and recrystallize in the same fashion as the anilide.

II. Preparation of Phenacyl and Benzyl Esters (By S_N2 reaction of the anion of the carboxylic acid on a Phenacyl or Benzyl Halide).

4. Preparation of a p-Bromophenacyl Ester

$$\text{Br-C}_6\text{H}_4\text{-C(O)-CH}_2\text{-Br} + \text{Na}^+ {}^-\text{O-C(O)-R} \xrightarrow{S_N2} \text{Br-C}_6\text{H}_4\text{-C(O)-CH}_2\text{-O-C(O)-R}$$

p-Bromophenacyl Bromide (2,4'-Dibromoacetophenone) → A p-Bromophenacyl Ester

General procedure:
Add **1.0 g of your unknown carboxylic acid and 1 drop of phenolphthalein solution** to **10 mL of water** in a 100 mL beaker and carefully neutralize the acid, with vigorous stirring to get the insoluble acid to react, to a faintly pink endpoint with **6N NaOH** solution. Avoid adding excess base (the hydroxide ion would be a much better nucleophile than the acid salt anion and would consume phenacyl bromide and create an impurity). Add one drop of **6N HCl** (more if the indicator is still pink) to make the solution just slightly acidic (and colorless), transfer this solution to a 50 mL round bottom flask, and then add **1 g of p-bromophenacyl bromide (2,4'-dibromoacetophenone)** and **10 mL of ethanol**. Heat the resulting mixture under gentle reflux for **30–60 minutes** (initially everything should dissolve at the boiling point of the solution, but eventually some p-bromophenacyl ester derivative may crystallize from the solution).

Cool the mixture, which should cause the crystallization of (or more of) the derivative, and remove the insoluble crude p-bromophenacyl ester by vacuum filtration. **Recrystallize** the crude p-bromophenacyl ester from **95% ethanol or an aqueous ethanol mixture**.

5. Preparation of a p-Nitrobenzyl Ester

$$\text{O}_2\text{N-C}_6\text{H}_4\text{-CH}_2\text{-Br} + \text{Na}^+ {}^-\text{O-C(O)-R} \xrightarrow{S_N2} \text{O}_2\text{N-C}_6\text{H}_4\text{-CH}_2\text{-O-C(O)-R}$$

p-Nitrobenzyl Bromide
α-Bromo-p-Nitrotoluene → A p-Nitrobenzyl Ester

General procedure:
Follow the same procedure as for the p-bromophenacyl ester derivative, substituting **4-nitrobenzyl bromide (α-Bromo-4-Nitrotoluene)** in place of p-bromophenacyl bromide. Work-up and recrystallize in the same fashion.

III. Neutralization Equivalent: Determination of the molecular weight of a carboxylic acid

A sample of your unknown acid (**0.2 g, weighed accurately to at least 3 decimal figures on an analytical balance**) is dissolved in **50 to 100 mL of water**. The mixture may be heated to dissolve all the compound. The solution is titrated with approximately **0.1000 N standardized NaOH solution,** using phenolphthalein as the indicator. The neutralization equivalent of the acid is calculated according to the formula:

$$\text{neutralization eq} = \frac{\text{weight of sample (in g)} \times 1000}{\text{volume of NaOH (in mL)} \times N}$$

Molecular Weight = (# of acidic groups) x (Neutralization Equivalent)

Unknown Amines - Test Tube Tests[1]

Hinsberg Test: Is an Unknown Amine a Primary, Secondary, or Tertiary Amine?

Procedure: Step 1 - In *Base*

Place **5 mL of 10% aq. NaOH** in a *large* (20 x 150 mm) test tube, add **0.3 mL (6-8 drops) of the amine** to be tested [Note 1], and then add **0.4 mL (8-10 drops) of benzenesulfonyl chloride [Caution!** Note 2]. Agitate the resulting mixture of materials vigorously with a glass rod or microkit spatula, by pumping it up and down in the test tube [Note 3].

If after **5 minutes** of vigorous agitation there is still *unreacted* benzenesulfonyl chloride [Note 4], **heat the mixture in a steam bath, with stirring, until all of the benzenesulfonyl chloride disappears** [Notes 5,6]. If in doubt about the presence or absence of benzenesulfonyl chloride, carry out the heating step for 2–3 minutes more just to be sure that all of it has been reacted.

Cool the reaction mixture in an ice bath, and check with litmus paper to make sure that the mixture is still basic (add more 10% NaOH, if necessary, to make it definitely basic to litmus paper). Scratch the sides of the test tube with a glass rod in an attempt to induce crystallization in the reaction mixture, *especially* if there is a water insoluble layer.

Interpretation of Results in *Base* [Note 7]:

A Primary Amine Should Form a *Single* Homogeneous Solution in *Base*: If at this point the reaction mixture is completely homogeneous (only one layer), the amine being tested is *probably* a primary amine (R–NH$_2$). However, for confirmation, one *must* see the results after the solution has been acidified (below).

$$R-NH_2 + Cl-SO_2-C_6H_5 \xrightarrow[H_2O]{10\% \text{ NaOH}} \left[R-N(H)-SO_2-C_6H_5 \right] \xrightarrow{OH^-} R-N(Na^+)^--SO_2-C_6H_5$$

Water Insoluble (More Dense than H$_2$O) [clear] → a Benzenesulfonamide (With an acidic N–H) → Water Soluble (Usually, but **Not** Always Completely)

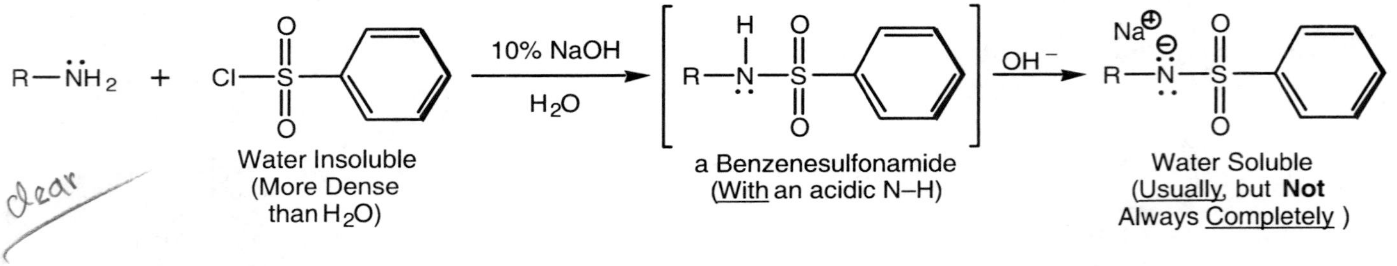

A Secondary Amine Should Form a Water Insoluble *Solid* in *Base*: If at this point there is a water insoluble solid in the reaction mixture, the amine being tested *MAY be* a secondary amine (R$_2$NH). However, for definitive confirmation, one *must* see the results after the solution has been acidified (below).

$$R-N(R')-H + Cl-SO_2-C_6H_5 \xrightarrow[H_2O]{10\% \text{ NaOH}} R-N(R')-SO_2-C_6H_5 \xrightarrow{OH^-} R-N(R')-SO_2-C_6H_5$$

Water Insoluble (More Dense than H$_2$O) [solid] → a Benzenesulfonamide (No acidic N–H) → **NOT** water soluble (No reaction with base)

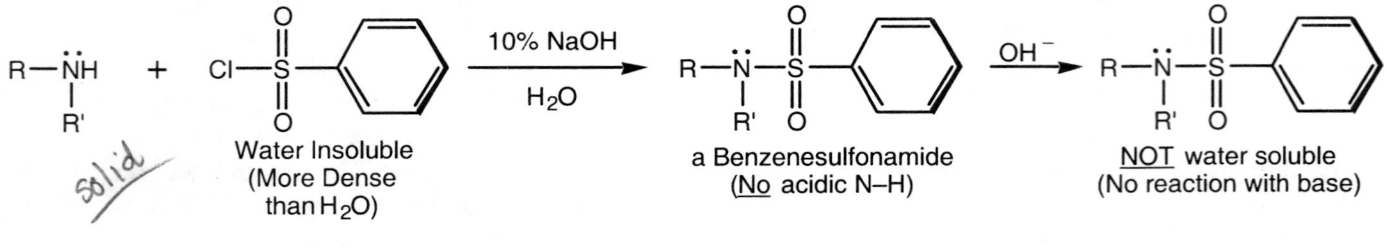

Warning: Sometimes (actually, quite frequently) a **primary amine** will form a benzenesulfonamide whose sodium salt is a solid that is not completely soluble in the **basic** solution, and the solution will therefore *appear* like that found with a **secondary amine**, whose benzenesulfonamide is *completely* INSOLUBLE in the basic solution. It is, however, possible to tell these visually similar results apart *by observing their behavior upon acidification*. The primary amine benzenesulfonamide **salt**, even though it may not be completely soluble in

---

[1] *From Joy of Organic Chemistry by Berton C. Weberg and John E. McCarty, Mankato State University, Mankato. Edited by M. J. Lusch*

base, is partially soluble and has some amount of the ionic sulfonamide salt dissolved in the solution [Note 8], whereas virtually none of the non-ionic secondary amine benzenesulfonamide is dissolved in the basic solution. Therefore, when the **dissolved primary amine sulfonamide salt is protonated** upon addition of acid and thus made water insoluble, it separates as a milkiness which eventually becomes the solidified benzenesulfonamide derivative, whereas the **secondary amine sulfonamide is unchanged by the addition of acid.**

A Liquid Tertiary Amine Should Form a Water-Insoluble *Liquid* in *Base*: If at this point there is a water insoluble liquid layer floating on top of (less dense than) the aqueous reaction mixture, the amine being tested MAY be a tertiary amine (R_3N). However, for definitive confirmation, one ***must*** see the results after the solution has been acidified (below).

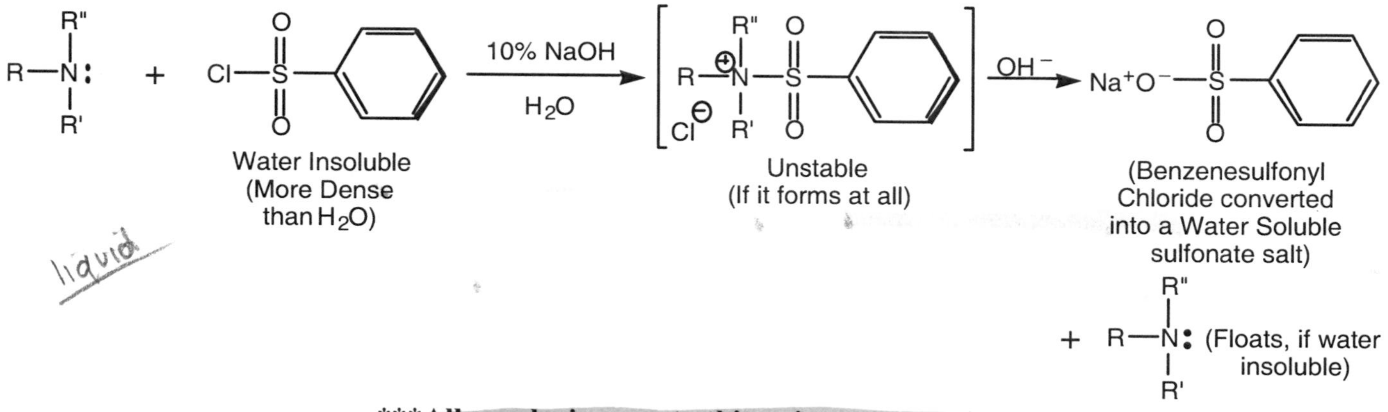

*All conclusions** up to this point are tentative, and *must* be verified by subsequent behavior/appearance after *acidification* of the reaction mixture.

Procedure: Step 2 - In *Acid*

Add **6 N aqueous HCl**, slowly and with thorough mixing, until the solution is distinctly acidic to litmus paper. In particular, watch to see if any temporary or permanent changes are taking place as you add acid and mix. **Save** any solid still present after acidification [Note 9].

Interpretation of Results in *Acid* [Note 7]:

A Primary Amine Forms a *Milkiness* Initially and Eventually a Water Insoluble *Solid* in *Acid*: If a milkiness forms in the homogeneous solution, as acid is added and the base becomes neutralized, and eventually a water insoluble ***solid*** forms when the solution becomes acidic, then the amine being tested is a **primary amine (R–NH$_2$)** [Note 9].

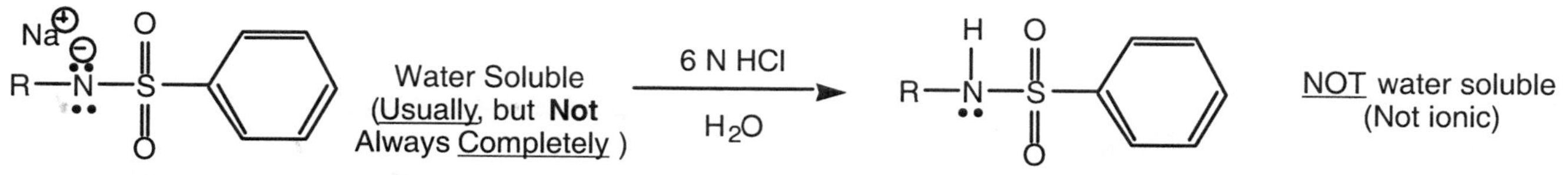

A Secondary Amine Should *Still* Have a Water-Insoluble *Solid* in *Acid*: If no **milkiness or additional solid** is formed when the solution becomes acidic, but the original water insoluble solid sulfonamide remains undissolved, then the amine being tested is a **secondary amine (R$_2$NH)** [Note 9].

A45

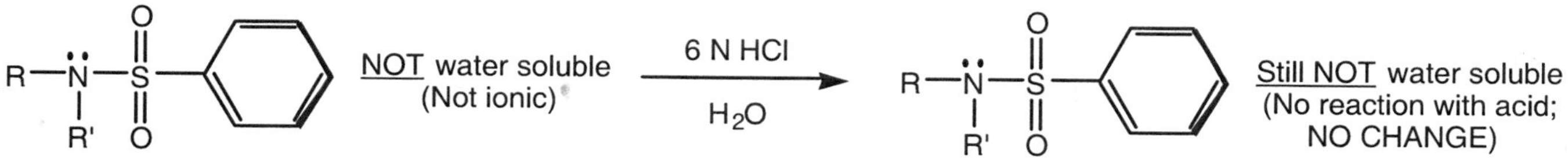

A Tertiary Amine Should Have Just One *Homogeneous* Solution in *Acid*: If the solution becomes homogeneous when it is made acidic, and any water insoluble liquid layer floating on top of the solution dissolves, then the amine being tested is a **tertiary amine (R_3N)**.

$$Na^+O^-\underset{\underset{O}{\|}}{\overset{\overset{O}{\|}}{S}}-C_6H_5 + R-\underset{R'}{\overset{R''}{N:}} \text{ Floats, if water insoluble} \xrightarrow[H_2O]{6\ N\ HCl} HO-\underset{\underset{O}{\|}}{\overset{\overset{O}{\|}}{S}}-C_6H_5 + R-\underset{R'}{\overset{R''}{\overset{\oplus}{N}}}-H\ Cl^\ominus$$

Water Soluble .... Benzenesulfonic Acid ... Ammonium Halide
.... (Water Soluble) .... Salt (Water Soluble)

ppt → clear

Everything Water Soluble – Clear Solution

NOTES

Note 1: Observe the behavior of the amine at this point – Is it soluble or insoluble in the aqueous solution? If it is insoluble, is it more (sinks) or less (floats) dense than the aqueous solution?

Note 2: Caution! Measure out Benzenesulfonyl Chloride in the hood! Benzenesulfonyl Chloride is a Lacrymator (causes eyes to water) and has a sharp, pungent, unpleasant odor. It is also slowly decomposed by moisture, so reseal the bottle *tightly* when finished removing the necessary amounts.

Note 3: Benzenesulfonyl Chloride is *insoluble in* and *more dense* than water, so it forms an oily globule on the bottom of the test tube. The **amines** being tested, on the other hand, if *not* water soluble, are usually **less dense** than water and float on the top of the aqueous solution. In order for the two to react, they must be physically mixed by some form of **agitation**, before the benzenesulfonyl chloride slowly, but inevitably, reacts with and is decomposed by the intervening water layer.

Note 4: The presence of benzenesulfonyl chloride can be detected by observing an oily, water insoluble globule on the bottom of the test tube, and by its strong, penetrating, distinctive odor.

Note 5: Benzenesulfonyl Chloride reacts with water slowly (because of its water insolubility), a process that is accelerated by heat, to form **Benzenesulfonic Acid**, which is *very* water soluble.

Note 6: Do not be fooled that benzenesulfonyl chloride is still present by the appearance of a larger volume of heavier-than-water oily liquid than the original amount of benzenesulfonyl chloride occupied, especially if it is dark in color. This may be **melted** benzenesulfonamide or benzenesulfonamide–salt product, especially in warm or hot solutions, which should solidify upon cooling and scratching with a glass rod to induce crystallization.

Note 7: To observe what the appearance of the test would be with typical primary, secondary, and tertiary amines, **before** running your unknown, perform the test separately on **sec-butyl amine** [primary amine, $CH_3CH_2CH(CH_3)NH_2$], **diethylamine** [secondary amine, $CH_3CH_2-NH-CH_2CH_3$], and triethyl amine [tertiary amine, $(CH_3CH_2)_3N$].

Note 8: Many times, as the basic solution is being neutralized with acid, and if the acid is added slowly and with *thorough stirring*, the *insoluble* portion of the primary amine sulfonamide *salt* will become more soluble in the less basic solution, and will either partially or completely dissolve as the solution approaches neutrality (and thus at *this* point will appear as a primary amine benzenesulfonamide salt theoretically should, i.e., soluble in (mildly) basic solution). When more acid is added beyond this point, and the solution finally becomes distinctly acidic, the protonated primary amine benzenesulfonamide then quite noticeably separates from the solution, first perhaps as a voluminous milkiness, and then eventually, especially with cooling and scratching to induce crystallization, as a solid derivative.

Note 9: Any water insoluble solid present after acidification is the **benzenesulfonamide derivative** of your primary or secondary amine. ***Save** any solid still present after acidification - It can be used as a **derivative**.* Since the melting point of this material can be used to identify the unknown amine, it is important that all benzenesulfonamide salt be completely protonated to the neutral benzenesulfonamide. This can be accomplished by thoroughly mixing the acidic solution and breaking up and powdering any solid that may have been present before the acidification began. The resulting solid should then be collected by vacuum filtration, washed with water, and then recrystallized (see derivation preparation procedures) to obtain purified derivative.

Chemistry 331- Organic Chemistry II Laboratory

Unknown Amine Derivatives
(see also pp 100–102, in "The Joy of Organic Chemistry Laboratory")

BenzeneSulfonamides:

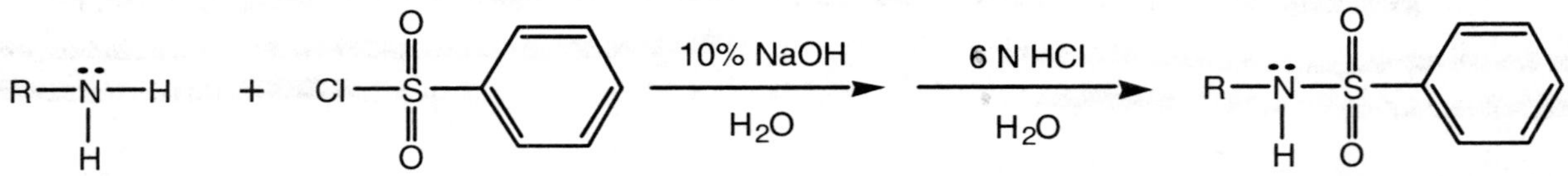

Primary (1°) Amine + Benzenesulfonyl Chloride $\xrightarrow{\text{10\% NaOH, H}_2\text{O}}$ $\xrightarrow{\text{6 N HCl, H}_2\text{O}}$ a Benzenesulfonamide

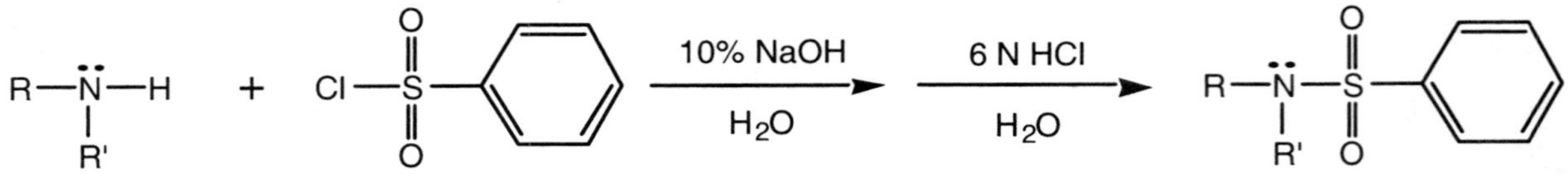

Secondary (2°) Amine + Benzenesulfonyl Chloride $\xrightarrow{\text{10\% NaOH, H}_2\text{O}}$ $\xrightarrow{\text{6 N HCl, H}_2\text{O}}$ a Benzenesulfonamide

The solid materials formed from *primary* and *secondary amines* and still present after acidification in the **Hinsberg Test** are the **benzenesulfonamide derivatives of these amines**. *Double or triple* the amounts described in the Hinsberg Test may be necessary to isolate sufficient amounts of the derivative.

Once the derivative has been **isolated by vacuum filtration and washed thoroughly with water**, its recrystallization should be attempted from **95% ethanol**, or from an ethanol (good solvent)–water (bad solvent) mixture.

p-TolueneSulfonamides:

Carry out the **Hinsberg Test** *procedure*, but use **0.5 gram** of **p-toluenesulfonyl chloride (a solid)** in place of the benzenesulfonyl chloride (a liquid), to produce the p-toluenesulfonamide derivatives of *primary and secondary amines*. Isolate and purify these derivatives in the same way as benzenesulfonamides.

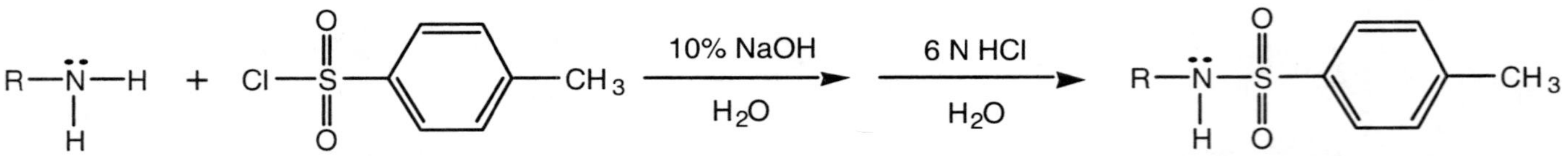

Primary (1°) Amine + p-Toluenesulfonyl Chloride (a solid) $\xrightarrow{\text{10\% NaOH, H}_2\text{O}}$ $\xrightarrow{\text{6 N HCl, H}_2\text{O}}$ a p-Toluenesulfonamide (of a 1° amine)

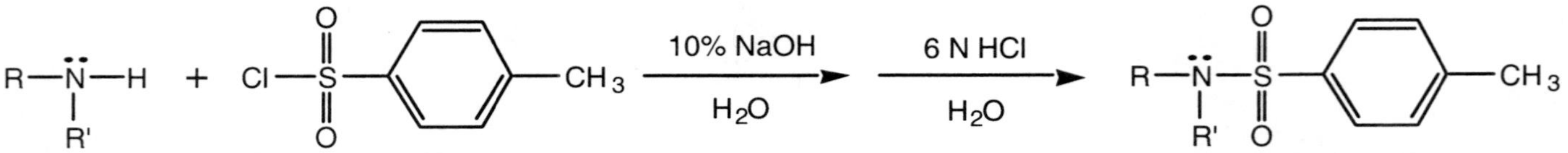

Secondary (2°) Amine + p-Toluenesulfonyl Chloride (a solid) $\xrightarrow{\text{10\% NaOH, H}_2\text{O}}$ $\xrightarrow{\text{6 N HCl, H}_2\text{O}}$ a p-Toluenesulfonamide (of a 2° amine)

Phenylthioureas:

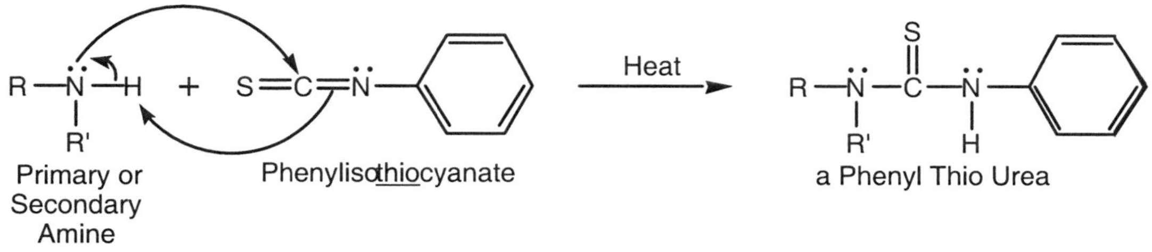

Place 0.5 gram (0.5 mL) of **PhenylisoTHIOcyanate** and 0.5 gram (0.5 mL) of **unknown amine** in a medium sized test tube, stir thoroughly, and heat in a steam bath (loosely stoppered) for at least **5 minutes**. Cool the reaction mixture in an ice bath and scratch with a glass rod (if necessary) to induce solidification. Stir thoroughly with **5 mL of hexanes** (to remove unreacted phenylisothiocyanate), breaking up the material into a fine powder, and decant off the hexanes solution; then stir the solid derivative thoroughly with a **50:50 mixture of 95% ethanol and water** (to remove unreacted unknown amine), and finally filter off the remaining solid and allow to dry thoroughly (it is usually not necessary to recrystallize these derivatives, although they can be purified further by recrystallization if necessary).

Benzamides: (See pg. 780, Williamson, 4th Edition)

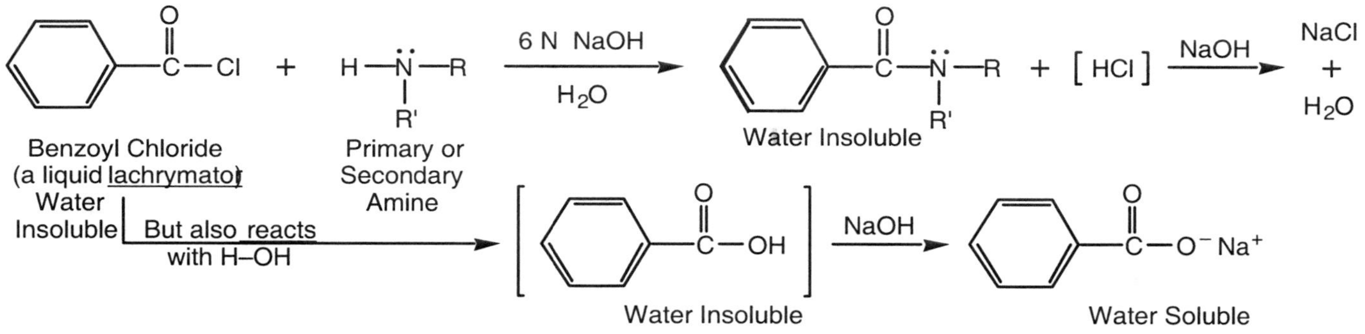

Place 0.5 gram (0.5 mL) of **unknown amine**, 10 mL of 6 N aq. **sodium hydroxide**, and 10 mL of **water** in a 50 mL **Erlenmeyer flask**, and then add, in portions of **0.1 mL** at a time, with vigorous stirring and mixing, a total of **1.0 mL of Benzoyl Chloride** (**NOT** BenzeneSulfonyl Chloride) [**Caution!** Benzoyl Chloride is a **lacrymator!** Use it *only* in the hood. It is also **more dense than** and **not very soluble in** *water*.] After all of the benzoyl chloride has been added, stir for an additional 10 minutes, and then check for the presence of unreacted benzoyl chloride (*odor* of benzoyl chloride, or an oily *liquid* layer on the bottom of the reaction mixture. If unreacted benzoyl chloride is still present, warm the mixture on the steam bath to speed up the hydrolysis of benzoyl chloride to benzoic acid.

Test the pH of the solution with pH paper, and adjust if necessary to **pH = 8** with either 3 M aq. HCl (if the pH is higher than 8), or with 5% aq. NaOH (if the pH is less than 8). The presence at this point of a water-insoluble **solid** indicates the formation of the benzamide derivative of the amine. Filter off the solid, wash it with cold water, and **recrystallize with 95% Ethanol** or ethanol/water mixtures.

Acetamides:

Heat a mixture of **0.5 gram or 0.5 mL** of **unknown amine** with **0.25 mL of acetic anhydride** in a large test tube on a sand bath for **5 minutes**. Cool in ice and add **2.5 mL of water** to the reaction mixture. Initiate crystallization by scratching, if necessary. Remove solid by filtration, and wash thoroughly with **5% aq. HCl** to remove unreacted unknown amine. Recrystallize the derivative from EtOH or other appropriate solvent.

A49

Picrates: (See pg. 780, Williamson, 4th Edition)

[Structure: Primary, Secondary, or Tertiary Amine R-N(R')(R''):H + Picric Acid (2,4,6-trinitrophenol, Not a Carboxylic Acid, but a very strongly Acidic Phenol) → 95% EtOH → Picric Acid Salt of an Amine (ionic, less soluble in EtOH than the Amine or Picric Acid), shown as ammonium cation with picrate anion]

Picrates can be prepared from all kinds of amines, i.e., *primary and secondary amines*, as well as *tertiary amines*, form picrate salt derivatives. **Caution!** Picric Acid is shock and friction sensitive and **explosive when completely dry** (not damp with some solvent). To prepare a picrate salt place **2.0 mL of a saturated solution of picric acid in 95% ethanol** (already prepared) into a medium-sized test tube, and add **0.5 grams (0.5 mL) of unknown amine**. Heat the resulting solution in a steam bath for at least **5 minutes**, taking care to avoid boiling out the ethanol solvent, and then cool the solution in an ice bath and scratch the sides of the test tube to induce crystallization of the picrate salt. Since many picrate salts seem to crystallized rather slowly, be patient when trying to induce crystallization, even to the point of stoppering the test tube tightly and letting it sit for a longer period of time (benign neglect).
Once crystallization has been achieved, vacuum filter the cold mixture, wash the collected yellow picrate crystals with pure cold 95% ethanol, and suction dry. Recrystallization is not usually necessary.

Methyl Ammonium Salts – Methiodides and Methyl p-Toluenesulfonates:
(use the procedures found on page 101 in the lab manual)

[Scheme: R-N(R')(R''): + CH₃—I → R-N⁺(CH₃)(R')(R'') I⁻ Methyl Iodide (a liquid) a Methiodide Salt of the Amine (ionic, less soluble in organic solvents than the reactants)]

[Scheme: Primary, Secondary, or Tertiary Amine R-N(R')(R''): + Methyl p-Toluenesulfonate CH₃-C₆H₄-SO₂-O-CH₃ (a liquid) → Toluene (NOT Benzene) → R-N⁺(CH₃)(R')(R'') ⁻O-SO₂-C₆H₄-CH₃ a Methyl p-Toluenesulfonate Salt of the Amine (ionic, less soluble in Toluene than the reactants)]

Methiodides:
Mix **0.5 g (0.5 mL) of the unknown amine** with **1.0 mL of Methyl Iodide** [**Caution! Harmful Vapor! Bp 45°C**]. Heat the mixture on a steam bath [**in the hood**] (but avoid boiling the methyl iodide vapor out into the air) for a few minutes, then cool in an ice bath.

Methylammonium p-Toluenesulfonate Salts: